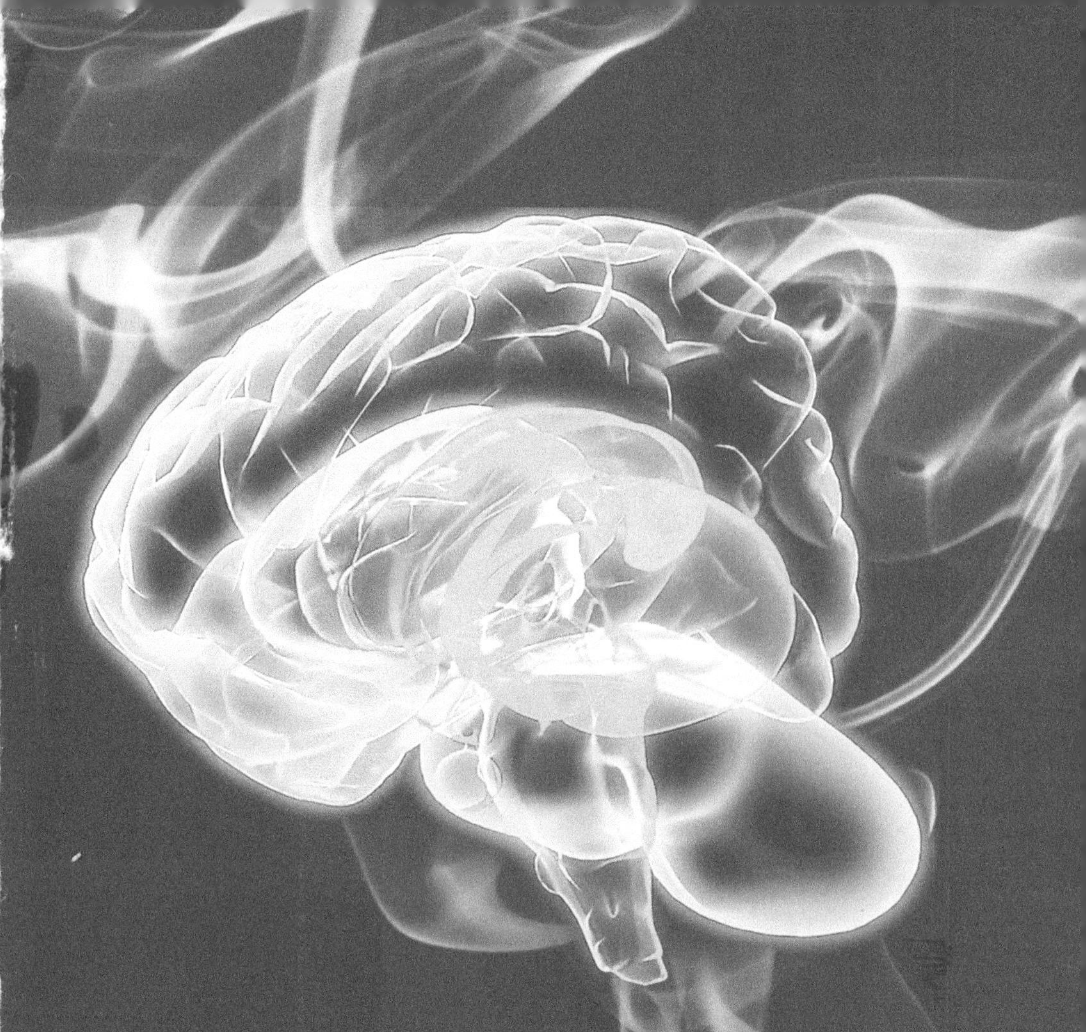

Sensing Biochemistry in the Brain

NEUROIMAGING

Sensing Biochemistry in the Brain

NEUROIMAGING

Patricia A. Broderick

JENNY STANFORD
PUBLISHING

Published by

Jenny Stanford Publishing Pte. Ltd.
101 Thomson Road
#06-01, United Square
Singapore 307591

Email: editorial@jennystanford.com
Web: www.jennystanford.com

British Library Cataloguing-in-Publication Data
A catalogue record for this book is available from the British Library.

Neuroimaging: Sensing Biochemistry in the Brain

ISBN 978-981-4613-49-1 (Hardcover)
ISBN 978-0-429-02974-5 (eBook)

Contents

Preface

From Singapore, is the author's authentic neuroimaging textbook, a monograph. The innovative scientific monograph explains with original experimental protocols in detail, the know-how of Neuromolecular Imaging (NMI) along with describing the discoveries made by the BRODERICK PROBE®. The author interjects her own person and her personality as she interacts with competitors and colleagues alike within her riveting travel to her original discoveries. Aggressively novel, her inventive art of sensing electrochemistry in the brain permits its energetic charge to flow *via* the power of her word and her magnificent raw data comes to the reader from the living subject LIVE. Right away, one sees that this is the book for the real thing—the LIVE art, *the very meaning of imagery that saves lives.*

This monograph focuses intensely on a multifunctional theragnostic brain imaging nanotechnology, Neuromolecular Imaging (NMI). In this exciting book, a new electrochemistry is born, bringing to the reader globally, a branch of analytical chemistry to the brain, biochemistry to avidly consume. The timeless sleek, ultra-small devices are many and are brought to the reader under the single umbrella of the BRODERICK PROBE®, named after the author's father, Patrick Broderick, an oxyacetylene torch welder, a blacksmith, an Irish immigrant to America. The products are patent protected and come with international trademarks that are separate for patients and for animals.

The book is the voice for nanoprobes of carbon, the molecule of life, whose allotropes are semiconductors that safely steer away from fossil fuels. The book is the voice for protein optical fluorescent nanoprobes that enlighten the imagery. These polymers are sensing devices that are much thinner, smaller, and more precise than molecular medical imaging equipment currently in use. The nanoprobes are biodegradable; the nanoprobes do not cause bacteria and do not scar. Valuation is predicated on a multi-million-dollar contractual offer that awaits the mature wisdom of *Eazysense Nanotechnologies Inc.*, which holds its equity. The University partners are in favor of investments to Eazysense and its non-profit arm, the *Broderick Brain Foundation,* for marketing the nanoprobes.

Patricia's voice uniquely delineates her inventive art of sensing and laser-driven voltaic photonic protein polymers in Chapter 1 and her original circuits to drive the sensors are drawn by her in Chapter 4 of the monograph. Quantum sensing enables precision at the nanoscale thus, making the surface a veritable haven of molecular recognition sites for Tau, the peptide, shown in Chapter 15, one likely solution to the dementias. The doctor shares her nanoprobes, carbon-based lipids and phosphatidyl's, that create algorithms for temporally synchronous walking during well-being in Chapter 10 while the biomechanics of the epilepsies are unearthed as never before in Chapters 12 and 13.

The electrocatalytic debacle of dopamine and ascorbic acid "mix-up." is abruptly aborted with Patricia's proof of concept, brilliantly demonstrated in Chapter 5. The author unabashedly delves into the *modus operandi* of epilepsy orgasms with the ease and agility of an Irish step-dancer as she writes Chapter 11; she respectfully includes notable practitioners in the field of sexual therapy. Orgasms are explained, enlightening the intellectual and the spiritual human mind/brain to the sublime and rhythmic artistry of the epilepsy orgasm within the confines of the temporally asynchronous *status epilepticus.*

The female psyche is discussed in Chapter 9, bringing to the monograph the field of endocrinology as this topic relates to hormones, the menstrual cycle and the estrous cycle. The inventor is pleased to bring more tech savvy to the nanotechnology of sex-specificity in gender medicine. Patricia brings her skill in inventing techs from her success in delving hands-on within the mechatronics of atomic absorption spectrophotometry and fluorescent spectroscopy during her studies about bipolarity. Too, the subject of caffeine comes to this disarming monograph In Chapters 8 and the polymer actually images caffeine in Chapter 9. The nanoprobe polymers are shown to be important in nutrition as the polymers come to a coffee shop near you in the form of middle chain triglycerides (MCT) products while the utterly beautiful jewel of perfume is imaged by the polymers in Chapter 6.

Also, demonstrated in Chapter 6 is major phosphorous signaling by the BRODERICK PROBE® polymer.

It seems like magic, but the formulae, waveforms, and images of peritumoral epilepsy, so closely allied with danger from gliomas in the white part of brain matter, are imaged with vivid online pictures to treat cancer patients clinically. There is a glaring contrast between the pictures imaged directly from gray versus white matter that highlight the versatility of this impressive nanotechnology, and the original data are presented in Chapter 15.

The book is for those who wish to learn; it serves the science of sensors, the basis of science, medicine, automotive, space, and the energy industries. The monograph brings together, students, professors, physicians and the lay person. This is our future. This is a book about a brand-new nanotechnology and everyone is the layperson. The world of polymers is for everyone. This provocative monograph is for everyone.

Within the precision of the art, the author shares her heartfelt passion for her inventive nanotechnology and for the people who have paved the way for her art to flourish. Mixing her life with science and medicine is the manner in which the author lives, and this becomes crystal clear in this charming monograph. Published by Jenny Stanford Publishing Pte. Ltd., the collaboration between the publisher and the author is well above the fray of all others. Indeed, the book is different from all others. We hope that you enjoy the wealth of the diversity within.

Introduction to Clinical Nanotechnology

Nitrous Oxide Is Electroactive: Biomarkers for NO versus N_2O

Introducing Nitrous Oxide: *The molecule, nitrous oxide, is not to be confused with nitric oxide, Indeed, nitrous oxide has two nitrogen atoms bonded with one oxygen atom whereas nitric oxide has one nitrogen atom bonded with one oxygen atom. Each molecular entity of nitrous oxide as opposed to that of nitric oxide has entirely different physiological functions albeit both are gaseous. Interestingly, we routinely speak of the electroactive nature of nitric oxide but not that of nitrous oxide. Ah but, we speak herein of the inherent electroactive property of the molecule, nitrous oxide, with empirical data imaged LIVE in the temporal lobe epilepsy patient, postoperatively, in situ, providing new dogma in the field of nanomedicine via imaging the molecule with the* BRODERICK PROBE® nanosensor. *Included in this Introduction, is an electroactive sensing distinction between these two gaseous molecules, nitric oxide (NO) versus nitrous oxide (N_2O).*

• What the Molecule Nitrous Oxide Does

Nitrous oxide (N_2O) appears as a structurally simple and small molecule that is comprised of two nitrogen atoms and one oxygen atom. Yet, its anesthetic, analgesic, and psychotropic properties

Neuroimaging: Sensing Biochemistry in the Brain
Patricia A. Broderick

ISBN 978-981-4613-49-1 (Hardcover), 978-0-429-02974-5 (eBook)
www.jennystanford.com

are indisputable [1, 13, 24]. Nitrous oxide is reported to act *via* opiate mechanisms because it induces met-enkephalin and β-endorphin release in murine animals and humans [25]; the sensory pain-relieving properties of nitrous oxide are reported empirically to be reversible by naloxone [11]. Also, but likely not exclusively, nitrous oxide may exert its effects *via* glutamate receptors, that is, administration of (80%) nitrous oxide to murine hippocampus has been shown to depress excitatory currents evoked by *N*-methyl-*D*-aspartate [15, 18].

The combination of nitrous oxide and oxygen has found its way into prehospital emergency treatment of pain as the combination is an effective therapeutic agent due to its release of endogenous morphine-like neurochemicals such as endorphin. Further, under the proprietary (brand) names, Entonox® and Dolonex®, the combination is administered in a 40–60% ratio; the dual treatment is used by paramedics when treating acute myocardial infarction [19]. In some areas of the world, it is used in emergency medicine in *lieu* of opioid analgesics for the management of painful injuries [16].

The BRODERICK PROBE® sensors successfully image cerebral cortex in neocortical temporal lobe epilepsy patients intraoperatively and nitrous oxide was administered as an anesthetic in combination with the general anesthetic, halothane [8]. In fact, nitrous oxide is used in adjunct and routinely with other general anesthetics such as halothane because nitrous oxide facilitates the reaction rate of "low blood/gas solubility partition coefficient" anesthetic gases; the phenomenon is known in anesthesiology as the "second gas effect" and facilitation is produced *via* its inherently high partial pressure. The BRODERICK PROBE® sensors successfully image cerebral cortex and hippocampus in neocortical and mesial temporal lobe epilepsy patients *in situ* [10, 20, 21].

• The Effect of Nitrous Oxide on Pain and Cerebral Blood Flow: PET Studies

Interestingly, in studies [14] used to map the effects of analgesics on pain, cerebral substrates for the nociceptive effects of nitrous oxide have been identified. A low concentrations (20%) of nitrous oxide was administered and the results were imaged using positron emission tomography (PET) scans of regional cerebral blood flow (rCBF). Inhalation of 20% nitrous oxide was found to be associated with enhanced rCBF in the anterior cingulate cortex in area 24, decreased rCBF in the hippocampus and posterior cingulate within areas 23 and 24 and finally there were decreased rates of regional cerebral blood flow occurring in the secondary visual cortices within areas 18 and 19.

Although Nitrous Oxide is known to cause abuse and misuse and deaths have been associated with the abuse and misuse of this atmospheric gas [2], the following excerpt from Dr. Gilman [13] is an interesting take on the abuse, misuse and addiction potential of Nitrous Oxide by professionals in the health care industry.

- **Dr Gilman, of Goodman and Gilman, *The Pharmacological Basis of Therapeutics*, Writes of the Abuse/Addictive Nature of Nitrous Oxide among Health Professionals.**

> "I present evidence that certain health professionals have a higher liability to drug abuse than does the general population. In view of this, data are presented in which the extremely limited contribution of nitrous oxide to this problem is assessed. I demonstrate on the basis of the limited published data now available that pure nitrous oxide addiction is very rare among health professionals, being the rarest drug of abuse among them. When used by the latter, it is most commonly part of a polydrug abuse pattern, not being a major component of the addiction in these cases. It is clear that nitrous oxide addiction is even less of a problem in the general population. Furthermore, after almost 200 years of use, its' very minor addictive potential should by now have clarified itself. It would appear that it is unlikely to become, has never been, nor is it the moment of any real significance as a drug of abuse when compared with the many currently available addictive drugs, which clearly pose a much more serious problem."

• Biomarkers for NO versus N_2O

The literature is replete with detailed and critical multifaceted studies of Nitric Oxide. Some topics take on a popularity in science and medicine that seems to be alright, "getting on the band wagon" so to speak, But, looking at life from the other side, it appears that Nitrous Oxide has not been extensively or even minimally studied. Nitrous oxide has been left out of the picture. Let us step back for a proverbial minute to discern the situation on the nanoscale at the quantum level.

Quantum cascade laser absorption spectroscopy (QCLAS) studies enable all of us to distinguish between nitrous oxide (N_2O) and nitric oxide (NO) [23]. The take-home message is clear; all organisms and pathways known to be involved in the catabolic branch of the microbial N-cycle have the potential to catalyze the reduction of NO_2^- to NO and then further to use electron transfer in redox to perform reduction of NO to N_2O. An urgent need to study both of these critical molecules is paramount to accuracy in detection and subsequent interpretation of the data. These aspects of human, mammalian and microbial redox biochemistry elucidate sensing mechanisms and provide an understanding to the professional and the lay person about the molecular basis of diseases and aging. Clarifying the molecule opens avenues for the development of preventive and therapeutic strategies in neuro-molecular medicine.

A quote from Schreiber et al., 2012 [23] follows:

> "Nitrous oxide (N_2O) is an environmentally important atmospheric trace gas because it is an effective greenhouse gas and it leads to ozone depletion through photo-chemical nitric oxide (NO) production in the stratosphere. Mitigating its steady increase in atmospheric concentration requires an understanding of the mechanisms that lead to its formation in natural and engineered microbial communities. N_2O is formed biologically from the oxidation of hydroxylamine (NH_2OH) or the reduction of nitrite (NO_2^-) to NO and further to N_2O. Our review of the biological pathways for N_2O production shows that apparently all organisms and

pathways known to be involved in the catabolic branch of microbial N-cycle have the potential to catalyze the reduction of NO_2^- to NO and the further reduction of NO to N_2O, while N_2O formation from NH_2OH is only performed by ammonia oxidizing bacteria (AOB)" [23].

Therefore, and at the nanoscale, the BRODERICK PROBE® nanosensor enables the distinction between Nitrous Oxide and Nitric Oxide by discovering a biomarker for each molecule. The biomarker for Nitrous Oxide is shown below with exciting operationally stable images derived LIVE, and i*n vitro* using the nanoprobe. New circuits and biosensors are invented and patented to image the brain after nanomedicine is administered and to perform these studies at the neuromolecular imaging (NMI) spectral analysis level is a significant contribution to the literature along with that of quantum cascade laser absorption spectroscopy. The Broderick nanoprobe sees Nitric Oxide; the report shows the distinct oxidation potentials during the actual redox reactions [22].

The importance of biomarking Nitrous Oxide in surgery, emergency medicine, pain, cardiovascular and psychiatry and dentistry needs no further explanation. There are virtually little or no direct techniques available to detect nitrous oxide unchanged in living tissue. The purpose here is to present such a technique using Neuromolecular Imaging (NMI) and carbon based nano- and microsensors.

The molecular structure of the nitrous molecule is shown in Fig. 1.

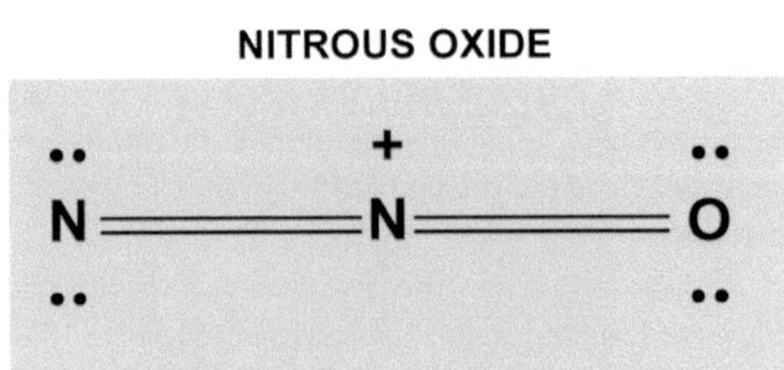

Figure 1 The nitrous oxide molecule.

• Experimental Design: Nitrous Oxide (N_2O)

The experimental design for and the results from *in vitro* assays, i.e., studies of nitrous oxide as N_2O is diffused into an electrochemical cell as well as those from *in vivo* imaging, i.e., studies of Nitrous Oxide which has stabilized in living tissue from N_2O infusion in the epilepsy patient.

This BRODERICK PROBE® NMI is a novel experimental assay for the gaseous solution, nitrous oxide *in vitro*. High-purity (99.9%) commercially available nitrous oxide (T.W. Smith, Brooklyn, NY) was diffused using a flowmeter, calibrated at 10 psi, into an electrochemical cell containing saline/phosphate buffer for 5 min to allow the gas to reach saturation at room temperature. The flowmeter was purchased from Fisher Scientific (Bridgewater, NJ). Nitrous Oxide concentrations in the approximate range of 10–100 μM were achieved.

Schematic designs as well as procedures for the signature images of neurotransmitters and neurochemicals by the BRODERICK PROBE® sensing medical devices are described [3–10].

• Real-Time Data-Nitrous Oxide-*In Vitro*

Figure 2 shows a representative recording of nitrous oxide detection *in vitro*. Nitrous oxide detection occurred at the oxidation (half-wave) potential of 0.53 ± 0.02 V. In addition, dopamine (DA) and serotonin (5-HT) signals are shown because increasing concentrations of DA and 5-HT were aliquoted into the electrochemical cell for use as standards. Thus, studies with the monoamines were conducted, which show the selective detection of nitrous oxide in the presence of the monoamine catecholamine, DA, imaged selectively, in real time, within seconds, within the semi-differentiated scan at a separate half wave/oxidation potential as the indoleamine, 5-HT, neurotransmitter.

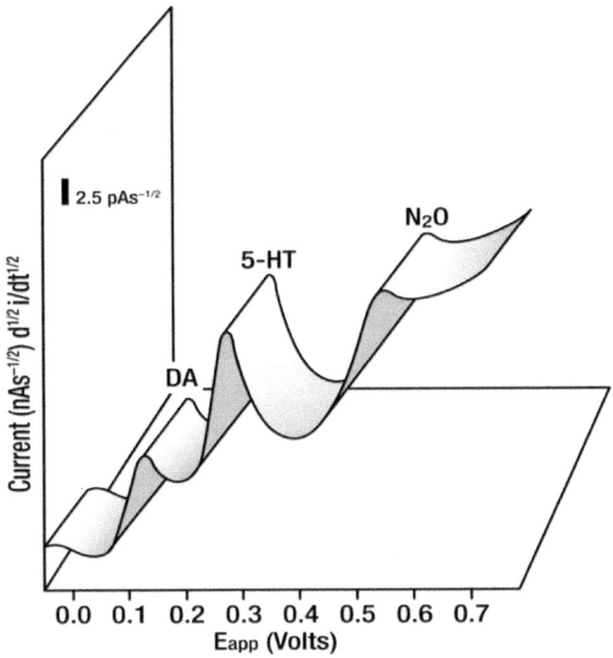

Figure 2 A representative recording showing the detection of DA, 5-HT, and nitrous oxide (N_2O) in saline/phosphate buffer by the BRODERICK PROBE® stearate sensor *in vitro*. The oxidation (half-wave) potential for N_2O is 0.53 V ± 0.02 V. Oxidation potentials for monoamines confirm previous reports [3–10]. Note that the sensitivity of the sensor for the indoleamine, 5-HT, is approximately two- to threefold greater than that for the monoamine DA *in vitro* as previously reported [3].

• Real-Time Data-Nitrous Oxide-*In Vivo*

Resected living tissue (hippocampal and neocortical) from temporal lobe epilepsy (TLE) patients was studied. Methods for patient classification and methods for delineating neurochemical profiles are published herein. Patients were administered nitrous oxide-oxygen anesthesia in a 40/60% concentration during intraoperative surgery. Figure 3 shows a representative recording from a TLE patient, *in vivo.* These images from TLE patients show reliable nitrous oxide signals that occurred at the oxidation (half-wave) potential of 0.53 ± 0.02 V, which is consistent with the detection of nitrous oxide at the same oxidation potential *in vitro*.

Figure 3 is shown below:

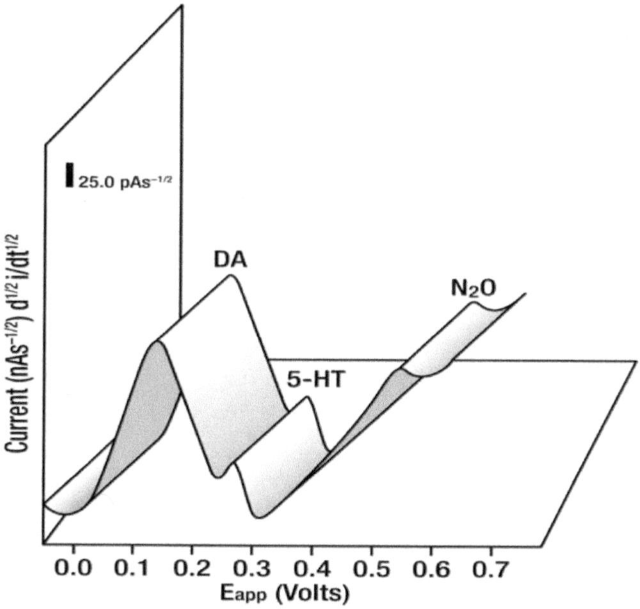

Figure 3 A representative recording showing the detection of DA, 5-HT, and nitrous oxide (N_2O) *in vivo* in a TLE patient, specifically an NTLE patient, a subtype of TLE that we found typically exhibits DA and 5-HT; the study was performed with a BRODERICK PROBE® stearate sensor. The oxidation (half-wave) potential for N_2O is 0.53 V ± 0.02 V. Oxidation potentials for biogenic amines confirm previous reports [3–10]. Note that *in vivo* data exhibit greater concentrations for catecholamines than for indoleamines.

• Selective Imaging of NO to N_2O using NMI

Therefore, further evidence for the reliable differentiated imaging of nitrous oxide from nitric oxide comes in this laboratory; the studies have shown the separate detection of nitric oxide (NO) at an approximate oxidation potential of 0.75 V [22]. These data are in general agreement with detection of NO using carbon fiber electrode [12]. Moreover, the reliable detection of nitrous oxide comes from the known predictive ability to detect oxygen at early negative potentials, again separating the detection of oxygen from that of nitrous oxide [26]. Finally, the stability of nitrous oxide at physiological temperatures is known [17]. Thus, these studies confirm the selective detection of nitrous oxide in the absence and presence of NO as well as in the presence of the monoamine neurotransmitters.

• Subtyping the Biomarker for Nitrous Oxide in Epilepsy Patients

Table 1 shows the presence of N_2O in resected temporal lobe tissue from epilepsy patients. *In Situ, en bloc,* in electroactive signals in the mesial temporal lobe patient (MTLE) as well as in the neocortical temporal lobe patient (NTLE), enabling the first subtypal neurochemical data of epilepsy patients according to the presence of the anesthetic neurochemical, nitrous oxide.

Table 1 Nitrous oxide (N_2O) in resected temporal tissues from human epilepsy patients

Patient number	Epilepsy type	N_2O signals imaged in:
2	MTLE	Neocortex (G)
3	NTLE	Neocortex (G,W); HPC (Granular cells)
4	NTLE	Neocortex (G,W); HPC (Polymorphic layer)
5	MTLE	Neocortex (G)
6	MTLE	HPC (Subiculum)
8	MTLE	HPC (Pyramidal layer)
9	NTLE	HPC (Pyramidal layer)
10	NTLE	Neocortex (G)
13	MTLE	Neocortex (G,W); HPC (Polymorphic, Pyramidal, Molecular Layers)
14	MTLE	Neocortex (G); HPC (Subiculum)
15	MTLE	Neocortex (G,W); HPC (Polymorphic layer)
16	MTLE	Neocortex (G,W)

Note: G is gray matter; W is white matter.

Table 1 shows the neuroanatomical location of nitrous oxide signals imaged in distinct neocortical neuroanatomic structures and hippocampal subparcellations. As expected, there was no apparent difference in the intensity of the signal imaging for nitrous oxide between NTLE and MTLE patients given the caveat that there were more MTLE patients than NTLE patients.

In MTLE patients 1, 7, 11, and 12, N_2O was below detection limits

Note: Figures and Table are previously deposited in the Prologue of Reference [20].

• Significance

In summary, the significance of imaging nitrous oxide anesthesia in living tissue is paramount in neurology, neurosurgery, emergency medicine, toxicology, substance abuse and dentistry [2]. Moreover, the **BRODERICK PROBE®** sensors image nitrous oxide signals on line yet separately from monoamines, metabolites, precursors. The data provide promise for optimizing such technology for selective nano- and micro-monitoring and measuring nitrous oxide intraoperatively.

This tiny sensing medical device, a carbon/lipid/phosphotidyl polymer, is experiencing quite the journey. From its acceptance into spectral analysis onto the pioneering nanotechnology of neuromolecular imaging, its presence in the nanobiotechnology world is a force with which to be reckoned. Then, from minimally invasive live imaging, onward to quantum mechanics in its non-invasive electroactive, photovoltaic, fiber optic, photosensitive, polymeric protein matrix, its promise for "personalized medicine" may well be unmatched. Stay tuned.

References

1. Archer WH (1944) Life and letters of Horace Wells, discoverer of anesthesia. *J Am Coll Dent* 11:81.
2. Bonson KR, Baggott M (2002) Emerging drugs of abuse: use patterns and clinical toxicity. In: Massaro, EJ, Broderick PA, Mattson JKL,

Schardein JL, Schlaepfer TE, editors. *Handbook of Neurotoxicology*, Vol. 2, Humana Press, Totowa, NJ, pp. 223–257.

3. Broderick PA (1987) Striatal neurochemistry of dynorphin-(1-13): in vivo electrochemical semidifferential analyses. *Neuropeptides* 10:369–386.
4. Broderick PA (1988) Distinguishing in vitro electrochemical signatures for norepinephrine and dopamine. *Neuro Lett* 95:275–280.
5. Broderick PA (1989) Characterizing stearate probes in vitro for the electrochemical detection of dopamine and serotonin. *Brain Res* 495:115–121.
6. Broderick PA (1995) Microelectrodes and their use in cathodic electrochemical arrangement with telemetric application. US Patent #5,433,710.1996; European Patent #90914306.7.
7. Broderick PA (1999) Microelectrodes and their use in an electrochemical arrangement with telemetric application. US Patent #5,938,903.
8. Broderick PA (2011) Introducing BRODERICK PROBE® biosensors clinically in epilepsy patients, *First International Congress in Molecular and Cell Biology, 4.1, Innovative Imaging Technologies*, Beijing, China.
9. Broderick PA, (2016) Noninvasive Electroactive Photonic Protein Sensor with Polymer Photovoltaic Optics for Memory Transduction Using Organic and Inorganic Elements as Platforms. *US Non–Provisional International Application No. PCT/US 16/68879.2020.*
10. Broderick PA, Pacia SV, Doyle WK, Devinsky O (2000) Monoamine neurotransmitters in resected hippocampal subparcellations from neocortical and mesial temporal lobe epilepsy patients: in situ microvoltammetric studies. *Brain Res* 878:49–63.
11. Chapman CR, Benidetti C (1979) Nitrous oxide effects on cerebral evoked potential to pain: partial reversal with a narcotic antagonist. *Anesthesiology* 51:135–138.
12. Crespi F, Campagnola M, Neudeck A, et al. (2001) Can voltammetry measure nitrogen monoxide (NO) and/or nitrates? *J Neurosci Methods* 109(1):59–70.
13. Gillman MA (1992) Nitrous oxide abuse in perspective. *Clin Neuropharmacol* 15:297.
14. Gyulai F, Firestone LL, Mintum M, Winter P (1996) In vivo imaging of human limbic responses to nitrous oxide inhalation. *Anesth Analg* 83:291–298.

15. Jevtovic-Todorovic V, Todorovic SM, et al. (1998) Nitrous oxide (laughing gas) is an NMDA antagonist, neuroprotectant and neurotoxin. *Nat Med* 4:460–463.

16. Malamed SF, Clark MS (2003) Nitrous oxide-oxygen: a new look at a very old technique. *J Calif Dent Assoc* 31:397–403.

17. May P, Cameron E (2017) Nitrous oxide: molecule of the month. June 1999 [Archived Version]. figshare.

18. Mennerick S, Jevtovic-Todorovic V, Todorovic SM, Shen W, Olney JW, Zorumski CF (1998) Effect of nitrous oxide on excitatory and inhibitory synaptic transmission in hippocampal cultures. *J Neurosci* 18:9716–9726.

19. O'Leary U, Puglia C, Friching TD, Kowey PR (1987) Nitrous oxide anesthesia in patients with ischemic chest discomfort: effect on beta-endorphins. *J Clin Pharmacol* 27:957.

20. Pacia SV, Broderick PA (2005) Bioimaging L-tryptophan in human hippocampus and neocortex: subtyping temporal lobe epilepsy. In: Broderick PA, Rahni DN, Kolodny EH, editors. *Bioimaging in Neurodegeneration*, Humana Press, Totowa, NJ, pp. 141–147.

21. Pacia SV, Doyle WK, Broderick PA (2001) Biogenic amines in the human neocortex in patients with neocortical and mesial temporal lobe epilepsy: identification with in situ microvoltammetry. *Brain Res* 899:106–111.

22. Rahni DN, Pacia SV, Broderick PA (2001) A novel microvoltammetric approach for the determination of nitrous and nitric oxides: human epilepsy. *Soc Neurosci Abstr*, Orlando, Fl.

23. Schreiber F, Wunderlin P, Udert KM, Wells GF (2012) Nitric oxide and nitrous oxide turnover in natural and engineered microbial communities: biological pathways, chemical reactions, and novel technologies. *Front Microbiol* 3:372. https://doi.org/10.3389/fmicb.2012.00372.

24. Thompson PL, Lown B (1976) Nitrous oxide as an analgesic in acute myocardial infarction. *J Am Med Assoc* 235:924.

25. Zuniga JR, Joseph SA, Knigge KM (1987) The effects of nitrous oxide on the central endogenous pro-opiomelanocortin system in the rat. *Brain Res* 420:57–65.

26. Radi R. (2018) Oxygen radicals, nitric oxide and peroxynitrite: redox pathways in molecular medicine. *PNAS*, 115(23) 5839–5848. https://doi.org/10.1073/pnas.1804932115.

Chapter 1

Laser Nanosensors for LIVE Imaging: A World of Polymers

From Restoring Eden, "the material creation is a fundamental of our worth." The creative mind never rests.

Nanotechnology is a futuristic creation, evolving in the present for relentless and significant place in the global market. Indeed, science and medicine may well eventually spin around the science of sensors and nanobiotechnology. The author is considered to be essentially the "one" that has pioneered the inventive art of nanobiotechnological devices within the mechanical and electrical engineering of the living human and animal mammalian brain while the subject is behaving in a variety of modalities such as walking. Actually, this inventive art has dramatically altered the way that science and medicine is viewed as such testimonials are published by formidable medical doctors and scientists. Creating within the realm of bionics and nanosensors linked to the electronic brain is rooted in the trademark, BRODERICK PROBE® *named after the author's father, Paddy Broderick. The generic nomenclature for the trademark is "microelectrode" and terminologies such as "sensor," "biosensor," "neurosensor," "photocell," "nanoprobe," "chip," and "nanochip" are used interchangeably throughout the monograph.*

Neuroimaging: Sensing Biochemistry in the Brain
Patricia A. Broderick

ISBN 978-981-4613-49-1 (Hardcover), 978-0-429-02974-5 (eBook)
www.jennystanford.com

1.1 The Background

The first chapter reflects a true discussion of several inventive patents [1–3, 5–8, 12] in order to enable a concerted view of these prodigious discoveries. Indeed, headway is made even through the historic pandemic of SARS-CoV-2 (COVID-19) and our attention to the science and medicine of (COVID-19) itself is clear as the tiny probe enables an interleukin-induced cytokine storm to be seen in its hippocampal havoc in walking subjects [11]. As has been noted throughout the globe, the work of the BRODERICK PROBE® represents the utter breaking of the mold of scientific design for science and medicine simply because the feat has never been accomplished heretofore. The applications are myriad! For example, the applications can be homeopathic, pharmaceutical, surgical, classically medical, novel medical or all of these and others. The integration of the patents delineating ground-breaking sensor and circuit designs include the inventive art of the BRODERICK PROBE® carbon allotropic polymers combined with photosensitive advanced materials such as fullerene [7] that facilitate other incredible inventions. For example, proton-driven Magnetic Resonance Imaging (MRI) technologies are supported by the nanoprobe [14]. Too, one of the author's latest patents includes the discovery of the living Tau peptide [8].

In the laboratory as well as in the clinical setting, images of waveforms from electrons transferred in redox reactions by the nanoprobe tell of the precise neurotransmitter milieu in the individual brain in its natural state, during its dysfunctional states and after therapeutic intervention in LIVE sequence. The Broderick laboratory video-tracks neuronal circuits within each healthy brain and brain disorder in a way never described heretofore. The emphasis is placed on the ability to see with the nanoprobe inside the healthy as well as the unhealthy living brain; real time same subject is used routinely as is not possible with previous technologies albeit their excellence.

The nanotechnology can be considered, in part, minimally invasive in regard to the instant surgery; it is not, however, invasive at all in regard to real time data processing which is LIVE and performed with tethered telemetry with formulas and applications thereof [1–3]. In the non-invasive design, moreover, the nanotechnology is entirely non-invasive. The sensor is a photocell external to the brain in the non-invasive design [5, 6]. Radio and/or optical frequencies of the electromagnetic spectrum enable human and veterinary use for the nanoprobe for personalized, point-of-care mechatronics. The nanosensor is simply placed near the skin external to the brain. Another patent in the portfolio [7] shows that the fullerene quantum nanoprobe is a semiconductor and fullerenes, as they embed in the fatty acids of the Broderick carbon/lipid phosphatidyl nanoprobe, are able to penetrate the lipid membranes in the skin surrounding the skull irrespective of lobe, whatever is needed. The Broderick photocell optically connects internal brain, skin and eye photocells for conversion of photons to electrons. The electrochemical potentiometric photonic probe, a polymer, works by the photoelectric effect.

In this chapter, in addition to covering all the patents, the ability of this nanotechnology to image the brain without opening the brain, that is, the noninvasive and contactless designs are highlighted. It is important to note that the noninvasive concepts and designs for the BRODERICK PROBE® microelectrodes/sensors for the noninvasive design are included in the Broderick 2016 non-provisional (Patent Cooperation Treaty (PCT) application [5]. It is noteworthy that neither the minimally nor the non-invasive sensor designs are invasive within the real time processing modality, Indeed, the sensor designs are known and accepted worldwide under the umbrella of the dual use of the BRODERICK PROBE® trademark that covers human and veterinary uses as separate entities.

The provisional application was filed on December 31, 2015. It was issued Serial No. 62/273,693 [4]. The non-provisional (PCT application), detailing this cutting-edge biosensor device, was filed on December 28, 2016, and was assigned International Application No PCT/US16/68879. The non-provisional (PCT application) is entitled "Noninvasive Electroactive Photonic Protein Sensor with Polymer Photovoltaic Optics for Memory Transduction Using Organic and Inorganic Elements as Platform" [6]. This application claims the benefit of U.S. Provisional Application No. 62/273,693, filed December 31, 2015, the disclosure of which is incorporated by reference herein in its entirety and is patented in the public domain, February 2017.

Thus, the nanoprobe continues to contribute to the way physicians and scientists think about the brain along with its diagnosis and treatment of disease. It is a nanobiotechnology that is already in marketing forecast and the tiny nanoprobe foresees the future in a novel LIVE Imaging molecular nanotechnology, named Neuro-molecular (NMI) in which electroactive neurotransmitters as well as enzymatic, non-enzymatic and non-electroactive neurotransmitters such as acetylcholine, aspartate and the like are handy for reading their signatures with this nanobiosensor.

Thus, the biosensor enters into a world of Emily Dickenson's, *Dream in Impossibility.* Applications reach far beyond this century and the lifetime of the author. The tiny probe may well be the dream of investors, philanthropists and humanitarians. Its timeless attributes for bringing beauty for our skin are discussed in Chapter 6 and sensing orgasmic auras with the nanoprobe are described in Chapter 11.

Take this journey with me.......

1.2 The Inventive Art: Potentiometric Polymers

The monograph brings to you the latest breakthroughs of the work; the marriage between the *electrochemical nature of the brain linking with the electrochemical nature of the tiny probe* is humanizing. The Latin language is a critical tool with which to study the sciences and medicine. Sensor comes from the Latin, *sentire* "to feel." Sensors have the attribute of feeling their surrounding environment to find things that are coupled, can be coupled or for that matter, should be coupled, In the field of electrochemistry, chemical, biological and physical sensors find use as transducers, meaning in Latin, "lead across," that is, at least, for one example, when an analyte is around. Responding to such things as heat, for example, taking one's hands off a hot stove involves sensors on a stove firstly to provide heat. Then, sensation of heat in the sensors of the hand signals sensors in the motor neurons of brain to respond; then signals are transduced reverting to muscles in the hand to move the hand as directed by sensors in the brain. Disaster is avoided by sensors. Sensors make sense.

Therefore, sensors are part of our daily lives and electron transfers from atoms (this is what sensors do) are natural, a part of nature and the fascinating part of this world of sensors is the fact that electrons are transferred back to the atom or to another atom. Nothing is lost! A valuable redox interaction, a redox reaction has taken place. The author sees that we are living redox reactions! The Broderick nanoprobes fix, ameliorate, see and feel, tell, and then put everything back in place, ever prepared for the next signaling event. We are left with the data we need in real time, online and LIVE.

In this inventive art, the potentiometric polymers respond to an *Electric Step Potential,* a change in the electric field, either manual or computerized and then, redox reactions in the brain from the Parkinson's medicine, L-DOPA, for example, change the plasticity of the Parkinson brain and neuronal motor signaling ceases trembling hands in place. "Relief" is the outcome. We are plastic; our brains are plastic/malleable. Dendrites on spiny interneurons disappear and regenerate and incredibly, proteins continue to fold and unfold unceasingly without regret and without our knowledge sometimes beneficially so and often to our detriment. Yes, the brain is plastic and the Broderick nanoprobe can and does image the complexities of multiple neuronal interactions; this nanoprobe is what we need to video-track neuronal integration in the natural physiological state as well as the neuronally degenerated state. The world is seeing the progress as the author is often interviewed internationally to tell of the nanoprobe. The sentiment is relayed among doctors, scientists, students and lay people alike. We need the probe in the market place; the healthy person also wishes to see inside their brains particularly in view of the upcoming wireless Broderick nanoprobe.

Further, the chapter includes critical theoretical and empirical inventive art of the resilient, original BRODERICK PROBE® carbon lipid phosphatidyl polymer in its journey from spectral analysis. The original is the electroactive voltaic polymer, a potentiometric polymer, a conductive polymer that has travelled smoothly from *in vivo* electrochemistry to neuromolecular imaging to the photovoltaic cell, a photovoltaic polymer. In its journey, it has been shown to have superior operational ability. Indeed, the nanoprobe promotes long-range electron transfer that can carry out redox chemistry at potentials near 1 V versus NHE. the normal (standard) hydrogen electrode. For example, the nanoprobe sees the peptide, Tau, separately and selectively at long range electron transfer with excellent spatial and temporal resolution [8–10].

The nanoprobe has many monikers, such as *potentiometric telemetric sensors*, organic and inorganic electrochemical transistors, polymeric nanoprobes, photocells, miniature optrodes, carbon sensors, protein sensors, carbohydrate sensors, chemiluminescent and photonic protein probes and rhodopsin fiber optical lasers. The Broderick sensor lives in the realm of polymer science which includes polymer chemistry, polymer physics, biophysics, material science and engineering. Products arising from the linkage of repeating units by covalent chemical bonds have been the primary focus of polymer science and the nanoprobe shares more than a similar focus. Carbon bonds with fluorine, oxygen and nitrogen (FON) in rhythmic numeric periodicity use and share equally and unequally their electrons in outer shells. However, it is the highly electronegative differences between carbon and FON that make for a dynamite nanoprobe.

Polymer science and molecular physics are part and parcel of the nanoprobe as carbon and silicon have four bonds to share and using opposite polarity for appeal, strong covalent bonds are forceful in changing the exciting likes and loves between the bonds that can change the nature of the human brain and hence, behavior. This is why we behave the way we do. The nanoprobe polymer of repeating units transduces a different brain into a different behavior. The nanochip can help the quadriplegic to walk, inserting the algorithm decoded for walking [4] with artificial intelligence (AI), machine learning, onto its interfacial surface on the nanoscale. The electroactive voltaic polymer gathers data telemetrically, catching points derived directly online, signaled by electron transfer to and from discrete brain synapses through a commutator while steps are tapped *via* carefully and strategically placed infrared photobeams or the like that map the motor pattern of the step.

The original polymeric carbon/lipid phosphatidyl nanoprobe, smaller than a human hair, is the inherent manufacturer of powerful inter- and intramolecular covalent bonds capable of changing the brain and its behavior without tampering with an action potential, that is, depolarizing a neuron unnecessarily. Such beneficial neuroplasticity is provided by the nanoprobe as the bionics of the nanoprobe polymer matches the bionic brain polymer, layer by layer in a scaffold-like and organized format. In other words, for one example, the nanoprobe images neuronal depolarization in the epilepsy person but does not cause neuronal depolarization in the epileptic brain.

Thus, both the original library of Broderick nanosensors and the new line of products are essentially not invasive, the new library of nanosensors are indeed non-invasive at the surgical level as well as the sensing level. Both generations of nanosensors are well suited to image neurotransmitter profiles LIVE with and without wires in healthy as well as neurodegenerative and psychiatric disorders in humans and animals such as in diseases of the motor and sensory systems, for example, the eye, brain, and skin. The sensing device enables the inside of the brain to be monitored without opening the brain as is conventionally required. In doing so, the inventive art provides substantial advancements in personalized medicine for diagnosis and treatment in, for example, disorders of Alzheimer and Parkinson with and without Lewy Body disease, epilepsy, stroke, prionopathies and other neurodegenerative and psychiatric brain diseases that occur and traumatize brain while inducing movement and cognitive decline in the process. This art enables LIVE imaging of neurotransmitters and by implication, the genes and electric field potentials associated with each neurotransmitter and neurochemical.

Another of the aspects of the instant inventive art relates to a photosensitive, potentiometric protein polymer sensing device that operates by converting photonic energy into electrochemical energy to image neurochemicals in real time. Let's add light as a stimulus for the nanoprobe!

1.2.1 How Does the Photonic Polymer Work?

Photons of light carrying the light-sensitive protein, rhodopsin, through ligand-gated channels in the skull and brain produce an electrochemical current based on the photonic light energy enabling the conversion of light energy to electrical energy and creating a photoelectrical current in units and subunits of voltage. This photoelectrochemical current is monitored using a potentiostat, spectrometer, and/or a spectroelectrochemical chemiluminesence device.

The inventive art relates to a miniature BRODERICK PROBE® carbon/lipid/phosphatidyl sensing polymeric chip or cell which is in another instant invention, a photocell, a photosensitive, electroactive polymer sensing device or nanosensor or nanoprobe. The patented polymeric photocell, a photovoltaic cell, operates by converting photonic energy into electrical and then electrochemical energy that generates a photocurrent in the brain without opening the brain. The output is provided in units of voltage. Laser diodes enable the electrochemical waveform to be seen as an electrochemical image in color separated into specific spectral frequencies. Thus, the photocurrent provides an image of neurochemicals in real time, noninvasively and/or minimally invasively. The embodiments provide a photoelectrochemical conductance device for nanodiagnostics, nanotherapies, and nanotheranostics.

The embodiments of this inventive art, whether with or without light, further relate to a memory transduction device, enabled by polymeric molecular bonding using what is aptly described as highly polar covalent hydrogen bonding, a covalent H-bond hybrid that forms intermolecularly as well as intramolecularly at least in part when the bonds are short and strong. In the example, $[F\text{-}H\text{-}Fl^{-}]$, when fluorine atoms are further away from the hydrogen atom, chemical hydrogen bonding takes over the competition.

Conventionally, one would not think of placing the word "covalent" alongside the word "hydrogen" in the armory of chemical bonds. However, current concepts of this bond, actually beginning about the time of the 1940's, are interesting. In fact, the lines between the covalent bond and the hydrogen bond are becoming less and less distinct. One becomes the other; hydrogen bonds become covalent bond as shown by spectroscopy and computer simulations; this phenomenon is shown in the following schematic, described in [13].

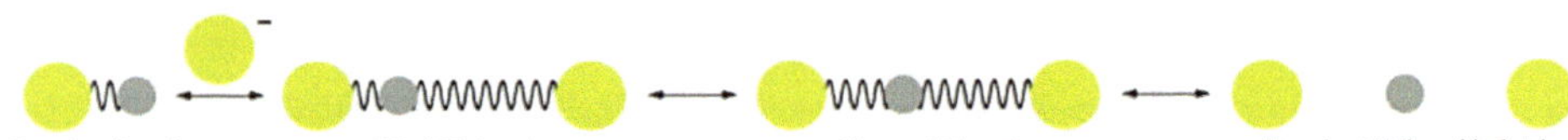

Figure 1.1 In introductory textbooks, hydrogen bonding is electrostatic, a special type of dipole-dipole attraction between molecules; it is not characterized as a covalent bond as bonded to a hydrogen atom but this new hybrid bond results from the attractive force between a hydrogen atom covalently bonded to a very electronegative atom such as carbon bonded to fluorine (F), oxygen (O_2) and/or nitrogen (N_2) (FON). In the current thinking, the terms covalent and hydrogen bonds can be spoken of as one entity when we think of highly electronegative atoms in the form of a covalent hydrogen bond hybrid [13]. The original carbon allotropic polymer is blessed with highly electronegative atoms throughout its molecular polymeric structure. *Symbols*: green, fluorine; gray, hydrogen.

Oh! The very thought of hydrogen actually bonding with two "other than" oxygen atoms and moving away from its electrostatic nature as its well-earned gold standard position in the molecule of H_2O may be anathema. But there was a student, named Huggins, who had some notes from his professor, Bray, and he thought this notion of a hydrogen bond forming from a covalent bond might be possible. Huggins even put a title at the top of his scribbling notes and Bray accepted the scrap paper as a term paper. Dr. Linus Pauling gave the notion a nod of assent while still going back to the molecular structure of the hydrogen bond and the covalent bond as separate. As the Nobel Laureate, Pauling, swayed back and forth around the conceptual dilemma, he wrote a double page letter to Huggins, gently warning Huggins to delay his publication about postulating a new hybrid hydrogen bond, just in case it was erroneous. There seemed to be no contention and Huggins published the work seemingly without recrimination from Dr. Linus. The author has read this letter from Pauling to Huggins with enjoyable interest; such is academia at its scholarly best in the era of the 1940's! Of course, the 1950's ushered in the intense competition concerning the nature of the chemical bonds of the deoxyribose nucleic acid molecule, DNA, Pauling was a part of this DNA endeavor as a master of the chemical bond but again, he appeared not to be cut-throat during the battle nor did the Nobel Laureate, Pauling, even attempt to learn X-ray crystallography.

But, hydrogen does, that is, bond as a hybrid, covalently to highly electronegative atoms such as, for one example, fluorine, and this likely operates, at least in part, within the Broderick nanoprobe formula in several places of its polymeric scaffold to enable the photocell to sense neurotransmitters LIVE. Relying on Dereka et al, [13] who have demonstrated the nature of short, strong H bonding using femtosecond, two-dimensional infrared spectroscopy in conjunction with quantum chemical calculations and their quantitative characterization of multiple coupled proton motions in the bifluoride anion [F-H-Fl$^-$], the nanoprobe unveils distinct prominent faces of this uncanny and unexpected transverse from the conventionally thought of electrostatic hydrogen bond to an unconventional short, bare and strong hybrid H-bond. Its fluorinated phospholipids and nitrogenous cerebrosides self-assemble into a potent probe that sees inside the brain even while the subject is walking.

The FON electronegativity with carbon is the operative concept. For example, PolyTetraFluoroEthylene (PTFE) (Teflon®)) is comprised of notable fluorine atoms-actually four. When Adams taught Broderick that Teflon® is inert, he was leaning toward referring to a complete lack of activity in the conductive capacity of Teflon® but Broderick thinks about those four fluorine atoms shared to complete the carbon four and the power of Teflon® cannot be denied. The nature of its inactivity lies in its insulative property of *poor* conductivity and *high* resistance to help provide a shell for the nanoprobe but the caveat that Broderick wants to share is that there is *some conductivity* in an insulator as opposed to the concept of a complete lack of electrical conduction. Therefore, electrical conductivity, for example, in the Broderick phosphatidyl voltaic polymer, semiconductor and photonic BRODERICK PROBE® nanoprobes, it is the nature of the fluorine bond in Teflon® that may elucidate electroactive signaling in its transition from insulator to conductor. The following is the author's rationale.

Since there's no or at least a low energy gap (Eg) between conduction band and the valence band in a conductor, no extra energy is used to transition between conduction band and valence band in electrical conduction states. The Eg is below 2 eV in a conductor. In the insulation mode, the eV is higher. Also, since the Eg is the difference between the bottom of the conduction band and the top of the valence band, the Eg in the semiconductor is higher than 2 eV, that is, the eV is higher but this extra energy that is needed for band transition in the semiconductor has the added advantage of better control of the electrical current at the p-n heterojunction. The electronegativity of the fluorine atom in Teflon® and PTFE as well as the O_2 and N_2 atoms can be used to manipulate energy to benefits needed for the Broderick carbon allotropic voltaic fullerene sensor and its photonic potentiometric polymer partner. In some cases, the electrical control enabled in the semiconductor state by FON bonding may occur within the electroactive voltaic nanoprobe.

Thus, chemical hydrogen bonds, hybrid H-bonds [F-H-Fl$^-$] and covalent bonds are highly likely to comprise the underlying mechanism of action of the Broderick polymer. Highly electronegative and polar properties are key. This advanced material in the world of sensors is endowed with both properties. It is further even more interesting when we think of the work cited in reference [13], that when the [F-H-Fl$^-$] hybrid is seen by vibrational potential in water, the hybrid disappears on H-bond bending whereas the hybrid begins with proton motion that is strongly coupled to the donor-acceptor vibrational stretching. The behavior is known as superharmonic vibrational frequency, likely non-linear.

Furthermore, unlike other spectral analysis and electrochemical art including what is called, *in vivo,* the embodiments include electrical devices as in mechatronics that enable the discovery of novel target compounds to diagnose and treat disease by using anion-cation exchange properties with electrochemical reduction voltammetry control. The embodiments also relate to brain sensing using optoelectronics, and producing novel nanomedicines and optogenetics to identify unique and/or continuous patterns of neurotransmitters and neurotransmitter-related genes for spinal and brain repairs that can be made luminescent.

Moreover, the art becomes brand new with the inventive art of the BRODERICK PROBE® fatty acid nanochips that are unsaturated. The teaching in the art pre-Broderick was to stay away from unsaturated fatty acids because of the nature of their chemical bonds. It was taught that unsaturated chemical bonds simply did not provide sufficient electroactivity as these fatty acids are endowed with double bonds. It was taught by Adams, Hubbard, Kissinger, Heinemann and more that unsaturated fatty acids are not viable electrodes for electrochemistry because of the nature of their bonds as these bonds are non-polar. Unsaturated fatty acids have some bonds to share, thus, disenabling their electroactive properties, it was taught. Broderick would like to share that it is the entire entity of the structural bonding of the molecule that matters. Broderick teaches that unsaturated fatty acids, albeit the bonds are non-polar, the bonds are electroactive and produce sensitive and operational stable sensors LIVE [1–3].

In fact, the art will utilize additional surface interfacial electroactive polymers such as proteins that are luminescent in addition to carbon allotropes that are photosensitive such as the "Bucky Ball" fullerene [7]. The disclosed art integrates cardiovascular systems with nanotechnology. For example, we can use inhibitors for serine proteases, such as thrombin, factor Xa and chymotrypsin because these enzymes that are proteinaceous that are light sensitive. In addition, this exciting art disclosed herein also includes a protein or protein induced photosensitive material made by hydrophobic bonds from phosphatidylcholine taken from the original carbon allotropic design incorporated into the protein, and by using hydrophobic bonds of the photolysed complex with protease from *Staphylococcus aureus* and trypsin, for example. The protein, albumin, can also be made photosensitive. Hence, the embodiments disclosed herein further include a method of making proteins luminescent and making other materials that are not proteins, luminescent. The inventive art disclosed herein includes chemiluminescent and materials that are or may not even be photonic but with the advanced materials made in our laboratory, light is emitted during catalysis when a second compound is catalyzed from a first compound during the reassembly of electrons.

1.3 The Work is Explained in Patent Language: A Pyramid Approach to Learning

The art comprises an electroactive photonic polymer sensing device, which includes an imaging device that converts photonic energy into electrochemical energy to generate a photocurrent that images a molecular parameter, wherein the imaging device includes at least one of a laser diode, photodiode, and photosensitive material; the art is a viable memory device that as each layer upon layer of the polymer intermolecularly bonds, a new molecular parameter is formed. The polymer is an advanced material that can comprise a vitamin, a lipid, a carbon allotrope and perhaps the highly electronegative fluorine in some of the designs that are comprised of carbon tetrafluoride and in its role as semiconductor, the advanced carbon tetrafluoride polymer further enables excited state proton transfer (ESPT).

The photosensitive material may include at least one of a protein, glycoprotein, protease, peptide, *amino acid, opsin, retinal, retinoic acid, retinol, rhodopsin, bathorhodopsin, lumirhodopsin,* metarhodopsin 1, metarhodopsin 11, carbon fullerene, lumirhodopsin, *trans*-retinal, purple membrane proton pump protein microbe from a class of helobacteria, eukaryote, prokaryote, carbon allotrope, and/or a vitamin. The memory device transduces a memory profile of a half wave of a molecule, and provides a peak signature associated with the molecule in ESPT. The electroactive photonic polymer sensing device includes a potentiostat monitoring photoelectrochemical current derived from the imaging device, and a spectrometer detecting and representing light energy derived from the imaging device. The laser diode may provide, for example, near-infrared photonic energy configured to extend the light through the brain and the imaging device includes at least one of a conducting and/or a semiconducting component. The conducting component includes at least a silica-containing material, piezoelectric material, and the semiconducting component could include tetrafluoromethane. The imaging device may include rhodopsin, the memory device may include carbon fullerene, and the rhodopsin and carbon fullerene may be disposed in a polymer shell [7].

1.3.1 The Optics

The BRODERICK PROBE® is a series of photonic, electroactive potentiometric sensing devices in accordance with embodiments disclosed herein and includes organic elements, such as, proteins, lipoproteins, glycoproteins, proteases, peptides, amino acids, and/or photosensitive biomaterials, such as, opsin, retinal, retinoic acid, retinol and rhodopsin, bathorhodopsin, lumirhodopsin, metarhodopsin, metarhodopsin and *trans*-retinal (free). Examples of organic photosensitive elements include purple membrane proton pump protein microbes used by archeae, from the class of helobacteria, as well as eukaryotes and prokaryotes, and organic photosensitive carbon allotropes. The sensing device can also include vitamins, such as, α-tocopherol, and/or lipids. In some embodiments, the sensing device can include organic conducting elements, such as carbon and/or microbes. In some embodiments of the art, the sensing device can include non-carbon conducting elements, such as silica containing materials, such as dioxosilane. In some embodiments, the sensing device can include inorganic semiconducting elements, such as tetrafluoromethane, and can further include organic and/or inorganic piezoelectric materials.

Photosensitive electroactive polymers of the disclosed embodiments enable molecular entities to be imaged through the skull and below the scalp without opening the skull. The polymer shell-based sensing device acts by using a series of intermolecular highly electronegative covalent, highly electronegative polar hydrogen/covalent hybrid bonds [F-H-Fl$^-$], chemical electrostatic, dipole-dipole hydrogen bonds and/or bonds that polymerize with thermal catalysis, albeit some non-polar bonds can factor into the equation. The sensing device tracks neurotransmitters, which are not limited to the movement of electrons as is detected conventionally in gels and petri dishes and the like and may use these but the BRODERICK PROBE® rather memorizes molecular entities while the subject moves, sleeps, and acts freely. The sensing device enables images to be generated of the natural, diseased, and surgically or medically treated states in real time of the same subject. The neurotransmitter that profiles the subject's outcome is seen by using a DC-to-AC converter or commutator coupled to a laptop, mobile phone, and/or computing device. The sensing device images anion-cation exchanges as electrically driven with a reduction circuit and can incorporate elements of near infrared, ultraviolet and ultrasound techniques using circuits.

1.3.2 Photovoltaics and Photodiodes

Photodiodes in the field of photovoltaics or electro-optics are important components of the BRODERICK PROBE® semiconductive potentiometric polymer, photonic potentiometric polymer and involve the production of volts from packets of photonic light energy.

Photovoltaics also involve the generation of a potential difference at the terminals of an unbiased semiconductor diode, which includes a p positive–n negative heterojunction as a result of illumination with light having energy higher than the Eg of the semiconductor. At the n junction of the semiconductor is the electron, proton sink, or quantum hole, in which the memory site is protonated after having received excess protons from the p juncture.

A schematic of a semiconductor design is shown in Figure 1.2. The schematic uses the semiconductor silicon for an example.

Semiconductor properties are known to behave as intermediates between conductors and insulators by using anode/cathode circuits that have an extra layer between the anode and the cathode. This extra layer is at an np heterojunction. A protonated part is the p layer or positively charged layer, which is also referred to as the protonated layer, and is formed by the addition of hydrogen bonds and protons. When photonic energy is present, the protonated positive layer passes protons to the extra layer between the anode and the cathode making the extra layer, the layer where the holes and spaces were, become positive. The anode is negative, the n layer, or nonprotonated layer, and the previous holes and spaces are positive and protonic. The hole comprises another charged particle as the electron is released. Charged particles can be electrons, holes, ions and electron-hole pairs. Electron-hole pairs are produced when the semiconductor absorbs energy enabling bands to undergo energy band transition. The circuit doesn't detect current when electron-hole pairs are compounded and insulation is the outcome. The peripheral circuit cannot see the current.

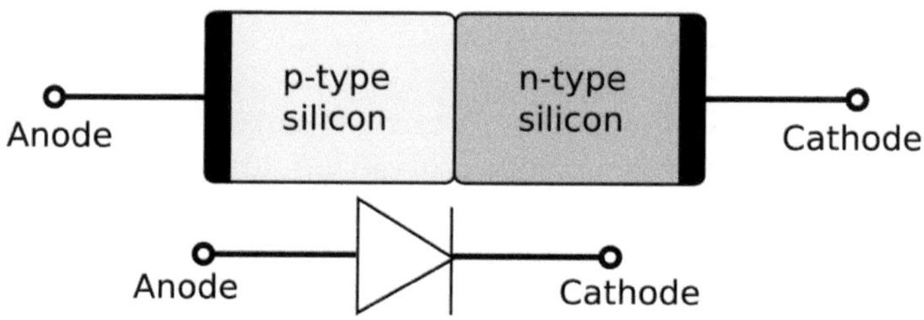

Figure 1.2 A schematic diagram showing the composition of a semiconductor. Two materials, p type and n type form a p–n junction inside the semiconductor wherein the p type holds an excess number of holes while the n type is comprised of excess electrons in the outer shells of electrically neutral atoms. One can manipulate a p–n junction to form a diode where the electronic action takes place. Credit: https://en.wikipedia.org.

The key to the BRODERICK PROBE® extra layer can be voltage-biased in either direction forward or reverse and is read as a photocurrent by using light pointed at any metric of the optical spectrum of the electromagnetic field including radio or radio telemetry described in the patent portfolio [1, 5, 6].

Thus, the BRODERICK PROBE® series of nanoprobes are semiconductor-based biochemical sensors with an electrolyte/insulator/semiconductor molecular structure. In the nanoprobe, nanosensor, nanochip, there is fullerene, a carbon allotropic semiconductor that is photoluminescent as well. Graphite is a semiconductor along with germanium and silicon also. Further, should the p-n heterojunction be biased on depletion, the width of the depletion layer is a function of the local value of the surface potential of the sensing device, the BRODERICK PROBE® neurosensor whether photoluminescent or not. When the nanochip is biased in reverse voltage, the depletion layer is enlarged. The local value of the bias voltage can be read out with AC photocurrent that is generated when a modulated light pointer is illuminated at any point along the carbon allotropic semiconductor. By illuminating parts of the surface of the nanochip with modulated light pointed at discrete parts of the semiconductor, additional charge carriers are generated and photocurrent forms; charge flows and current does not flow.

1.3.3 But, When Does the Photon Convert to an Electron?

Light produces additional charge carriers and these arrange and re-arrange as photons of light energy are turned on and off. ***It is the arrangement of the charge carriers that depends on voltage. The process is voltage-dependent! The photon converts to an electron; it is at this point and a photocurrent is created. The photoelectric effect happens in the depletion layer in the semiconductive* BRODERICK PROBE® *chip.*** The released alternating current (AC) is a photocurrent, due to a rearrangement of charge carriers in the depletion layer of the semiconductor, the nanoprobe and while the photonic light energy is turned on and off more and less conduction and its sequel, less and more resistance becomes its operative words! The resulting voltage and photonic energies are seen as coupled and yet, quantum calculated as correlated separate entities. The principles of holes and spaces and charge carriers are used to their utmost in the Broderick power-wireless radio optics inclusive or not using transition bands in PTFE as insulator, semiconductor and/or conductor.

More accessible power for stable operative ability is available with the nanoprobe without unnecessary interruption that occurs in high frequency harmonics. The Broderick nanochip is low resistance enabling photoelectric effect with fast band transition. Graphite/Carbon is a semiconductor along with silicon and germanium. The **electron–hole pair** is the fundamental unit of generation and recombination in inorganic semiconductors which comprise some of formulations of the BRODERICK PROBE®. In the case of inorganic semiconductors, an electron transitioning between the valence band and the conduction band where generation of electron is a transition from the valence band to the conduction band and recombination leads to a reverse transition.

What this means for the rhodopsin retinal sensor or nanosensor is that the part that the rhodinone catechol protein plays in the sensor represents another advancement for the nanoprobe's advantage over conventional techniques as it relates to ESPT for its optical voltaic memory properties. The advances provided by ESPT are faster by a factor of 1 over the factor of atomic observation for electron transfer without light ranging in the approximate range of 10^{-4} to 10^{-8} of a joule in the anode cathode semiconductor in contrast with non ESPT-based conventional techniques. Accordingly, the disclosed embodiments enable critical improvement over previous electrochemical and spectrophotometric devices, which include live imaging without opening the skull and the use of ESPT.

The neuroprobe/nanosensor signals from the brain are remembered by phototransduction of light photons, conversion to photovoltaic energy, which is further converted to photoelectrochemical imaging, and visualized by collecting photonic waveforms by a spectrometer, and collecting electronic waveforms potentiometrically or by galvanic means. Online half waveform signaling, the LIVE visualization of molecular memory can be rectified from charge flowing from voltaic optical currents in one direction to alternating current for reading profiles.

1.4 Placing Avant-Garde Inventive Art Patents Together

1.4.1 The BRODERICK PROBE® is a Voltaic Polymeric Cell

The following drawing is provided by way of example only and without limitation, wherein like reference numerals (when used) indicate corresponding elements throughout the several views.

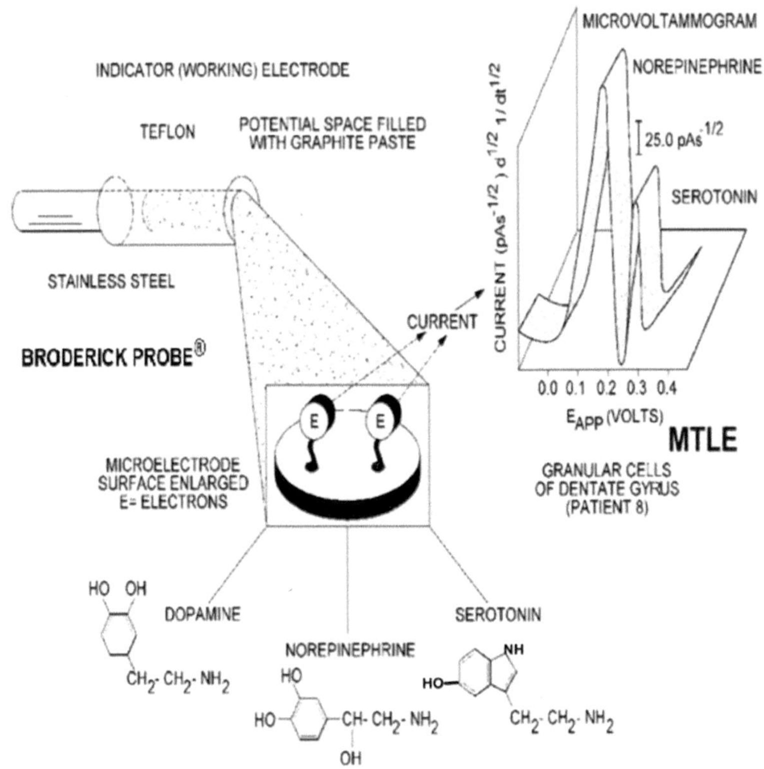

Figure 1.3 Operation of a Broderick probe in detecting neurotransmitters and a scan resulting therefrom. Results obtained with the BRODERICK PROBE® carbon-lipid-phosphatidyl polymer were in excellent agreement with those obtained by standard gas chromatography [1].

NMI scans can be called microvoltammograms.

1.4.2 The BRODERICK PROBE® is a Photovoltaic Polymeric Cell

The following drawing is provided by way of example only and without limitation, wherein like reference numerals (when used) indicate corresponding elements throughout the several views.

Figure 1.4 Shows an embodiment of the BRODERICK PROBE® wherein the semiconductor carbon allotrope, fullerene is embedded in its center [7].

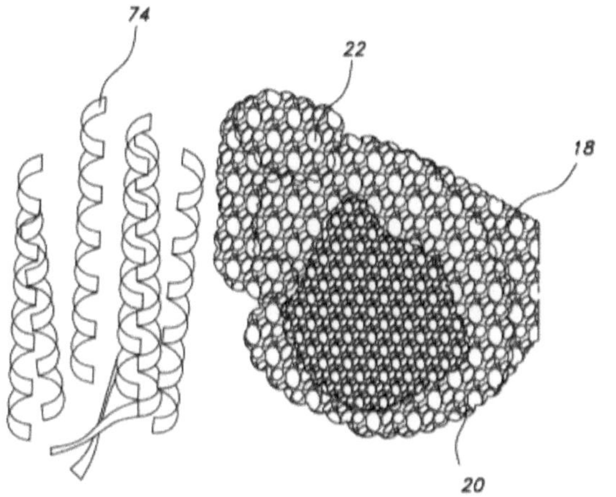

Figure 1.4 Shows a neuroprobe or nanosensor in accordance with one or more of the disclosed embodiments. In some embodiments of the art, the photocell which comprises a sensing device, includes a polymer shell 20 shown in Fig. 1.4 above. For example, the sensing device can include a carbon fullerene (18) and an opsin (22) (e.g., rhodopsin) (74) in the polymer shell (20). The polymer shell (20) can include several lipids and/or oils. The sensing device transduces a memory profile of the half wave of a molecule, which produces a peak signature for each target molecule in an excited state proton transfer (ESPT). The memory of the target molecule is embedded within the spaces/holes of the depletion layer in the semiconductor sensing device and this template is trained to memorize anions and cations for diagnosis and treatment.

Memory in the BRODERICK PROBE® polymeric photocell, a protein rhodopsin sensor uses the photoelectric effect, which activates the photovoltaic effect by translating photonic potential energy to voltage. Rhodonine, a constituent of rhodopsin, is a known semiconductor, and may not be photosensitive, but photoreceptors in rhodopsin can be used to image photonic light. (Near infrared may be safer than ultraviolet for humans and animals). Rhodopsin, when activated by near-infrared or photonic light, acts as a semiconductor by ESPT. The color change in rhodopsin shows that when subtypes of rhodopsin convert, rhodopsin bleaches and changes color from its various states. For example, rhodopsin can change from being bleached to a green, to a red, and then to a black when the molecule is no longer in an active state. In this way, signaling occurs and memory sites, such as catecholamines and glutamates, for example, glutamines, are detected in the brain, on-line, or *in vivo* within nanoseconds.

ESPT, which provides an improvement over conventional techniques, enables the embodiments of the neuroprobe/nanosensor to operate by reassembling highly polar hydrogen bonds to substantially improve the imaging of targets in the brain to diagnose patients and animals. The brain includes sites for charges, and neurons are charged cells; much like an electronic device using anode/cathode circuits, the neurons in the brain, body, and blood are electronic devices using anode/cathode circuits. The neuroprobe nanosensor and a neuron act as a lock and key to anion cation exchanges using semiconductor properties. Thus, light enables semiconductor properties of the rhodopsin rhodonine nanosensor to chemically and magnetically bond with the semiconductor properties of the neuron.

1.4.3 The BRODERICK PROBE® Laser Diode Circuit-Quantum Mechanics

One can manipulate a p-n junction to form a diode where the electronic action takes place such as what the Broderick probe does.

It is well known that both the forward-bias and the reverse-bias properties of the p–n junction demonstrate the p–n heterojunction as a diode. A p–n junction diode allows electric charges to flow in one direction, but not in the opposite direction; negative charges (electrons) can easily flow through the junction from n to p but not from p–n, and the reverse is true for holes. When the p–n junction is forward-biased, electric charge flows freely due to reduced resistance of the p–n junction. When the p–n junction is reverse-biased, however, the junction barrier (and therefore resistance) increases and current is blocked but not likely entirely.

The Broderick laser diode provides near-infrared photonic energy configured to extend the light, for example, Angstroms through the brain and the imaging device includes at least one of a conducting and/or a semiconducting component. The conducting component includes at least one of a silica-containing material, piezoelectric material, and the semiconducting component could include tetrafluoromethane. The imaging device may include rhodopsin, the memory device may include carbon fullerene, and the rhodopsin and carbon fullerene may be disposed in a polymer shell [5–7].

The Physics of Semiconductor Devices is described by Professor M. Fischetti 201D Marcus Hall, Department of Electrical and Computer Engineering-University of Massachusetts Amherst, MA 01003 Spring 2010.

> "Classical Mechanics: F = ma from an unusual perspective: Quantum Mechanics is the background of semiconductor physics: Semiconductors operate thanks to the existence of 'energy gaps' whose existence stems from the wavelike nature of electrons; many modern devices, such as junction (injection) lasers, quantum cascade lasers, resonant tunnel diodes principles of operation are based on basic quantum properties of electrons" [15].

The nanoprobe uses but is not limited to a laser diode circuit.

The following drawing is provided by way of example only and without limitation, wherein like reference numerals (when used) indicate corresponding elements throughout the several views.

Figure 1.5 is a block diagram of a miniaturized near-infrared laser diode and photodiode circuit for use in accordance with one or more of the disclosed embodiments. An Optical spectral analyzer with photodiode or photodetector is a non-limiting embodiment.

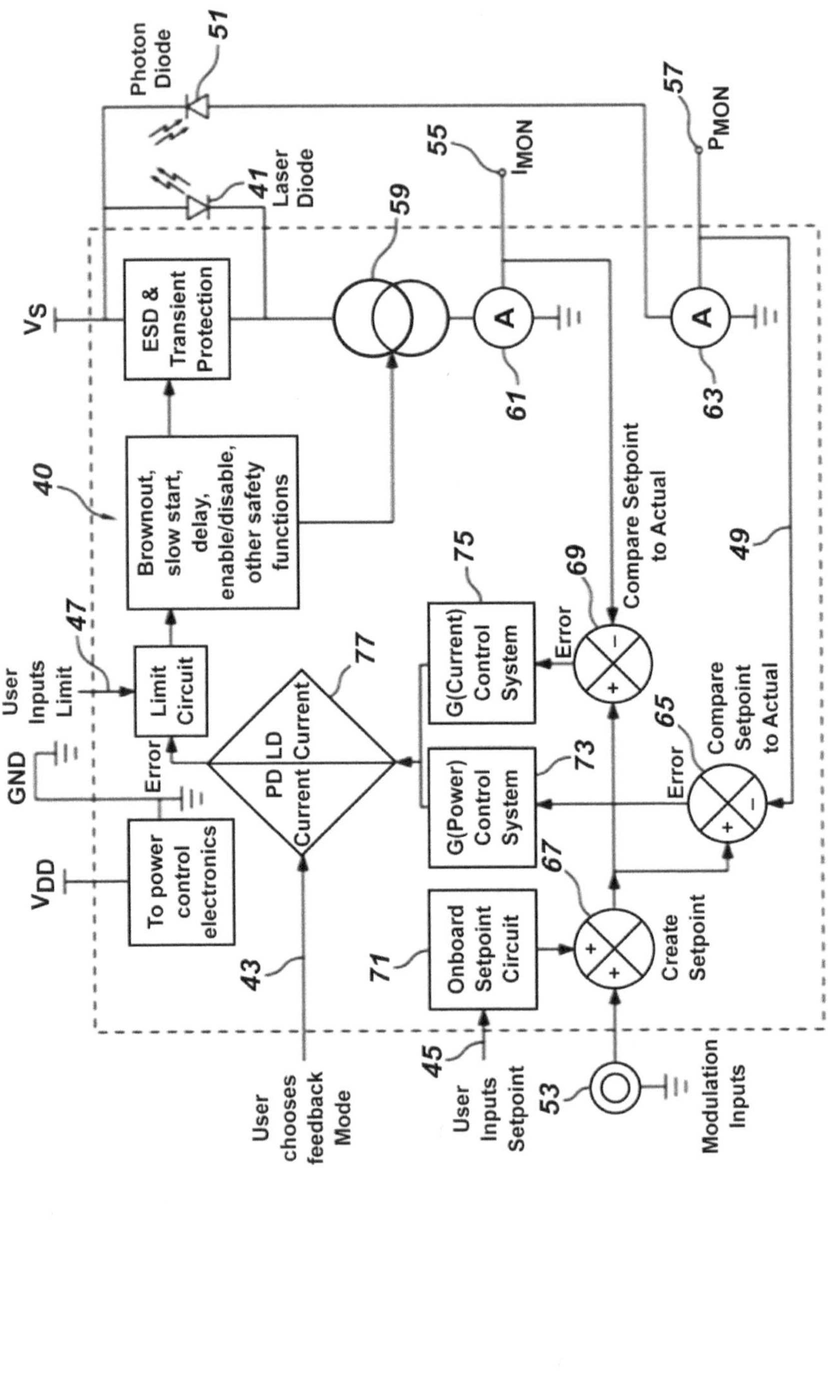

Figure 1.5 Shows the dual laser circuit for at least one use of the nanoprobe.

1.5 The Neuroelectronic Devices Are the BRAIN and the BRODERICK PROBE®

Memorization of molecular entities and concentrations takes place by one of a covalent H-bond-hybrid for one example and the memory of the half wave potential is further embedded within the neuron by ESPT. A newly formed sp^3 valence layer enables movement and reassembling of hydrogen bonding, which is highly polar covalent to recognize and signal to the anode cathode counterparts in the neuron. Each reassembly of bonds forms a memory of the molecule under study, and with continuing pulses of light, the resulting half wave signal is memorized and memorized data is embedded in the sensor by continuing pulses of light, thereby forming long term memory.

Memory is enabled in the Broderick neuroprobe/nanosensor by photonic reassembly of carbon, oxygen, fluoride (such as carbon tetra fluoride), and nitrogen bonds in amino acids, peptides and proteins embedded in cationic and anionic sites in the protein rhodopsin sensor of the fiber optic neuroprobe operatively coupled with the laser diode. Charge sites in rhodopsin are identified and sites are memorized by repeated and/or continuous pulses of photonic light from the near-infrared laser. The memory of charge is embedded in rhodopsin of the nanoprobe and the nanoprobe links to the brain neurotransmitters, such as dopamine, serotonin, aspartate, and glutamate.

The disclosed embodiments of this inventive art relate to live imaging of neuromolecules directly related to brain, body and blood disorders, preferentially in mammalian patients and animals, through the periosteum covering the scalp below the skull that surrounds the brain's cranium. These embodiments are noninvasive, as molecules are imaged live without opening the skull. The inventive art uses a series of polymeric protein sensors that exhibit photosensitive and electroactive properties. A photosensitive and electroactive protein, such as rhodopsin, is used with and without its chromophore, retinal, and includes a photovoltaic semiconductor, which is a rhodopsin complex catechol protein, referred to herein as rhodonine, which is derived from opsin as is rhodopsin.

Photons of light, in the near-infrared spectral band, pass through fiber optic materials of glass or plastic in the form of photonic bundles of quantum light energy. These fiber optic cables contain quanta photons of energy, and are electrically connected to a series of laser diodes, which are powered by a low wattage designed to reach into the neuroanatomy of the brain, such as the hippocampus and cortex of, for example, an epilepsy patient, which may be intractable to conventional drug treatments. Rhodopsin is in the core of the fiber optic material connected to the laser diodes.

Neurons are dynamically polarized cells that are responsible for electrochemically transmitting information throughout the nervous system. Scientists have characterized hundreds of neuron types based on location, morphology, and gene expression. In general, there are three main types of neurons: motor neurons, sensory neurons, and interneurons. Despite great variability in size and shape, most neurons share common morphological features including a cell body, dendrites, axon, and axon terminals.

Near-infrared energy [5, 6] activates photoreceptors in the rhodopsin retinal sensor, which is in the core of the fiber optic of the laser module. When excited by light, polar covalent hydrogen bonds of carboxylic acid and amine moieties in rhodopsin retinal react to produce either positive or negative neuron proteins, thereby forming memory in the nanosensor and the target, for example, the neuron, enabling the photoelectromechanics, a memory chip that mimics the function of memory in animals and humans in the neuronal circuit of the memory neuroanatomic circuit.

Molecules correlate with frequency, which is the number of cycles of photonic energy absorbed during a given period of time. In rhodonine complexes of rhodopsin, chemical bond transformations occur in a donor acceptor region of the semiconductor, rhodonine. In both examples, light induced chemical bond transformation creates, rearranges, polymerizes, catalyzes, and cleaves covalent hydrogen hybrid and hydrogen chemical bonds for one example and covalent hybrid hydrogen fluorine nitrogen bonds to manipulate molecular signals to one or more protonated or deprotonated hydrogen bond memory sites. The brain and behavior change according to bonds in brain polymers and the BRODERICK PROBE®.

Memory sites in the neuron formed by near-infrared include protonated electrons in the rhodopsin retinal sensor, such as in metarhodopsin 1 formed from rhodopsin and converted to metarhodopsin II, creating embedded cationic sites, whereas, in halorhodopsin, in its chemical transformation to lumirhodopsin, anionic memory sites are formed and memorized by a quantum of photon packets induced by light. The neuroprobe provides memory by bleaching or changing color.

Rhodopsin is a protein polymer that absorbs blue green light and appears reddish purple during chemical transformation of memory sites embedded in the protein matrix of the neuroprobe. Photosensitive electroactive polymers with blue light emitting laser/rhodopsin/rhodonine photodiodes using one or more photon lasers in the near-infrared spectrum are described herein, but are not limited to the specific examples described herein. Rhodopsin is bonded covalently to lysine amino acid of 11 *cis* retinal, preferably over a *trans* isomer of retinal in the rhodopsin retinal complex. Phototransduction of the memory function through light induced chemical bond transformation signals specific spectral and electrochemical waveforms at specific wavelengths and half wave potentials by anion/cation exchange. The generation of potentials is visualized by color changes in the fluorescent protein. Waveforms and redox waveforms are seen on a spectrometer and a potentiostat on-line via, for example, a personal computer or cell phone.

Thus, the neuroprobe remembers by performing the function of memorizing the signaling process through autotrain with continuing flashes of light. Although each autotrain may include a small amount of photonic energy, that is, high voltage energy at a number of volts through the photovoltaic effect, the final memory process in terms of quantum packets of energy is substantially more effective than each flash of light.

Keys to the memory function in the nanoprobe are the polymer bonds formed during photon absorption and electron adsorption between and within the **electronic devices, the nanoprobe and the brain**.

The following figures further detail the inventive art of non-invasive NMI. Figures 1.6 and 1.7 are provided by way of example only and without limitation wherein like reference numerals (when used) indicate corresponding elements throughout the several views.

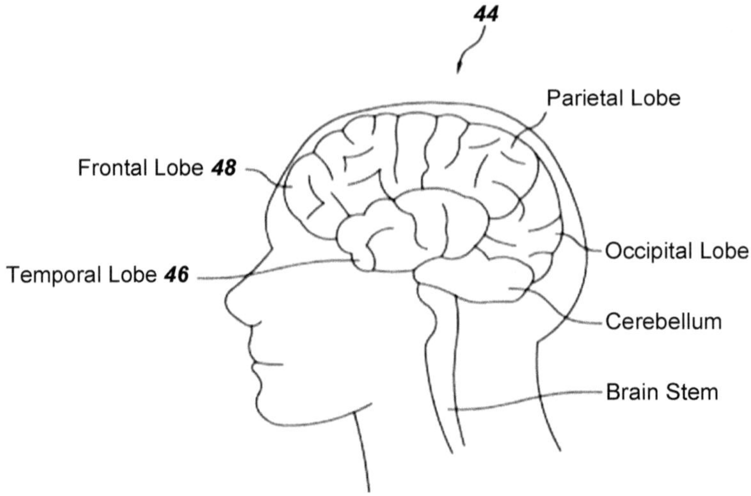

Figure 1.6 Shows a patient's skull and internal channels in the brain.

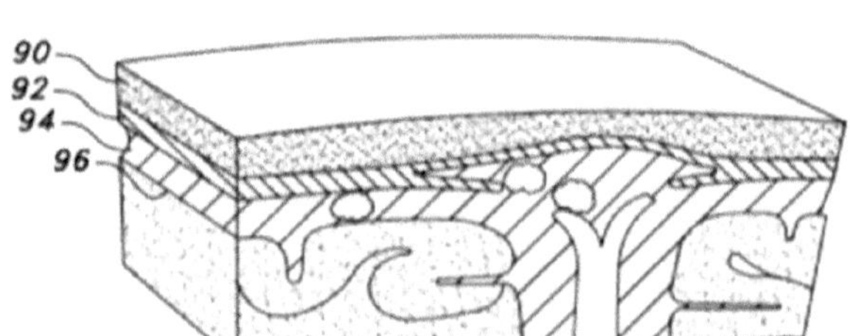

Figure 1.7 The following drawing is provided by way of example only and without limitation, wherein like reference numerals (when used) indicate corresponding elements throughout the several views shows a cross-section of a portion of the patient's skull and brain.

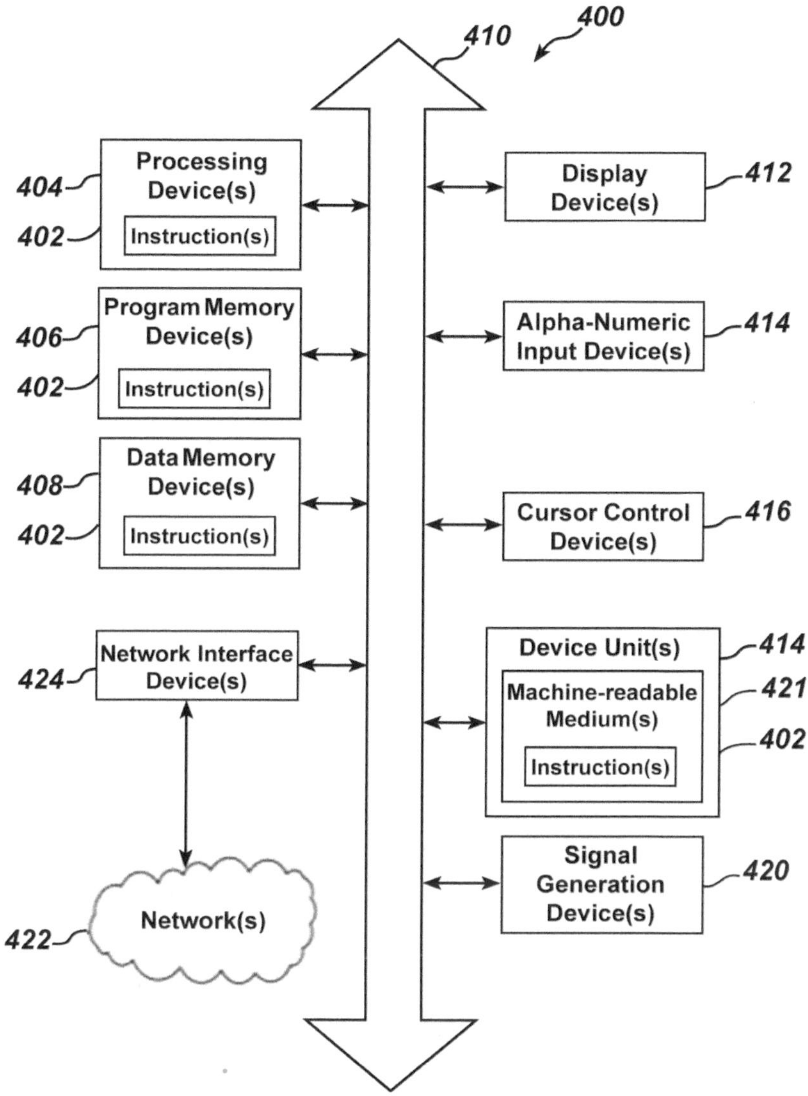

Figure 1.8 Shows elements of a non-limiting brain-computer interface to be used with the Broderick nanoprobe. The figure is a block diagram of at least a portion of an exemplary machine in the form of a computing system that performs methods according to one or more embodiments disclosed herein.

1.6 Photoacoustic Piezoelectric Innovations in the Nanoprobe: BRODERICK PROBE®

A photoacoustic piezoelectric device acting on a temporal bone or tympanic membrane is activated to record the response of a photonic sound pulse in decibels. A miniaturized speaker is mounted above and behind the ear of the patient, which provides the background noise and/or stimuli. Responses are recorded at, for example, 200 ms intervals, after presentation of the stimulus. The movement responses are transduced by a piezoelectric accelerometer. The patient's seizure movement from the decibel stimulus generates an electrical current in the piezoelectric nanoprobe that distinguishes firing from non-firing neurons. The nanoprobe detects the pressure pushing down on the patient's ear by photonic pulses on the piezoelectric ceramic platform due to the eliciting stimulus in decibels. Some of the disclosed embodiments are designed to minimize extraneous noise and vibrations to attenuate influences from external stimuli.

Photoacoustic responses to auditory stimuli *via* muscle movements transduced electrically use piezoelectric materials in epilepsy patients and patients with neurodegenerative diseases and/or psychiatric disorders and drug abuse. Data is recorded as a potential in voltage and concatenated for compatibility using, for example, spreadsheet software. The maximum voltage, V_{max}, is used in calculations, which represents the highest voltage detected during the response window, which represents the peak value of the response.

The following drawing is provided by way of example only and without limitation, wherein like reference numerals (when used) indicate corresponding elements throughout the several views.

Figure 1.9 shows the operation of the nanoprobe in accordance with one or more of its disclosed embodiments. Note that the electroacoustic application of the BRODERICK PROBE® is shown at the bottom of the figure and is marked with the number 32. The BRODERICK PROBE® fullerene polymer encased in another carbon allotrope is shown at the top of the diagram. A schematic of rhodopsin for the neuroprobe is numbered 74. Online real time scans imaged from this novel voltaic polymer during a LIVE series of nimodipine studies for post-operative memory loss are shown in number 30.

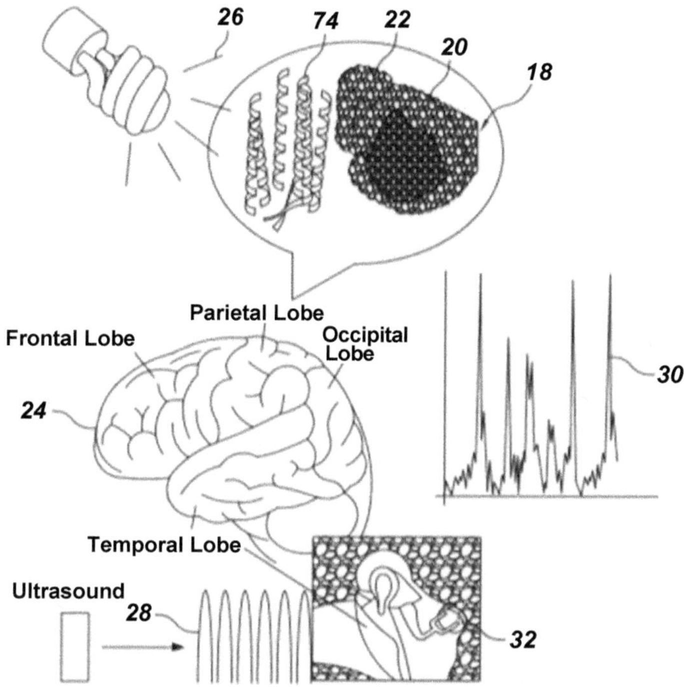

Figure 1.9 The operation of the nanoprobe in accordance with one or more of its disclosed embodiments.

The reader is advised and indeed encouraged to view all the patents in the Broderick patent portfolio and indeed the reader is encouraged to network and participate in investments for these needed medical advances. The reader is indeed encouraged to contact the author at broderick@med.cuny.edu.

1.6.1 Principles of Electroacoustics and Piezo Resonance Effects

The basic principles of acoustics measurement and piezo effect are as follows. The conversion of electrical pulses to mechanical vibrations and the conversion of the returned mechanical vibrations back into electrical energy is the basis for ultrasound testing. The transducer converts the electrical energy into acoustic energy. Piezo material is polarized material (i.e., some parts of the molecule are positively charged, while other parts of the molecule are negatively charged) with electrodes attached to two of opposing faces of the piezo material. When an electric field is applied across the material, the polarized molecules align themselves with the electric field, resulting in induced dipoles within the molecular or crystal structure of the material. In piezophotonics, semiconductors are important and relevant to the disclosed embodiments, as is photoacoustics in conjunction with piezo semiconductor properties.

The active element of acoustic transducers is typically ceramic, but can include other materials, such as cadmium and titanium. Piezoelectronics include semiconductors for imaging anions and cations in patients and animal brains. Pressure of the movement from the seizure in the epilepsy patient, for example, changes ultrasound photonic energy as changes from firing to non-firing neurons transduces acoustics in decibels to volts enabling sleep materials and alert systems for treatment of epilepsy patients because the tympanic membrane is a muscle and thus, mechanical energy is provided by movement of the membrane with and without seizure.

Therefore, the disclosed embodiments can distinguish between epilepsy and non-epilepsy while also providing an alert system for the patient and animal during aura period, usually by the smell of burning rubber. Sensor photonics with the BRODERICK PROBE® provide a substantially advantageous technique to study, diagnose, and avoid danger during seizure and aura including non-genital orgasmic aura Disclosed embodiments include pattern profiling of a patient's need for personalized medicine, brain printing or an imprinting system for video-tracking neurotransmitters in real time and providing diagnosis, strategies for treatment, and solutions to underlying disorders previously unable to be reached or ascertained. Disclosed embodiments also provide a danger-free zone for epilepsy patients and animals.

Piezoactuators are electromechanical transducers that convert electrical signals into a mechanical displacement to regulate control systems and vice versa. The conversion process is the focus of actuator technology. Electric field strengths of about 2000 V/mm are used for an actuator deflection of 1.4 to 1.7, for example.

Electroacoustic scans are called sensograms.

1.7 Telemetric *In Vivo* Electrochemistry: Real Time Data Processing: The Path to Successful Patenting of Sensing Devices: Approved-Assigned-The Albert Einstein College of Medicine-The City University of New York

Let's look at what happened around the time of Patricia's first patent [1]. These are the facts, the public records. ***The Ironic Twist!***

Patricia's first patent application was approved by Dr. Dominick P. Purpura, MD, Dean, the Albert Einstein College of Medicine, Montefiore Hospital, Yeshiva University. The letter of approval reached the author, dated July 3rd, 1985, from the Chair of the Patent Committee, Dr. John P. Wexler, MD, PhD. The letter reads:

> "The Dean has approved that the college assume the responsibilities for patenting your disclosure." The patent disclosure number is P-124; the patent is entitled, "Novel method for measuring biogenic chemicals using in vivo electrochemical means."

The letter was addressed to the author at the Department of Psychiatry, the Albert Einstein College of Medicine.

Somehow, somebody at the Albert Einstein College of Medicine had Patricia's back! You see, Patricia had already accepted a position at the Cornell University Medical College, The Burke Rehabilitation, Department of Neurology, Dementia Research Center.

Dr. Peter Davies, The Albert Einstein College of Medicine (AECOM), facilitated this new opportunity for the author's work. Patricia set up her new laboratory and continued to work diligently on the neuronal brain, intravascular, pulmonary and cardiovascular applications of the Broderick probe. Drs. Gibson and Blass served as the principle scientists at the site. Patricia served as Post-Doctoral Research Associate.

The author's response to Dr. Wexler, as paraphrased was, "I will send all requested materials needed to move this approval forward right away." Patricia did so. Patricia speaks respectfully of Drs. Purpura and Davies *in memoriam*.

On October 3rd, 1985, a letter from Dr. Eugene Moroz, PhD, Esq., *Morgan, Finnegan, Pine, Foley and Lee* was forwarded to the author from Morton Amster, Esq., of Rothstein and Engleberg, copied on Dr. Wexler's letter.

The letter reads:

> "I am presently reviewing the agreement sent with your letter of September 18th, 1985. Since reference is made in the agreement to AECOM's Official Policy on Patents and to the Broderick Disclosure Data Sheet, neither of which have I seen, will you please forward copies of those documents to assist me in my review."

Once again, Patricia said yes and performed the necessary tasks to be done.

Relevant to the conversation: Dr. Hadassah Tamir, PhD, *Columbia University College of Physicians and Surgeons,* had contacted Patricia and Dr. Moroz to set up the initial meeting that was to change Patricia's life and the world. Hadassah protected the BRODERICK PROBE®.

John Dugan Esq, *Chairman of the Patent Committee, The City University of New York,* met with Patricia to discuss the transfer of her patent from AECOM to CUNY. Of course, with my Associate Professor position on tenure track, it made sense to approve such a transfer. On September 2, 1988, Patricia met with Lawyer, Dugan and was asked by John Dugan, Esq to send all relevant materials to him right away. Patricia did so. The author's first patent was issued in 1989 [1].

1.8 Summarizing

Quantum mechanics, about which, it is said, "angels fear to tread," comes to the forefront of the penultimate and ultimate discoveries in the Broderick brain/nanoprobing set of systematic formulas and devices. Who benefits? Applications begin with the natural brain and spinal cord and proceed to diagnostic therapy in brain, body and blood disorders in animals in the veterinary world, patients in the clinical world and yes, plant physiology in the realm of the plant kingdom as well. Nothing can be left out! It is here for you to imagine, dream, act on and use these patents and publications for your springboard. Let us imagine, dream and discover. Let us image.

But what happened early on in the advent of electrochemistry into spectral analysis caused a tech enmity unmatched in the history of scientific inventive art. Should this have happened in today's world, financial investments and legal ease would have flooded the multisided fierce battle and both sides would have their say despite the press releases that spoke of the inventions as a dueling episode. At the time spoken of herein, though, it caused, instead, suffering and pain for the author scientist. With finances funneled unilaterally in the direction of the most vocal, the author quietly invented and invented and proved and proved.

Thus, Patricia moved forward with gusto to the proof of concept stages for the BRODERICK PROBE®.

References

1. Broderick, PA (1989) Cathodic Electrochemical Current Arrangement with Telemetric Application. *US Patent Serial No. 4, 883,057.* Issued.
2. Broderick PA (1995) Microelectrodes and their Use in Cathodic Electrochemical Current Arrangement with Telemetric Application. *US Patent Serial No. 5,433,710.* Issued.
3. Broderick PA (1999) Microelectrodes and their use in an Electrochemical Arrangement with Telemetric Application. *US Patent Serial No. 5,938,903*, Issued.

4. Broderick PA (2013) Neuromolecular imaging shows temporal synchrony patterns between serotonin and movement within neuronal motor circuits in brain. *Brain Sciences* 3:992–1012.
5. Broderick PA (2015) Noninvasive Electroactive Photonic Sensor with Polymer Memory Transduction Using Organic and Inorganic Elements as Platforms. *US Provisional Application No. 62/273,693.* 2017.
6. Broderick PA (2016) Noninvasive Electroactive Photonic Protein Sensor with Polymer Photovoltaic Optics for Memory Transduction Using Organic and Inorganic Elements as Platforms. *US Non-Provisional International Application No. PCT/US 16/68879.2020.*
7. Broderick PA (2021) Identification, Diagnosis, and Treatment of Neuropathologies, Neurotoxicities, Tumors, and Brain and Spinal Cord Injuries Using Electrodes with Microvoltammetry. *US Patent Serial No. 10,980,460*, Issued.
8. Broderick PA (2020) Identification, Diagnosis, and Treatment of Tau-Associated Pathologies Using Electrodes. *US Provisional Application No.* 63/130,453. December 24th.
9. Broderick PA (2021) Tau peptide signals are seen live in parkinson subjects by Broderick Probe® sensors. *MOJ Clin Med Case Rep.* 11(2):46–50. DOI: 10.15406/mojcr.2021.11.00380.
10. Broderick PA (2021) Sensing polymers for parkinson tau signaling: late stage nanomedicine. *Medical Research Archives (in progress).*
11. Broderick PA, Cofresi SL (2021) Neuroimaging a cytokine storm by transducing IL-1α to hippocampal cornu ammonis: COVID-19 SARS-CoV-2. In: Preedy VR, Patel VB (eds.) *Compendium on The Neuroscience of Depression: Genetics, Cell Biology, Neurology, Behavior and Diet.* Chapter 11, pp: 107–117 Elsevier Press, NY and London.
12. Broderick PA, Pacia SV (2006) Identification, Diagnosis, and Treatment of Neuropathologies, Neurotoxicities, Tumors, and Brain and Spinal Cord injuries using Microelectrodes with Microvoltammetry. U.S. Patent Serial No. 7,112,319, Issued.
13. Dereka B , Yu Q, Lewis NHC, Carpenter WB, Bowman, JM, Tokmakoff A (2021) Crossover from hydrogen to chemical bonding. *Science* 371(6525):160-164. doi: 10.1126/science.abe.
14. Fatouros, P, Dorn H (2006) Fullerenes yield stable, powerful MR imaging agent. *National Cancer Institute*, July 26.

15. Fischetti M (2010) Classical Mechanics: F = ma from an unusual per-spective. Lecture: *The Physics of Semiconductor Devices.* Department of Electrical and Computer Engineering-University of Massachusetts Amherst, MA 01003 Spring 2010.

Chapter 2

A Remarkable Happening in the Advent of *In Vivo* Electrochemistry into Spectral Analysis

2.1 Battles over Technologies: The Story behind the Story!

Yes, there were "battles". The battles were between scientists who were previously immersed and dedicated to technologies such as "push pull cannula" that evolved into microdialysis (having nothing to do with the kidney but having to do with a dialysis tube) and those who were pioneering electrochemistry to the point of making electrochemistry LIVE. The conflicts went on over decades and, therefore, interrupted the advent of in vivo electrochemistry into its rightful place in spectral analysis. Oh, they were heated! These heated debates have made, truthfully, life more miserable than it should have been. Battling over devices does not make sense.

Conversations ensued between the *father* of *in vivo* electrochemistry, *in memoriam*, Dr. Ralph Adams, PhD, *University of Kansas* at Lawrence and the author, Dr. Patricia A. Broderick, OP, PhD, Medical Professor, the *City University of New York School of Medicine, The City College of New York*; her legacy is known worldwide as the inventor of LIVE electrochemistry, the nanotechnology called, Neuromolecular Imaging (NMI).

Neuroimaging: Sensing Biochemistry in the Brain
Patricia A. Broderick

ISBN 978-981-4613-49-1 (Hardcover), 978-0-429-02974-5 (eBook)
www.jennystanford.com

2.2 Adams and Broderick Began a Dialogue: And—in the Midst—an Ironic Twist

Adams was an off-campus mentor to Broderick from time to time. Each considered the other a professional friend. Dr. Ralph Adams of Kansas, known as "the father" of *in vivo* electrochemistry, was interested in studying the brain that was beset with schizophrenia; his interest centered on electrode brain implants in murine models; his publications showed the effects of antipsychotic medications on the neurotransmitter, dopamine since schizophrenia was considered to be fully a dopamine disease at the time. Adams and Broderick began a dialogue since Patricia wished to study *in vivo* electrochemistry for application to neurodegenerative and psychiatric brain functionalities in patients as well as in murine models. *In vivo* electrochemistry was the rational next step for Patricia. Her expertise in atomic absorption spectrometry to detect medicines such as lithium and her proficiency in techniques of fluorescent spectrophotometric analysis of biogenic amines in addition to her knowledgeable facility in brain mechanics and redox reactions placed her squarely in the perfect position for an adept grasp of the essentials. Patricia gained expertise in spectral analysis as this field relates to brain and behavioral symptomatology in manic-depression and bipolar types of mental illness. The author served as predoctoral fellow while achieving her doctorate in Pharmacology, *St. John's Graduate School of Arts and Sciences* under Dr. Vincent de Paul Lynch, PhD.

A priori, Patricia, as an Einstein postdoctoral fellow, initiated a meeting with Dean Dominick Purpura, MD, then Chairperson, *Department of Neuroscience* at the *Albert Einstein College of Medicine.* Patricia mentioned to the Dean's secretary that the purpose of the meeting was to request permission to pursue a cryptic visiting scientist role in the laboratory of Dr. Adams while still serving in her Einstein postdoctoral position. Patricia wished to become ensconced in her newly desired field of study that she named "electrochemical neuroscience" and Patricia mentioned that she needed to meet with Dean Purpura to request such permission as soon as possible. Dean Purpura agreed to the meeting with gusto. The meeting was set up right away. During the meeting with the Dean, Patricia requested a three week stay in Kansas to work one-on-one with Dr Adams. Dean Purpura received Patricia's request to study with Adams exuberantly. Dean Purpura gave to Patricia readily his full and enthusiastic support; the Dean mentioned with certainty that Patricia would be assiduous in her task of originating new diagnostic technical devices and therapies for brain disorders. Indeed, he was sure that she would be a zealous and original contributor to the electrochemical field as pioneer in cutting-edge inventive art for Einstein's celebrated patent portfolio. The Dean asked Patricia whether or not five hundred dollars would take care of her travel and living expenses to which Patricia answered with a resounding "Yes". Purpura called Adams and Broderick's visiting scholarship was set in motion.

Too, Dr. Wagner Bridger, then, Chairman of the *Department of Psychiatry* at the *Albert Einstein College of Medicine*, was an ardent supporter of Patricia's work in brain and behavior and had indeed personally invited her to become a member of his department in 1982. Dr. Bridger had heard of Patricia's developing work in the evolving field of brain and behavior. This is another amazing part of Patricia's professional life. Indeed, it had always been a dream of Patricia's to be a part of the "Einstein" community and the realization that Dr. Bridger liked her idea of studying "electrochemical neuroscience" thrilled her intensely. Wagner immediately gave his assent to her short visiting scholarship in Kansas. Dr. Bridger wanted Patricia right back to her postdoctoral fellowship in Einstein which is exactly what Patricia wanted as well.

However, another ironic twist occurred. The twist was to bring Patricia instead to the laboratory of Dr. Ross F. Lane at the University of Oregon at Eugene. You see, Dr. Adams felt that his laboratory was not at the level of organization required to provide Patricia her beginning existential dialogue and formal training in the field of *in vivo* electrochemistry with himself and his team. Then, Patricia received a phone call from Adams himself, honestly lending an explanation to his situation. Quickly, Patricia rerouted the boxes of a few extra jackets and other necessities that she had already sent to Kansas on over to the west coast, to Eugene, Oregon.

2.2.1 The Oregonian Experience

The Oregonian experience was not quite the learning experience that Patricia had expected because there was not an electrode or sensor manufactured nor was there an experimental assay or surgery done in her presence. The surgery was performed by Dr. Lane's assistant. Whether or not there was data gathered post-surgery is not clear because these experiments were performed in a room directly across from the office of Dr. Lane that was separated from the laboratory proper. As Patricia sat outside in the laboratory proper, she is wondering right at this precise moment in time, whether or not surgery was actually done at all. She was told that surgery to implant an electrode in the brain was in progress and that it had failed.

Nonetheless, Ross and his assistant, Chuck (now, Dr. Charles Blaha) did share with Patricia, the visiting scientist at Eugene, some of their past electrochemical data and these interactions were helpful. Reference electrodes particularly peaked the author's interest as Patricia had taught electrochemical principles to her students in Cathedral and Msgr. Scanlan High Schools in the New York area. Redox was, even then, no stranger to Patricia's mind-set and she avidly read and re-read hundreds of electrochemical papers each and every day during her visit to the Lane Laboratory. Patricia continued to read, self-study and analyze on her own all aspects of electrodes and circuits as this is her custom. She continued to read and re-read every weekend of her three weeks spent in Oregon.

Oh, there was much comradery. In the evenings, the suppers with the laboratory and with the family of Dr. Ross Lane were so very pleasant; living with the Lane's was also engaging since the family was most welcoming. There was a mutual "like" among us. Patricia and Margo were both easy-going and Ross was a fun guy, a good-looking guy who resembled John Denver and we loved teasing him about this and listening to *Annie's song* by the late John Denver. On each morning during the week, Patricia jumped into Dr. Lane's car as the radio played, "Morning has

Broken", still a popular song which is really a religious hymn that Patricia enjoys to the present day. You see, Patricia patiently awaited some hands-on experience each day or perhaps she waited to enjoy seminars in the laboratory or seminars given by other Professors in the University from other departments. Apparently, the stay needed to be longer but Patricia had to return home to her work in Einstein and her religious life with Sr. Bridget O'Sullivan, OP, MA, Order of St. Dominic. Nevertheless, the ascent from her spectrometry world to her "electrochemical neuroscience" world congealed in Patricia's brain.

Ross, Chuck and Patricia met often at the *Society for Neuroscience* meetings and these meetings were productive and wonderfully collegial. Patricia shared co-authorship with Ross and Chuck on two of her publications derived from her work at Einstein albeit neither Ross nor Chuck had contributed neither in the writing nor in the empirical gleaning of Patricia's data. Patricia shared these publications *via* co-authorship with Ross and Chuck since the electrochemists had come to Eugene, previously trained at and working in the laboratory of Dr. Arthur T. Hubbard, PhD at the University of California, at La Jolla. This group of scientists were considered the pioneers within the group of electrochemists interested in becoming neuroscientists. Therefore, the motivation for Patricia to share with Drs. Lane and Blaha was to encourage both scientists in their academic careers as she did the same with so many of her colleagues and students. It's intriguing, though, that there was no contact among Ross, Chuck and/or the author all through the years after the short time Patricia spent in Oregon, that is, with the exception of emailing the publications co-authored and conviviality with all during the *Society for Neuroscience* meetings.

2.2.2 Adams to New York

Dr. Adams and Dr. Broderick remained in contact. Dr. Adams mentioned to Dr. Broderick a particular experience he had while grant writing. Ralph had applied for a grant involving the use of electrochemistry to study schizophrenia. Adams was considered to have some "nerve" to make such a proposal and he was advised to "stick" to his own field. The New Yorker, Dr. Broderick, mentioned to Dr. Adams that she had a similar, but beguiling enough, the same experience with the exception that the fields were reversed. Patricia was told, in writing in a grant summation that she had some, to paraphrase, "nerve" working with electrochemistry when she is a neuroscientist; she was further told to "stick" to her own fields of pharmacology and neuroscience and not to meddle in electrochemistry. But Adams and Broderick forged ahead in their endeavors and Broderick followed in the footsteps of Adams to become an expert in both nanobiotechnology and the synaptic workings of the elusive brain using *in vivo* electrochemistry.

Patricia invited Ralph Adams to New York to visit the Broderick laboratory at the *Albert Einstein College of Medicine* to discuss their work. In the process, Ralph Adams presented an in-depth seminar on his seminal electrochemical work for Dean Purpura, the Professors, post-doctoral students, technicians and staff. After the seminar, the two scientists, Adams and Broderick enjoyed a cheese burger together at the neighborhood diner on the corner near the Forchheimer Building at Einstein. Dr. Adams encouraged Patricia, a postdoctoral fellow in the *Department of Psychiatry* as well as in the *Department of Neuroscience* at Einstein to move forward with neuroscience and advance her work in electrochemistry LIVE. Broderick took Adams' advice and she continued her legacy work of wedding nanotechnology with the brain, sensing the brain into a LIVE molecular technology. Now, known as NMI, there is much written about this nanobiotechnology. The medical doctors have written "the BRODERICK PROBE® will change the way that physicians view the brain" [4].

In fact, Adams, who was nicknamed "Buzz" for his adeptness in piloting aircraft, sent Broderick some equipment from time to time and on the notes, there was scribbled "figure this out Pat, Buzz" and Broderick got the job done!

2.2.3 Professor Dr. Urban Ungerstedt of the Karolinska Institute: The Dialysis Tube

Thus, be that as it may, here is a brief and true historical picture of how the Broderick Research Laboratory has made its *mark* in the sojourn of the legacy of sensing devices in spectral analysis. Importantly, the true story tells how duels were pitched between a nanotechnology and a "push pull cannula" evolution combined with a well-known instrument in the market known as High Performance Liquid Chromatography (HPLC). The concept of dialysis, further considered, micro, was born in the early 1970's. It so happened serendipitously as discoveries often do, that while examining the cross-section of a blood vessel among fluorescent nerve endings, Professor Dr. Urban Ungerstedt, PhD, of the *Karolinska Institute* in Stockholm, Sweden, had an idea. The Professor thought of using a dialysis tube as "an artificial blood capillary" in order to monitor chemical events in tissue. The work of Ungerstedt is exemplary and has born a new technology which has blossomed into effective therapies for brain disorders to improve the health of human beings and animals to this day and beyond [21].

But, the battles between and among the scientists, the technicians and students were fierce and the contest had to be fought and then, shall we say, "won" with empirical data. Indeed, *in vivo* electrochemistry due to the work performed by this author in human epilepsy patients in real time [5] and work done by others such as Gerhardt and team [14] *in vivo* electrochemistry has made its place in spectral analysis as a reliable and viable electrochemical technology. With the work done by the author, the work has reached the nanobiotechnology and nanotechnology status [6]. The battles among the scientists remain a shocking surprise for the author as the Machiavellian strategies that were on display from some of the educated populace were unexpected insofar as the author's predilection of the academia was and is concerned.

2.3 A Conference Held by the New York Academy of Science, Manhattan, NY, USA, 1983

This debate between technologies, a rather major conflict, came to a head during a conference held by the *New York Academy of Science* in Manhattan, NY, in 1983. The conference was called

"Neurochemical Analysis of the Conscious Brain: Voltammetry and Push-Pull Perfusion"; the results were published in 1986 [1]. Dr. Broderick, presently full and tenured Medical Professor at the *City University of New York School of Medicine, the City College of New York,* was a Postdoctoral Fellow at the *Albert Einstein College of Medicine/Yeshiva University/Montefiore Medical Center* at the time of the Conference. The author's work fully embraced and focused on a new type of neuronal molecular imaging that the author invented and pioneered, abbreviated NMI.

The work that the author originated and presented at this inaugural meeting was live imaging distinguishing presynaptic and postsynaptic dopamine release mechanisms within the dorsal striatal movement substrate in the brain in an anesthetized mammalian animal model. The studies were performed online and in real time during the administration of the biphasic partial agonist dopamine medication, apomorphine, used today as indicated for advanced Parkinson's disease [13] and used for quite some time as indicated for erectile dysfunction [16].

Of note is the terminology used for this conference. The terms "Voltammetry" and "Push-Pull Perfusion" are in use today. Voltammetry at that time was used to denote an electrochemical reaction since the units, volts, or voltage subunits were required to perform the electrochemical reaction. Voltammetry referred to "electrical" circuits used as well. We used phrases such as microvoltammogram for the recordings/scans. Voltammetry records signals for neurotransmitters, neuromolecules and the like; the molecules are detected *via* electron transfer and scans are taken at the time of recording. The technology includes such advantages as "no temporal lag" between scans and "no tedious perfusion, centrifugation and radioactive labeling assays". The Broderick *in vivo* electrochemistry technology enables same-subject studies. This is recognized universally as a critically important advancement in the LIVE electrochemical field as well as in the brain and behavior field in science and medicine.

The term "Push-Pull Perfusion" has been, for the most part, replaced with the terminology, dialysis, called, "Microdialysis." Essentially, dialysis is an extension of "Push-Pull Perfusion" and it requires high-performance liquid chromatography (HPLC) into which electrochemical detection has been produced. The electrochemical detection is performed "after the fact", *post factum,* whether or not the neuromolecules in the brain of the murine animal model are assayed during life or not. The extracellular fluid, that is, the exudate from brain that is removed and placed in centrifuge tubes, are saved for later analyses such as in the centrifugation part. *This means that the next set of data for the same study are actually taken from a different animal or subject.* The dialysis techniques are currently described by experts as a process that creates an area around the membrane in which all solutes capable of crossing the membrane are depleted. Furthermore, there are other limitations described, including ascorbic acid interference and misinterpretation of the data [10].

Nonetheless, the battle over which device was better in operational stability, for example, ensued. This part of the argument was and is valid. However, some of the debates related to unimportant things such as the popularity of the scientist who performed a particular method. Which would be better, an electrochemical one or a liquid chromatography/push-pull/dialysis one? Which device would be better at assaying neuromolecules in the living brain? Broderick kept thinking "Why cannot both technologies be called valid since both technologies really are valid!" But, instead, the battle was intense on that day as well as it was all the days before and after. In fact, the battle maintained its intensity for many years; the battle has now abated somewhat as liquid chromatography/push-pull/dialysis companies are including sensors into their product lines, enabling sensors to take the place of dialysis or enabling sensors to work within this established market, speaking of which, the BRODERICK PROBE®, biosensors are currently poised for a place in this pre-clinical market.

Electrochemistry, with NMI and the BRODERICK PROBE®, has gained considerable status simply due to the fact that NMI provides an online, real-time nanotechnological pre-clinical and clinical device that has proven to be operationally stable and to exhibit superior temporal and spatial resolution. This nanotechnology provides excellent and reliable electroactive signals with LIVE imaging applications in virtually every field of neuroscience, neurodegeneration, psychiatry, psychology, psychoimaging, drug delivery systems, the neurobiology of brain reward, reinforcement and addiction processes as well as vascular; cardiovascular and pulmonary medicine. In addition, the nanoprobe provides novel surgical diagnostics and therapeutics and many novel applications in sub-fields of imaging technologies including specific technical devices at the nanoscale.

2.3.1 What Exactly Happened on that Dreadful Day!

On that day, in 1983, insults were flying between the opposing scientific schools of thought. "You cannot do what you are saying you can do!" There were nasty comments, hurling one to another. Outside the conference hall, the battle remained incorrigible. Now, one heard, "Did you hear this" and "Did you hear that? "Can you believe this" and "Can you believe that?

All left the conference, no matter which side or camp one was on, with the distinct thought that the push-pull cannula device, then beginning to be known as dialysis, had won the battle of the senses. "Microdialysis made sense and Electrochemistry did not." That was the quote of the day. During the conference, the push-pull/dialysis group pronounced the invincible precision of their assay while, on the other hand, the *in vivo* electrochemists told of the true struggles, the advantages, and the limitations of their electrochemical devices.

There were only a few types of sensors at that time. The push-pull/dialysis assay consisted of only one of a few types as well and there were actually limitations to this assay but these limitations were somewhat downplayed. For example, in many instances, the neurotransmitter, itself, is not directly monitored with microdialysis; what is monitored is the by-product that results from the assay. Sometimes, the molecule assayed is hydrogen peroxide and not the neurotransmitter, acetylcholine, for one example. The group is to be congratulated, nonetheless, as the success in the market is clear. Professor Dr. Urban Ungerstedt, PhD, was and is the leader in this group and he is rightly credited with the successful invention of the microdialysis methodologies [21]. Another distinguished member of this group is Professor Dr. Kjell Fuxe, PhD, also a Professor at the Karolinska Institute, Stockholm, Sweden [12].

2.3.2 A New Top Tier-Established in the Neuroscientist-Electrochemist Camp

Therefore, on that fateful day at the *New York Academy of Sciences* in the year 1983, the neuroscientists/ electrochemists, were downtrodden but some were still determined to continue the work even though there was a sense of disheartenment.

Patricia noted that the critical importance of *in vivo* electrochemistry in the brain was not conveyed adequately or strongly enough at this inaugural meeting, *"Neurochemical Analysis of the Conscious Brain: Voltammetry and Push-Pull Perfusion".* The leaders in the neuroscientist-electrochemist group could not communicate with the general audience, assurance of a reliable future for *in vivo* electrochemistry in the brain sciences or in the market. Something was missing! Oh, the author whispered! "the leaders of this neuroscientist-electrochemist group are failing to get to the point". It seemed that they could not answer the questions thrown at them and the questions were undeniably "thrown" at them. Adams fumbled somewhat; his demeanor unfortunately bespoke of being tired of the unending criticism of his fine work that was undoubtedly tainted with the dreaded "electrochemical electrocatalysis". The nexus of the electrocatalytic thorn was the lack of the ability to separate the electroactive signal for dopamine from that of ascorbic acid. Adams, the very father of *in vivo* electrochemistry, could not adequately assuage the dialysis camp that he and his team had everything under control, that it was safe to go to market with his electrode!

To compound the felony, Dr. Lane seemed unsure of the nature of his work insofar as electrocatalysis was concerned. It seemed that he could not either cope with or solve the electrocatalysis problem. Lane's demeanor, the author is sorry to say, seemed to show a real lack of confidence in his data. One could feel his annoyance directed at his attackers. The author felt sympathy for him. Lane, who was the second major leader in this camp, presented his work at the academy from the stage. Lane's abstract [15] for his presentation was:

> "The development of electrodes with improved selectivity for the catecholamines has been an active area of research interest, and changes in ascorbate and DOPAC concentrations are sufficiently large so as to obscure the much smaller changes in extracellular DA, confusing the analysis of drug-induced changes in DA levels."

Lane was attempting to discuss different methods that may solve the nexus to this electrocatalytic thorn but there was no data that proved that Lane had the solution. Then, to clinch the nemesis of the attributes of his *in vivo* electrochemical data, his initial sentence engulfed his downfall. Lane's introduction to the subject belied the meaning of his presentation!

Be it known on that day and ever after, the push-pull/dialysis camp was ever more confident that they had won. Perhaps even a certain arrogance appeared in a few of the disciples of that particular school of thought! The animosity that had entered into the equation on that dreadful day was exacerbated and it permeated among and between the members of the different camps. One scientist in any group was against any scientist in the other group and the hostility was obvious amidst the scientists of the same group/camp. The battles were intense, at times leading to "cruel" behavior.

Some electrochemists-turned neuroscientists made a few calls to publishers to block the work of Broderick! She heard about it personally and she heard what was said to whom and where it was said. The words that simultaneously hurt and flattered were: *"block Broderick-she's coming up the line-she's good-she will make a difference"!* Cliques developed! Patricia handled the situation tranquilly-her faith in God prevailed and her inner physiological being took the trauma. Patricia remained performing her research and to her credit, she, who serves on multiple editorial boards globally and routinely as she did so during those turbulent times, accepts manuscripts in micro-dialysis and any technology or nanotechnology involving and engaging in any aspect of the brain or spinal cord that suit a particular journal's guidelines.

Thus, the scientific argument over the two competing technologies rested on the shoulders of the leaders in these technologies at first and then it filtered down to their disciples. Neuro-electrochemistry leaders in the *in vivo* electrochemistry group and the leaders in the technology, called microdialysis, were formidable in their arguments and there were ardent followers in each camp who stood by their leaders. But some leaders were diminished during this conference and others, some disciples, became leaders. A new top tier was established in the neuro-electrochemist camp of scientific thought. The new top tier continued the assaults on Patricia, for example, who was in the same camp. How could this happen? The author is learning that one would ordinarily see such dastardly experiences in the academe but the author thinks that this should not be so.

Finally, it was strange that the comradery between Lane and Broderick was gone on that day. It seemed to have disappeared and a nasty remark was made by Ross to Patricia; the remark was a big surprise to Patricia and she believes that Ross was suffering from an unfortunate circumstance in the academe as he had made another fatal mistake. His talk had not gone over well and Patricia's ability to focus and configure circuits and sensor design on her own seemed to be a source of irritation for the former leader of the neuroscientist-electrochemist camp. So, a milestone was met on that day in 1983.

Perhaps that terrifically unfortunate misfortune by Lane instigated a prejudice against carbon paste materials. This is the pity of misplaced priorities. The science of advanced materials must not be altered in this unprofessional and unethical way. After all, capacitors are made of carbon paste and scientists all over the world work wonders with carbon paste as does the author.

2.3.3 The BRODERICK PROBE® Separates from the Rest

But, the discontent on the neuroscientist/electrochemist camp was not felt by Patricia. After all, she had presented her own original work on electrochemistry and neuroscience at the prestigious *Annals of the New York Academy of Science*, inaugural conference! Patricia was thinking that, although she was a postdoctoral fellow, her novel work was *on par* with the professionals! Patricia was certain of her data; you see, she had already begun patenting her data showing a clear distinction between the two molecules, dopamine and ascorbic acid even in the early days and indeed before the *online* proof of concept shown here in Chapter 5. Broderick had already started her trademarking prominence, naming her sensor after her father, Patrick Broderick, from County Clare, Ireland.

Furthermore, Patricia was honored to attend the Conference with Dr. Montford Piercey, PhD (Monty) with whom she worked at the time of the conference. The author was ever so proud to attend the conference with Monty since she served as consultant for the *Upjohn Pharmacia Company* in Kalamazoo, Michigan before and after the conference, actually for seventeen wonderful years. Monty traveled to New York for the meeting and was delighted to accompany Patricia to the meeting. In fact, the round trip between Kalamazoo and New York was continual, productive, enlightening during these years of learning the pharmaceutical industry while in academia.

But, another part of Patricia's delight on this day was the simple joy that Dr. Piercey felt as recipient of the warm Irish-American hospitality of Sister, Bridget O'Sullivan. Sr. Bridget was particularly fond of Monty as she was of so many doctors who honored and graced our home throughout the decades of years. Sr. Bridget, without a blink of an eye, connected and interweaved with brilliant world scientists. Broderick and Piercey went on to publish prolifically through their seventeen scholarly years until his untimely death in 1998. Drs. Piercey and Broderick [8, 17] are known for their LIVE imaging work, integrating electrophysiology and electrochemistry into the neuroscience of neurodegenerative disorders and Carlsson's autoreceptors.

2.3.4 The Calm after the Electrochemical Electrocatalytic Storm

Presently there is a truly bright light at the end of the tunnel. There is respect among the different schools of thought. In fact, as mentioned, many dialysis companies are adding electrodes and sensor devices to their product lines essentially admitting that the electrochemistry school of thought merited more recognition than was previously given. Indeed, many microdialysis scientists are now collaborating with *in vivo* electrochemists and neuroscientists in multiple possible roles. Indeed, there are many microdialysis scientists who approach this author continually wishing to use microdialysis with her sensors or to replace the dialysis technique with her sensors. Thus, microdialysis people are speaking with neuro-electrochemists and electrochemists who became neuroscientists. Important in this progress is a 1986 Conference initiated by Dr. Peter Kissinger, PhD, founder of *Bioanalytical Systems*, Lafayette, Indiana. The Conference took place in Cherry Hill, New Jersey; Pete invited Patricia to speak [2]. Of course, the answer was in the affirmative and the conference was stimulatingly refreshing.

A good deal of frustration has been spent in the lack of a distinction of a sharp, distinct electroactive signal for DA from and away from the electroactive signal for AA. Just five years ago, a covalently anchored p-aminobenzene sulfonate multilayer on a graphite pencil lead electrode was used to attempt the separation of AA from DA; the attempt is fair enough but the separation appears as not optimal albeit the attempt is novel [11]. On the other hand, in this nanobionics, proof of concept chapter, we contribute the full stellar data about the electroactive delineation of the neurotransmitter DA from AA, the latter being the critical cofactor (vitamin) in the biochemical dopaminergic pathway leading to the amino acid L-3,4-dihdroxyphenylalanine (L-DOPA) to the neurotransmitter DA. Imaging each of these neuromolecules while in the process of separating these neuromolecules *at the same time*, while still these molecules are unfolding *on line* and *in vivo*, is shown in the following data herein In Chapter 5.

2.4 The Interpretation of the Data Was Based on the Technology Instead of the Science

There was more to this no sense/nonsense about sensors and the difference between sensors and dialysis, and that difference was the scientific concepts of synaptic physiology, neurochemistry, biochemistry, biomedical engineering and/or molecular/cell biology. In other words, the debate between *in vivo* electrochemistry and microdialysis took on meaningful neuroscientific interpretations concerning the distinction between synaptic release and/or reuptake of neurotransmitters. You see, Broderick showed release mechanics took precedency in her NMI publication in *Pharmacology, Biochemistry and Behavior* reporting the interaction between cocaine and gamma butyrolactone [3] but a precedency of synaptic reuptake inhibition mechanics was favored by the dialysis scientists; an example of reuptake inhibitory mechanics is given in [22]. Takahashi et al., shows release mechanisms as intraneuronal [20]. Rice et al. favors release mechanisms as well [19].

An interesting side-note: Dr. Rice (the Adams lab) and Dr. Broderick (off campus, the Adams lab) were each separately awarded the *Irish America* Prize, September 12, 2018. The celebration took place at the famous Metropolitan House in New York. The doctors were applauded for their work in *in vivo* electrochemistry, (having nothing to do with release versus reuptake mechanisms).

The author feels that the senseless part of these distinctions insofar as the pettiness was concerned, resides in placing judgements on one or the other dependent on the technology of choice. The critical and valid synaptic release mechanism were diminished because reuptake inhibition of neurotransmitters at the neuronal extracellular fluid in the interstitial space took priority. The latter interpretation assumed importance for being associated with dialysis. In the end, the direct release of neurotransmitters using the NMI technology has proven to be substantially valid in fact and NMI strongly holds its theoretical and empirical validity. Still, to attach such critical scientific discoveries about the properties of the neuron to the differences between the importance of dialysis to viz a viz electrochemistry and NMI, in truth, does not make sense.

2.4.1 The Happiness Abounded from the Beginning

Nonetheless, some of the debates were entirely friendly and provided "food for thought" for such illustrious scientists as those seeking the truth, for example, Drs. Monty Piercey, Patricia Broderick, and Nobel Laureate, Dr. Arvid Carlsson worked together. Indeed, yes, the author has had the privilege of working one on one with the Nobel Laureate, Dr. Arvid Carlsson [18]. Patricia has fond memories of Drs. Piercey and Carlsson, attempting to make her distracted during her presentation; oh, she just smiled. Another fond memory is that of Drs. Broderick and Carlsson having a frankfurter side by side while sitting on the edge of a stage in the *Society of Neuroscience* as there were no seats left. Another fond memory is that of calling the home of Dr. Arvid Carlsson on the day his Nobel was announced to congratulate him personally. What about the time that Dr. Piercey treated Dr. Carlsson and a bevy of very bright scientists from the National Institute of Health to a Sunday's sandwich. Another fond memory is that of Patricia having dinner at his home with his family and the events please her even this moment.

The outcome of these mind-blending associations encumbered long-time sound friendships with such prestigious minds and the results of the meeting of the minds enhanced the intensely focused productive debates. It was interesting and exciting to debate the scientific and medical concepts along with the merits of each of the technologies associated with the microdialysis versus the NMI and/or *vice versa*. It was amazing participating at the *Society for Neuroscience* conferences held in the United States, attended internationally, along with other prominent international meetings. Neuroscientists profoundly yet respectfully still became downright passionate during the senseless battles of one school of thought over the other.

Thus, such an example of a true scientific debate can be shown by the scientific and real friendship between Drs. Piercey and Broderick as humorously shown *in vivo* as well as in the cartoon in Fig. 2.1.

Figure 2.1 (Left) A photo taken of Drs. Piercey and Broderick, *Society of Neuroscience,* showing how the dopamine (DA) release versus reuptake inhibition debates may be linked to the treatment of psychosis. The photo shows friendship built on true sharing of strategic novel treatments using live imaging, NMI, and electrophysiology online. We typically looked into several technologies and nanotechnologies to seek the advantages and limitations of each one. (Right) Humorous cartoon depicting, perhaps, what neuroscientists Drs. Broderick and Piercey may well have looked like during the more friendly, non-competitive, yet, aggressive discussions/debates. The surrounding scientists enjoyed these debates between Piercey and Broderick. In fact, Dr. Broderick was invited to work with the *Pharmacia & Upjohn Company* in Kalamazoo as a consultant, bringing the BRODERICK PROBE® to industry. Patricia said yes to the invite and the healthy debates continued as Monty and Patricia traveled between Kalamazoo and New York and further internationally to pursue the underlying science of neurodegenerative and psychiatric diseases. The Broderick team published strikingly innovative data in the LIVE Parkinson brain using the BRODERICK PROBE® library of sensors [7, 9]. Drs. Piercey and Broderick had the privilege of working one on one with Nobel Laureate, Dr. Arvid Carlsson, MD using the BRODERICK PROBE® in Parkinson therapy [18]. *Calvin and Hobbes* is a daily American comic strip created by cartoonist Bill Watterson that was_syndicated from November 18, 1985, to December 31, 1995. https://en.wikipedia.org/wiki/Calvin_and_Hobbes.

2.4.2 Not the Final Ironic Twist by any Means

In a speech that has inspired Patricia, Professor Dr. Ungerstedt presented the Nobel prize in physiology and medicine to Professor Dr. Carlsson in 2000 for the discovery of dopamine as a neurotransmitter *in its own identity* instead of as a precursor to a neurotransmitter. (Award ceremony speech. NobelPrize.org. Nobel Prize Outreach AB 2021. Tue. 28 Sep 2021. *<https://www.nobelprize.org/prizes/medicine/2000/ceremony-speech/>*).

2.4.3 The Conclusion

The penultimate is filled with bittersweet success, so much suffering and so much success, a veritable roller coaster ride it was and is! The feeling now, though, is followed by the ultimate, which is this book, the telling of the story. The bittersweet element is diminished and is now replaced with a feeling of accomplishment, a quite modest sense, if you will, of achievement together with a certain respectful awe of the sublime nature of the humble beginning and the maturity of the BRODERICK PROBE® NMI nanotechnology wed with the brain. The totality of this effort, made by the author and her team and other teams with her, which appears effortless to many, is made worthwhile. The progress made during this marriage between nanotechnology and the brain brings a realization that NMI and live quantum mechanics with *Electroactive Photovoltaic Fiber Optics* have dramatically advanced diagnosis and treatment for a myriad of brain disorders.

Such are the sentiments of a brain that fills the senses and savors the nuances of the brain. Realizing the neuron as an electronic and electrochemical device brings an immense fulfillment to the authors' own brain and being. No longer bittersweet is the experience derived from the inventions of the sensors and the circuits. The contents of the book just shows how a unique brain works simultaneously on both sides through each hemisphere to fathom such a *phenomenal phenomenon* of a brain sensor nanobiotechnology.

...and with this book, a new and consolidated expertise is given birth; one will readily see how this sensor nanobiotechnology, wedded to the brain and the spinal cord, has already and continues to be poised to further science and medicine and may, in fact, change the face of science and medicine. There is a lot written between the lines of the book, the minutes, days, hours, months, and years of study and testing, the years of pain and sorrow alongside inexplicable joy and the satisfaction of "finding solutions'.

Socrates has left us no written words. The thoughts of Socrates are written in Plato's works and are termed Platonic Socrates. Many forms of a "genius" defined, have cropped up from these Platonic Socrates works as well as from other Greek philosophers and modern philosophers. One of these phrases is "It is the genius who sees a target that no one else sees". Patricia did see a target that only a few could see and she did bring it to fruition and she continues to do more and more......

Bringing the brain sensing target to success and continuing to work diligently and unceasingly to solve the problems embedded within this highly complex integrated target to help alleviate brain suffering is overwhelmingly satisfying. The cost of caring and believing in a just cause is enormous. The price paid is worth its weight in carbon, even more precious than gold!

This was the most difficult chapter to write...

References

1. Broderick PA (1986) Delineation of striatal dopaminergic autoreceptor agonist properties of (-)-apomorphine with *in vivo* electrochemistry. In: *Neurochemical Analysis of the Conscious Brain: Voltammetry and Push-Pull Perfusion*, Myers RD, Knott PJ, editors. Annals of the New York Academy of Sciences. 478: 506–508. https://doi.org/10.1111/j.1749-6632.1986.tb23603.x.
2. Broderick PA. (1986) *In vivo* electrochemistry in biomedical research. *International Electroanalytical Symposium* 13:42–50.
3. Broderick PA (1991) *In vivo* voltammetric studies on release mechanisms for cocaine with gamma-butyrolactone. *Pharmacol Biochem Behav* 40:969–975.
4. Broderick, PA (2010) BRODERICK PROBE® Biotechnology for epilepsy patients, Press Release, NYU Langone Med Ctr.

5. Broderick PA (2018) Sensing the epilepsy brain: a unique nano-surgical biomedical device to treat epilepsy patient. *Med Case Rep MOJ Clin* 8(20):60–65.

6. Broderick PA (2018) A Live Imaging Nanotechnology: Sensing the Brain from Spectral Analysis to Neuromolecular Imaging to Voltaic Photonics. *30th Annual Congress on Nanotechnology and Nanomaterials,* Stockholm, Sweden, Sept. 10–11.

7. Broderick PA, Jacoby JH (2017) Nanomedicine in brain repair: Alzheimer's Parkinson's and the Proteome. *e-pub,* December 2016; *Journal of Neural Transmission* 124(**1**). DOI:10.1007/s00702-016-1660-0.

8. Broderick PA, Piercey MF (1990) Regulation of nerve impulse frequency and transmitter release by serotonergic autoreceptor agonists. In: *Presynaptic Receptors and the Question of Autoregulation of Neurotransmitter Release.* (Eds. Stanley Kalsner and Thomas C. Westfall) *Ann. N.Y. Acad. Sci.* 604: 596–597.

9. Broderick, PA, Wenning L, Li Y-S (2017) *epub* (2016) Neuromolecular imaging, a nanobiotechnology for Parkinson's disease: advancing pharmacotherapy for personalized medicine. *Journal of Neurotransmission* 124: 57–78.

10. Chefer VI, Thompson AC, Zapata A, Shippenberg TS (2009) Overview of Brain Micro dialysis. *Curr Protoc Neurosci.* Apr; CHAPTER:Unit7.1.doi:10.1002/0471142301.ns0701s47, PMID: 19340812, NIHMSID: NIHMS236586, PMCID: PMC2953244.

11. Devaramani S, Sreeramareddygari M, Radhakrishna Reddy, M, Thippeswamy, R (2017) Covalently Anchored p-Aminobenzene Sulfonate Multilayer on a Graphite Pencil Lead Electrode: A Highly Selective Electrochemical Sensor for Dopamine. *Electroanalysis* 29(5):1410–1417. https://doi.org/10.1002/elan.201600627.

12. Fuxe K, Hamberger B, Malmfors T (1967) The effect of drugs on accumulation of monoamines in tubero-infundibular dopamine neurons. *Eur J Pharmacol* 1(4): 334–341.

13. Hegell P, Odin P (2001) Apomorphine in the treatment of Parkinson's disease. *J Neurosci Nurs* 33(1): 21–34, 37–8.

14. Humeiden ML, Quinter JE, Slevin JT, Gerhardt GA (2018) Fast analytical sensing technology: microelectrode-based recordings of tonic and phasic neurotransmitter signaling in the mammalian brain. In: Lhatoo SD, Kahane P, Luders HO, editors, *Invasive*

Studies of the Human Epileptic Brain: Principles and Practice. Oxford University Press pp. 500–510.

15. Lane RF, Blaha CD (1986) Electrochemistry In Vivo: application to CNS Pharmacology. In: *Neurochemical Analysis of the Conscious Brain: Voltammetry and Push-Pull Perfusion*, Myers RD, Knott PJ, editors. Annals of the New York Academy of Sciences. 478. https://doi.org/10.1111/j.1749-6632.1986.tb23603.x.
16. Montors F (2002) The role of apomorphine SL in the treatment of erectile dysfunction. Eur Urol Suppl 1(3): 4–11.
17. Piercey MF, Broderick PA, Hoffman WE Vogelsang GD (1990) Neurotransmitter release and impulse frequency studies of dopaminergic autoreceptor agonists. In: *Presynaptic Receptors and the Question of Autoregulation of Neurotransmitter Release.* (Eds. Stanley Kalsner and Thomas C. Westfall) *Ann. N.Y. Acad. Sci.* 604: 637–638.
18. Piercey MF, Svensson KA, Lin CH, Haadsman-Svensson SR, McCall RB, Smith MW, Broderick PA, Carlsson A (1993) U-93385, a high intrinsic activity 5-HT1A agonist with good oral activity. *Soc Neurosci Abstr* 19:1866.
19. Rice M E, Patel J. C, Cragg S J (2011). Dopamine release in the basal ganglia. *Neuroscience* 198, 112–137. 10.1016/j.neuroscience.2011.08.066.
20. Takahashi, A, Ikarashi Y, Ishimaru H, Maruyama Y (1993) Dopamine output upon death reflects intraneuronal aspects while alive: accumulation of releasable dopamine during tetrodotoxin perfusion. *Neuroscience Research* 18:45–51. https://doi,org/10.1016/0168-0102(93)90104-x.
21. Ungerstedt U, Pycock C (1974) Functional correlates of dopamine neurotransmission. *Bull. Schweitz. Akad. Med. Wiss.* 1278:1–13.
22. Wimalasena K. (2011). Vesicular monoamine transporters: structure-function, pharmacology, and medicinal chemistry. *Med. Res. Rev.* 31, 483–519. 10.1002/med.20187.

Chapter 3

Oxidative Stress: A Time-Sensitive Detection Proxy for Electrocatalysis of Ascorbate

Nathaniel Hawthorne wrote: "Time flies over us, but leaves its shadow behind" Twenty-five years—one quarter of one century, a century of shadows, of joyful shadows, of sorrowful shadows, each and every nanosecond remembered. Each and every scientist, colleague, family, friend and acquaintance, remembered. Each and every student remembered. Each and every discovery remembered. Each and every kind greeting, encouragement and each and every hurtful word and action remembered. Each and every supporter and each and every detractor and/demoralizer, all limbic-related in emotion and cortically analytically ordered to help Patricia's mind/brain/soul grow in depth, mentally and spiritually, is deeply remembered. William Shakespeare wrote: "Time is very slow for those who wait, very fast for those who are scared, very long for those who lament, very short for those who celebrate"—But for those who love, time is eternal. For this author, time is eternal.

The problem of electrocatalysis, that is, the mix-up between the detection of dopamine (DA) and ascorbic acid (AA) electrochemically, was indeed a problem. This problem was the very bane of the existence of the neuroscientist who wished to be the "one" to be the expert "in vivo electrochemist" and vice versa. But these scientists, neuroscientist or electrochemist apparently could not do it! Each and every scientist

Neuroimaging: Sensing Biochemistry in the Brain
Patricia A. Broderick

ISBN 978-981-4613-49-1 (Hardcover), 978-0-429-02974-5 (eBook)
www.jennystanford.com

involved began to author manuscripts in which the authors could detect ascorbic acid or dopamine but none could detect both molecules separately and clearly. What could be done? Someone had to do it! Then, Broderick stepped up to the plate, so to speak. What did Broderick do? Broderick accomplished the necessary! What was accomplished between 1983 and 2008 was the saving grace for in vivo electrochemistry for growth in the electrochemistry field in vivo and preparation for the advent of neuromolecular imaging and cerebral polymeric, photosensitive, photoelectric and piezoelectric live imaging. The future of in vivo electrochemistry in spectral analysis now has its place, humbly, via the contribution of the neuroscientist electrochemist, Broderick and her team.

This chapter begins the dialogue to discern the solution to the issue of electrochemical "electrocatalysis." There did not seem to be a rationale underlying this concept of "electrocatalysis" as an issue in the first place. The reason that electrochemical electrocatalysis became an issue has its' basis in the scientists' lack of understanding of the issue. The issue was misunderstood and then politicized. It seemed that due to the grievous and fatal error that had been made by some anxiety-laden electrochemists desiring to convert to neuroscience, the author had to suffer from the prejudice of their errors. Her work, however, within *in vivo* electrochemistry, neuromolecular imaging and live quantum mechanics flourished. The suffering was and is horrendous but still, not a real problem as she is known as a *tour de force,* her strength of mind, photographic and analytical memory with her team persona override the minutia.

What has always concerned the author, though, was the fact that in the absence of the correct knowledge and sensor insight into the electrocatalysis of AA from AA, there was created inadvertently an inherently flawed analysis of brain disease and hence its progress in science and hence, medical treatment. One will see throughout this imaging monograph that the author's empirical solution to this electrocatalytic complexity has enabled her to pioneer the LIVE and continuous image of electroactive signals in normal (natural) physiological states as well as altered and/or diseased states in brain and spinal cord, even in different modalities within the central and peripheral nervous organ systems such as sensing the brain while walking.

3.1 Why?

Why did the Broderick lab take on this feat that others could not accomplish? After all, the study is complex. It was not easy to achieve success in this regard since AA exists in one redox form in the body and is transported into the brain in a different redox form while AA is a cofactor on one hand in the synthesis of DA from L-3,4-dihydroxyphenylalanine which is L-DOPA and on the other side of the equation, AA works again as a cofactor in the synthesis of norepinephrine from dopamine. The importance of achieving the electrochemical separation of DA from AA has never been underestimated! Thus, the Broderick lab took on the ultimate challenge and separated not only a biogenic amine, DA, from AA electrochemically and also separated an indoleamine, serotonin (5-HT) from AA using neuromolecular imaging.

3.2 How?

- A paradigm of oxidative stress was selected for study using neuronal molecular imaging (NMI) since oxidative stress is a condition that amply produces free radicals and free radicals under-pin brain disease such as those in the fields of neurodegenerative and psychiatric dysfunctions.
- The paradigm had to include key players such as ascorbate, dehydroascorbic acid, dopamine, morphine, naloxone, opioids, oxidative stress and the BRODERICK PROBE®.
- Early on, the paradigm was designed and the empirical studies were carefully performed with live imaging, BRODERICK PROBE® biosensors, on line, and in real time to separate DA from AA. This process took quite a number of dedicated and disciplined years.
- As mentioned, the LIVE imaging is named, Neuromolecular Imaging, abbreviated, NMI.
- With confidence, the PI, the author, then submitted her manuscript to Prof. Dr. Peter Riederer, Editor-in-Chief, Chemist and Successor to Dr. Arvid Carlsson, *Journal of Neural Transmission.* The Co-Editor-In-Chief is Prof. Dr.

Etienne Hirsch, PhD presently. The author believes that the superb background of Dr. Riederer as a Chemist enabled him to understand redox reaction to a much greater degree than did others.

- *The manuscript underwent severe, impassioned and profound scrutiny for one full year.*
- Concepts of sensing mechanisms in the brain were feared. But, astute editors together with the author persevered, answering all positive and negative critiques accurately and with strong scientific evidence. The paper is reprinted here with permission from *Springer Publications*, Berlin/Germany. London, UK, New York, NY, USA.

3.3 Background

Previous studies had shown that morphine increased extracellular DA in nigrostriatal and mesolimbic pathways in rodent brain [23]. These morphine effects have been postulated to be a factor in the reinforcing effect of morphine and the known increased locomotor effects produced by morphine [39]. Another study using a microdialysis technique has shown that morphine increased extracellular DA specifically in the NAcc [37]. Two other reports have provided indirect evidence that morphine reduces pain by increasing extracellular DA in the NAcc. This evidence is derived from research with a new opioid peptide, nociceptin, which produces hyperalgesia [29, 30] and decreases morphine-induced extracellular DA in the NAcc shell [18].

Moreover, the antioxidant, ascorbate, when administered orally in high doses, has been shown to relieve pain, reduce opioid use, inhibit endogenous opioid degrading metalloenzymes, increase endorphin levels, and ameliorate withdrawal symptoms from heroin addiction [21]. Chronically, but not acutely administered ascorbate has been shown to inhibit morphine withdrawal [26]. Dehydroascorbate (DHA), the oxidized form of ascorbate, enters the brain by facilitative transport [3]. DHA increases ascorbate in brain and increased DHA and ascorbate in brain provides neuroprotection during cerebral ischemia and stroke [15, 25].

Serotonin (5-HT) has been implicated in morphine's pain-relieving effects because 5-HT reduces hyperalgesia *via* L-tryptophan, the precursor to 5-HT [1]. Morphine increased striatal extracellular 5-HT levels as well as DA levels in striatum of laboratory rats [19, 20]. In these previous studies, morphine was also reported to increase the oxidative metabolites of monoamines by showing increased striatal ascorbate and glutamate. The authors suggested that oxidative stress is mediated by the u-opioid receptor at nigrostriatal sites and that ascorbate and glutamate release may be tonically controlled by this same mechanism. Therefore, this animal model is important because ascorbate release provides a valid measure of oxidative stress in specific neuroanatomic substrates of the brain. In the present chapter, the oxidative stress animal model was tested by studying morphine/ascorbate/monoamine oxidative reactions at a mesolimbic site, the vlNAcc, in the anesthetized, male, Sprague-Dawley laboratory mammalian animal model.

3.3.1 General Overview: Conventional Microvoltammetry

Conventional microvoltammetry involves the implantation of an indicator electrode and a reference electrode in specified regions of brain, the application of a potential to an indicator electrode, the oxidation or reduction (redox) reaction of selected neurochemicals, and the recording of current derived from that redox reaction. In essence, the potential is applied between the indicator and the reference electrode wherein the indicator electrode reads current and the reference electrode provides a relative zero potential. Information about analytes is provided as a function of the potential difference, as this potential difference activates the surface of the indicator electrode. Dr. Ralph Adams, Lawrence, Kansas *in memoriam*, has been named, the father of conventional *in vivo* electrochemistry [2].

3.3.2 Conventional Microvoltammetry

1. Involves measurement of current in an electrochemical cell as a function of applied potential.

2. Electroactive species undergo redox reactions at a characteristic redox potential.
3. The formula is: O + ne− ⇔ R, wherein, ne = number of electrons, O = oxidation, R = reduction.
4. The amount of current that is produced is proportional to concentration according to the Cottrell equation [27].

Methods based on conventional voltammetry and micro-voltammetry, as pioneered beginning in the 1970s, have validated that the flow of charge, i.e., amount of current in amperes, which passes through the surface of an indicator electrode is proportional to the concentration of the electroactive species studied. The following formulas describe this relationship in terms of charge, electron transfer, current, diffusion layer, time, Faraday's constant, size of the indicator electrode and concentration (mass) of electroactive species:

$$Q = nFVC_{oR}$$

$$i = dQ/dt$$

$$i = nFV\, dC_{R,t}/dt,$$

where V is the volume of the diffusion layer on the electrode where the measurement is being made, n is the number of electrons transferred, F is the Faraday constant, and C_o denotes initial concentration. The Cottrell equation is derived from the formulas written above and demonstrates that current, i.e., charge, and mass, i.e., concentration, are proportional.

3.3.3 General Overview: Neuromolecular Imaging

This is a novel biotechnology which uses conventional microvoltammetry as its basis. In this paper, as described under "Recordings and Circuits," neuromolecular imaging (NMI) employs a semiderivative voltammetric circuit to apply a potential difference to a sensor. The term voltammetry denotes a specific electrical circuit, and there are several different types of circuits, by which a potential difference can be applied to a sensor. Voltammetry does not define neurotransmitter selective sensors nor does voltammetry denote the detection or imaging of neurotransmitters *per se*. NMI, on the other hand, includes detection and imaging of neuromolecules. Thus, a significant difference between conventional microvoltammetry and NMI is the development of new sensors to image neurotransmitters within seconds over long periods of time, *in vivo*, *in vitro*, and *in situ*.

The BRODERICK PROBE® sensors relate to several formulations of unique, patented, and trademarked miniature carbon sensors comprising a series of compounds which include, among others, classes of molecules in the biochemical categories of lipid, glycolipid, lipoprotein, saturated and unsaturated fatty acid. These inventions are able to detect electrochemical signals for a vast number of neurotransmitters, neuromodulators, and metabolites, including neuropeptides [7, 8, 10, 11, 13]. NMI primarily but not exclusively detects release mechanisms for neurotransmitters as shown by γ-butyrolactone studies [9]. NMI has shown, for the first time, a distinct difference, in tissue resected from epilepsy patients, between hippocampal white versus gray matter [12]. A schematic generic diagram of the BRODERICK PROBE® sensor is shown in Figure 3.1. In this chapter, results from new BRODERICK PROBE® *N*-stearoylcerebroside and laurate sensors are presented.

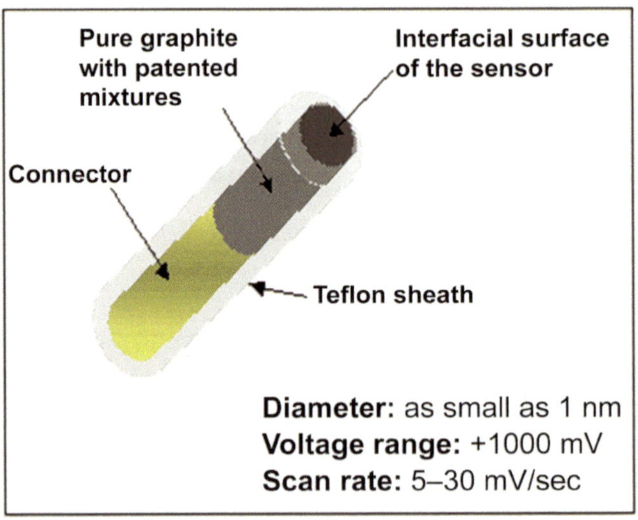

Figure 3.1 Schematic drawing of the BRODERICK PROBE® sensor.

3.3.4 Manufacture of BRODERICK PROBE® *N*-Stearoylcerebroside Sensors

N-Stearoylcerebroside sensors were manufactured on site. *N*-Stearoylcerebroside is a complex lipid wherein the cerebroside portion is derived from bovine brain (formula: ($C_{42}H_{81}NO_8$). *N*-Stearoylcerebroside sensors, for these studies, comprised <1 mg aliquot of stock, a mixture of ultra-pure carbon (0.75 g), oil (Nujol) containing vitamin E (0.062 g), and *N*-stearoylcerebroside (5.0 mg). Details of the construction of the *N*-stearoylcerebroside sensor have been published [10, 11].

- **NMI properties of BRODERICK PROBE® *N*-stearoylcerebroside sensors**

Oxidation potentials for these sensors, *in vitro*, are, in the order of detection from lowest to highest potential, ascorbate [0.07 V]; DA [0.12 V]; 5-HT [0.27 V]. Oxidation potentials for these sensors, *in vivo*, are, in the order of detection from lowest to highest oxidation potential, ascorbate [0.08 V], DA [0.14 V], 5-HT [0.28 V].

3.3.5 Manufacture of BRODERICK PROBE® Laurate Sensors

Laurate sensors were manufactured on site. Lauric acid is a saturated fatty acid (formula: $CH_3(CH_2)_{10}COOH$). Laurate sensors, for these studies, comprised <1 mg aliquot of stock, a mixture of ultra-pure carbon (1.5 g), oil (Nujol) containing vitamin E (1.24 g), and lauric acid (100 mg). Details of the construction of the laurate sensor have been published [10, 11].

- **NMI properties of BRODERICK PROBE® laurate sensors**

Oxidation Potentials for these sensors, *in vitro*, are, in the order of detection from lowest to highest potential, ascorbate [0.08 V]; DA [0.13 V]; 5-HT [0.29 V]. Oxidation Potentials for these sensors, *in vivo*, are, in the order of detection from lowest to highest oxidation potential, ascorbate [0.09 V], DA [0.15 V], 5-HT [0.3 V].

3.3.6 BRODERICK PROBE® *Recordings and Circuit*

Within seconds, ascorbate, DA, and 5-HT, each at distinct and separate signature oxidation potentials, were selectively recorded. Each neurochemical was detected in 10 sec. Repetitive scanning took place every 2 min for 1 to 2 h in an electrochemical cell for *in vitro* studies. For *in vivo* studies in the animal brain (the vlNAcc), repetitive scanning took place for one and one-half hour after drug(s) administration. Charging (background) current was recorded and eliminated in the first 20 sec of each recording.

For these studies, a semiderivative circuit was used; this circuit provides the first half-derivative of the analog linear signal. A semiderivative circuit combines an additional series of resistors and capacitors, called a "ladder network" with linear scanning circuits and the addition of this "ladder network" to the linear scanning circuit records sharper, more defined peaks [31]. Presently, computerization and remote control features for NMI imaging are performed with an Autolab (Brinkmann Inc., Westbury, NY).

3.4 Protocol: *In Vitro* Studies

Ascorbate, DA and 5-HT in crystalline soluble form and buffer reagents were purchased from Sigma/Aldrich of St. Louis, MO. Sensors were treated in buffer by a medium exchange technique [7, 10, 11]. This technique conditioned sensors repetitively with increasing concentrations of ascorbate and each monoamine over a period of three days. Aliquots of DA and 5-HT were pipetted into the electrochemical cell containing saline (0.16 M) phosphate (0.01M) buffer (pH 7.4). Concentrations of ascorbate were 3–300 times greater than those of monoamines.

3.4.1 Pretreatment

N-Stearoylcerebroside and laurate BRODERICK PROBE® sensors were pretreated with phosphotidylethanolamine (PEA) (10% w/v) and bovine serum albumin (BSA) (1% w/v) for 30 min, then immersed in saline (0.16 M) for 45 min, and further immersed in saline (0.16 M) phosphate (0.01M) buffer (pH 7.4).

3.4.2 Rationale

The rationale for the PEA/BSA pretreatment of sensors is based on data from this lab showing that BRODERICK PROBE® sensors produced significantly enhanced images for ascorbate and monoamines when pretreated with the lipid, PEA (10% w/v) and the protein, bovine serum albumin (BSA) (1% w/v) [22]. This pretreatment provides an optimal "simulated brain in the beaker" and further provides optimal electrochemical electron transfer signaling processes for NMI. With PEA/BSA pretreatment, neurotransmitter and neuromodulator images are significantly enhanced over those produced when sensors are immersed in saline/phosphate buffer without PEA/BSA pretreatment. Moreover, *in vivo* images, pre- and post-calibration data are better reflected in enhanced images from sensors pretreated with PEA/BSA [14, 22, 32].

Since the nanoprobe, laurate sensor was selected for the *in vivo* part of this study, this sensor was additionally tested for performance without pretreatment with PEA/BSA. Laurate was selected for the *in vivo* studies because the laurate sensor exhibited higher surfactant properties and hence greater electron-transfer capabilities than the nanoprobe *N*-stearoylcerebroside sensor, albeit both sensors exhibit operational stability, sensitivity, and selectivity [10, 11].

3.4.3 *In Vitro* Experimental Specifications

The semiderivative circuit was scanned from –0.2 V to +0.45 V; sensitivities for pre-calibration studies were in the range of 0.1 nA/V to 0.5 nA/V, at a time constant of 1 sec or 5 sec. Scan rates were in the range of 10–20 mV/sec, for one to three conditionings in the medium exchange technique. Sensitivities for post-calibration studies ranged from 2.0 nA/V to 10.0 nA/V. Calibration curves have been published [14, 32]. Time constants were set at 1 or 5 sec. Scan rates were in the range of 10–20 mV/sec. Oxygen was removed from buffer with pre-purified nitrogen gas (N_2) extra dry grade 5 psi (T. W. Smith Co., Brooklyn, NY). N_2 gas was perfused through the buffer, which was continually regulated by a variable area flowmeter

(Purgemaster, Fisher Scientific Fadem, NJ). The variable area flowmeter was calibrated to 20 cc/min. Perfusion of N_2 took place for 1 min before each cell deposition was activated and before scanning images began at the 0.2 V setting on the potentiostat.

3.5 Protocol: *In Vivo* Studies

Male Sprague-Dawley murine animals were purchased from Charles River Laboratories, Kingston, NY. Animals were housed in the animal care facility at The City College of New York (CCNY) for one to two weeks before surgery was performed. The animals were housed in pairs before surgery and housed individually after surgery. Animals were fed Purina Chow and water *ad libitum*. A 12 h dark-light cycle was maintained both in the housing of the animals and throughout the experimental studies. *The Animal Care Facility operates under the auspices of the CCNY, Institutional Animal Care and Use Committee (IACUC), in compliance with National Institute of Health (NIH) guidelines.*

Pentobarbital Na, ascorbate, and DHA were purchased from Sigma/Aldrich (St. Louis, MO). Morphine sulfate and naloxone were purchased from Cardinal Health (Dublin, OH). All drugs were stored at room temperature. Pentobarbital Na, morphine sulfate, and naloxone were stored in a locked safe, approved by the Drug Enforcement Administration (DEA), under approved and current NY State and Federal DEA licenses.

Surgical procedures began with an intraperitoneal (ip) injection of the anesthetic, pentobarbital Na (50 mg/kg in a dilute 6% solution). Then, BRODERICK PROBE® laurate sensors were inserted stereotaxically (Kopf Stereotaxic, Tujunga, CA) within the NAcc (coordinates: AP = +2.5; ML = +2.6; DV = −7.3) [33] into the vlNAcc of each subject. An Ag/AgCl microreference, which provided a zero point for current measurement and a stainless steel microauxiliary, which provided an additional ground, were placed in contact with the cortex.

Sensors were held in place with Splintline Acrylic (Lang Dental Mfg. Co., Inc., Wheeling, Il.). Body temperature was continuously monitored with a rectal probe and thermometer (Fisher Sci., Fadem,

NJ). The body temperature was maintained at 37.5±0.5°C with an Aquamatic K module heating pad (Amer. Hosp. Supply, Edison, NJ). Supplemental doses of pentobarbital Na were administered to maintain an adequate level of anesthesia throughout surgery. Pinnal, corneal, and leg flexion responses were monitored throughout surgery. The total time for surgery was approximately 2 h. Stable and selective NMI of ascorbate, DA, and 5-HT was achieved before either ascorbate, DHA, morphine, or naloxone was injected. Scans were recorded for 1.5 h after injection of each drug(s).

3.5.1 *In Vivo* Experimental Specifications

The semiderivative circuit was scanned from −0.2 V to 0.45 V; sensitivity ranged from 2.0 nA/V to 10.0 nA/V. Time constants were set at 1 or 5 sec. Scan rates were in the range of 10–20 mV/sec. Cell deposition was 2 min; scans were recorded every 5 min which includes a 2 min cell deposition time.

3.5.2 The Electrocatalytic Reaction: Selective Detection of DA in the Presence of AA

Ascorbate has been studied in terms of the methods used to determine neurotransmitters in its presence [5]. The principal reason for determining neurotransmitters in the presence of ascorbate is that signals for DA in the presence of ascorbate may not have appeared distinct in data from certain previous conventional voltammetric electrodes. Called an electrocatalytic reaction, the process involves an initial electro-oxidation of DA to free radicals such as orthoquinones, followed by the oxidation of ascorbate by orthoquinones; some of which may regenerate the DA signal [16]. Inventive miniature sensors, used in this study, diminished electrocatalysis [7, 10, 11]. Slow scanning techniques also diminish electrocatalysis [38].

It is noteworthy that generally, higher relative concentrations of ascorbate compared with each monoamine are required for ascorbate to be recorded within similar electrochemical parameters. Detection of ascorbate is enabled by the presence of higher concentrations of the vitamin both *in vitro* and *in vivo.*

Detection of ascorbate is also enabled by increased electrochemical parameters such as sensitivity and time constants at the oxidation potential of ascorbate.

3.5.3 Facilitative Transport into Brain: Ascorbate

Ascorbate is present primarily in the reduced form. Agus et al. (1997) [3], using Fisher 344 rodents and Balb C mice in radioactive [^{14}C] ascorbic acid transport studies, identified the chemical form of ascorbate that readily crosses the blood–brain barrier and this is the oxidized form of ascorbate, DHA. Ascorbate, in the reduced form, cannot cross the blood–brain barrier. In addition, the transport of DHA into the brain requires L-glucose. The facilitative glucose transporter, GLUT1, expressed on the endothelial cells of the blood–brain barrier, is responsible for glucose entry into brain.

Therefore, GLUT1 transports DHA into the brain where reduced ascorbate is once again formed. Thus, DHA increases ascorbate in brain, providing neuroprotection against free radicals. This antioxidant activity of ascorbate has been found to be protective against stroke [15]. The concentration of ascorbate released is indicative of the relative degree of oxidative stress present. In this paper, studies of injectable ascorbate and DHA were performed *in vitro* and *in vivo* for additional controls.

3.5.4 Sensor Placement

Placement within vlNAcc was confirmed by the potassium ferrocyanide blue dot method (specifications: current 30 mA, time in sec, 40). Virtually no damage to brain tissue occurred.

3.5.5 Statistical Analysis

One-way analysis of variance (ANOVA) was used to determine statistical differences in (A) DA release in the vlNAcc among groups studied. Statistical analysis within each group was further determined by Tukey's *post hoc* test. Statistical differences were determined according to first hour and second hour morphine sulfate effects followed by third hour naloxone effects. One-way ANOVA was then subsequently performed, in this same manner, to determine statistical differences in (B) 5-HT release in the vlNAcc among groups. Significance was determined at the $p < 0.05$ level.

3.6 What Were the Results?

Figure 3.1 shows a schematic diagram of the BRODERICK PROBE® sensor. The interfacial surface of the sensor is the detecting device which attracts electrons from neurotransmitters and neuromodulators to produce current, proportional to neuromolecular concentration, according to the Cottrell equation. The part of the schematic labeled "pure graphite with patented mixtures" and the interfacial surface of the sensor, comprises a mixture of pure graphite, Nujol, in addition to, in this case, either the *N*-stearoylcerebroside or laurate components [10, 11]. Clinical applications of these and other sensors to neurodegenerative diseases are described in [13, 14, 32].

3.6.1 *In Vitro* Studies

Controlled tests, systemized experimental designs comprised of different concentrations of ascorbate, DA, and 5-HT in saline (0.16 M) phosphate (0.01 M) buffer (pH 7.4) were performed for the purpose of delineating NMI signaling *in vitro.* One series of tests used this saline/phosphate buffer. Another series used a protocol wherein sensors first underwent a pretreatment in optimal concentrations of a critical brain-derived lipid and protein, i.e., PEA [10% w/v] and BSA [1%w/v] before they were tested in saline/phosphate buffer. Pretreatment with this mixture provided enhanced NMI electrochemical signals for ascorbate, DA and 5-HT compared with NMI signaling when sensors were immersed in conventional saline/phosphate buffer alone. The BRODERICK PROBE® *N*-stearoylcerebroside and the BRODERICK PROBE® laurate sensors exhibited operational stability, sensitivity, and selectivity. See Figures 3.2A and 3.2B.

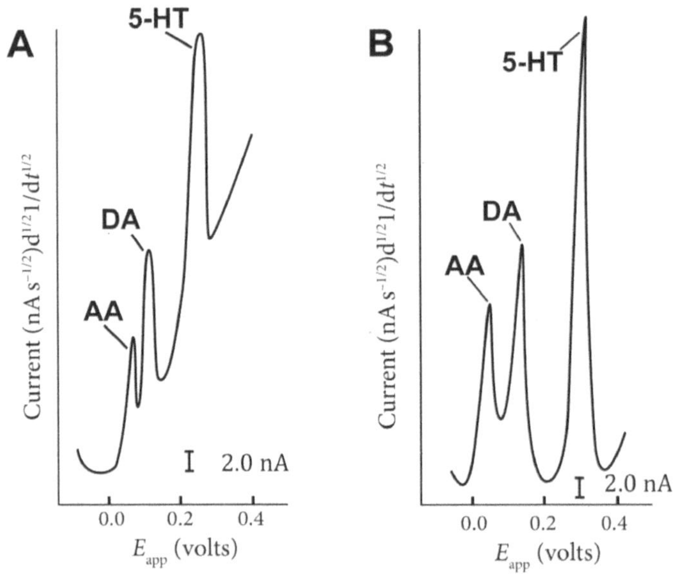

Figure 3.2 *In vitro* electrochemical images (NMI) of ascorbate, DA and 5-HT in saline/phosphate buffer derived from the BRODERICK PROBE® *N*-stearoylcerebroside sensor pretreated with PEA/BSA **(A)** and from the BRODERICK PROBE® laurate sensor **(B)**. The signature for each neurochemical, defined as the oxidation potential in mV, is shown on the *x* axis. Current is shown on the *y* axis; concentration of each neurochemical is proportional to current according to the Cottrell equation. Several and repetitive studies were performed. Three distinct and separate cathodic oxidation peaks were produced by the pretreated BRODERICK PROBE® *N*-stearoylcerebroside and the laurate sensors. The *N*-stearoylcerebroside and the laurate sensors exhibited operational stability, sensitivity, and selectivity.

Figure 3.2A shows a representative *in vitro* recording, derived from the BRODERICK PROBE® *N*-stearoylcerebroside sensor. The sensor was pretreated with the PEA/BSA mixture. Three well-defined cathodic peaks are shown. In order, from the lowest oxidation potential recorded to the highest, the NMI electrochemical signal for ascorbate was first, the DA signal was second and the 5-HT signal was third. Recording specifications for this image were: sensitivity [2 nA/V]; time constant [1 sec]; scan rate [10 mV/sec]. Charging current (−0.2±0.0 V) is not shown.

The recording, Figure 3.2A, shows results from the pretreated BRODERICK PROBE® *N*-stearoylcerebroside. The recording was derived from the oxidation of ascorbate, DA and 5-HT at the following concentrations: ascorbate = 11.2 uM, DA = 150 nM and 5-HT = 150 nM. Therefore, ascorbate concentration was approximately 75 times higher than that of either monoamine. Currents produced by each neurochemical at the concentrations noted were ascorbate = 6 nA, DA = 10 nA and 5-HT = 18 nA. Thus, all three neurochemicals were detected at the same electrochemical parameters of sensitivity = 2.0 nA/V, 1 sec time constant and 10 mV/sec. Thus, greater concentrations of ascorbate were used for all three neurochemicals to be seen within the detection ranges of monoamines noted. The data also show that detection of 5-HT produced higher current values than did DA at equivalent concentrations.

In this same study using the PEA/BSA pretreated *N*-stearoylcerebroside sensor, ascorbate was first detected at 2.2 uM when each monoamine concentration was 50 nM. Recording specifications for these images were: sensitivity [2 nA/V]; time Constant [1 sec]; scan rate [10 mV/sec] (recordings not shown).

In an additional study, using the PEA/BSA pretreated *N*-stearoylcerebroside sensor, ascorbate was detected at 0.07 uM when the concentration of each monoamine was 20 nM. Moreover, continuous and ascorbate images occurred with increasing amplitude, as increasing concentration of ascorbate were added into the electrochemical cell, up to a final concentrations of ascorbate of 0.07 uM. Dopamine and 5-HT images were unchanged from the signals, derived from 20 nM, when no additional monoamine was added to the electrochemical cell. When DA was added to the electrochemical cell, only the DA image increased in amplitude. Ascorbate images were observed at the same electrochemical parameters as was used for the monoamines. Recording specifications for these images were: sensitivity [2 nA/V]; time constant [1 sec]; scan rate [10 mV/sec] (recordings not shown).

In further studies using the PEA/BSA pretreated BRODERICK PROBE® *N*-stearoylcerebroside sensor, ascorbate was pipetted into the electrochemical cell at concentrations of 0.07 uM 1.5 uM, 2.2 uM, 3.7 uM, and a final 3.7 uM. Monoamine concentrations were sequentially 10 nM, 10 nM and 20 nM. Ascorbate was first observed at a concentration of 1.5 uM when the concentration of each monoamine was 40 nM. At the final concentration, ascorbate concentration (11.2 uM) was 275 times greater in concentration than DA (40 nM); this proportionality of ascorbate and DA produced an ascorbate current which was higher than, and still separated from, the DA current. The concentration of monoamines recorded, remained the same at 40 nM while clear and separate electrochemical signals for ascorbate increased with increasing concentrations of ascorbate. All three neurochemicals were detected at the same electrochemical parameters. Recording specifications for these images were: sensitivity [2 nA/V]; time constant [1 sec]; scan rate [10 mV/sec] (recordings not shown).

Figure 3.2B shows a representative *in vitro* recording, derived from the BRODERICK PROBE® laurate sensor pretreated with the PEA/BSA mixture. Three well-defined cathodic peaks were delineated. In order of lowest to highest oxidation potential, the NMI electrochemical signal for ascorbate was the first to be recorded, the DA signal was second and the 5-HT signal was third. Recording specifications, specifically for this imaging, were: sensitivity [2 nA/V]; time constant [1 sec]; scan rate [10 mV/sec]. Charging current (−0.2±0.0 V) is not shown.

The recording, shown in Figure 3.2B, from the pretreated BRODERICK PROBE® laurate sensor, was derived from the sequential oxidation of ascorbate at concentrations of 0.07 uM, 1.5 uM, 2.2 uM, 3.7 uM and a final 3.7 uM and also from the concomitant and sequential oxidation of DA and 5-HT each at concentrations of 10 nM, 10 nM and 20 nM. Therefore, the recording shows images for ascorbate, DA and 5-HT at the following concentrations: ascorbate = 11.2 uM, DA = 40 nM and 5-HT = 40 nM. It is important to point out though that the distinct image for ascorbate began to appear at a concentration of 0.07 uM when the concentration of the monoamines was 40 nM. Corresponding currents produced by each neurochemical at the final concentrations were: ascorbate = 9 nA, DA = 12 nA and 5-HT = 24 nA. All three neurochemicals were detected at the same electrochemical parameters of 2.0 nA/V, 1 sec time constant and 10 mV/sec. Higher concentrations of ascorbate were needed for all three neurochemicals to be seen simultaneously within the detection ranges of monoamines noted.

Moreover, detection of ascorbate was tested in a series of concentrations from 0.07 uM to 11.2 uM when each monoamine concentration was 40 nM when using the pretreated laurate sensor. In this study, ascorbate first appeared at 1.5 uM when each monoamine concentration was 40 nM, clear and separate electrochemical signals for ascorbate, DA and 5-HT were observed at the same electrochemical parameters as were used for the monoamines. Recording specifications for these images were: sensitivity [2 nA/V]; time constant [1 sec]; scan rate [10 mV/sec] (recordings not shown).

In addition, since the BRODERICK PROBE® laurate sensor was selected for *in vivo* studies of oxidative stress, studies were also performed on the laurate sensor, *in vitro*, when this formulation of sensor was not pretreated with the PEA/BSA mixture. The results showed that three well-defined cathodic peaks were delineated. In order of lowest oxidation potential to the highest, the NMI electrochemical signal for ascorbate was the first to be recorded, the DA signal was recorded second and the 5-HT signal was recorded third. Recording specifications, specifically for this imaging, were: sensitivity [0.2 nA/V]; time constant [1 sec]; scan rate [10 mV/sec] (data not shown).

The results from this study, in which the BRODERICK PROBE® laurate sensor was not pretreated with PEA/BSA, showed that when ascorbate was added to the electrochemical cell to a concentration of 5 uM and each monoamine was aliquoted into the electrochemical at 100 nM, the approximate currents produced were: ascorbate = 7.0 nA, DA = 10.0 nA and 5-HT = 40.0 nA. Comparable current ranges for recording specifications at electrochemical sensitivities of 2.0 nA/V, such as detection ranges for sensors pretreated with PEA/BSA, are: ascorbate = 0.7 nA, DA = 1.0 nA and 5-HT = 4.0 nA.

3.6.2 *In Vivo* Studies: DHA

Figure 3.3A represents baseline data as this novel biotechnology uses same animal control and baseline data represent 100% (control values). Therefore, Figure 3.3B shows that there were no significant *differences in DA or 5-HT release as measured from baseline after DHA was administered at any time during injections of DHA (20–100 mg/kg, ip).*

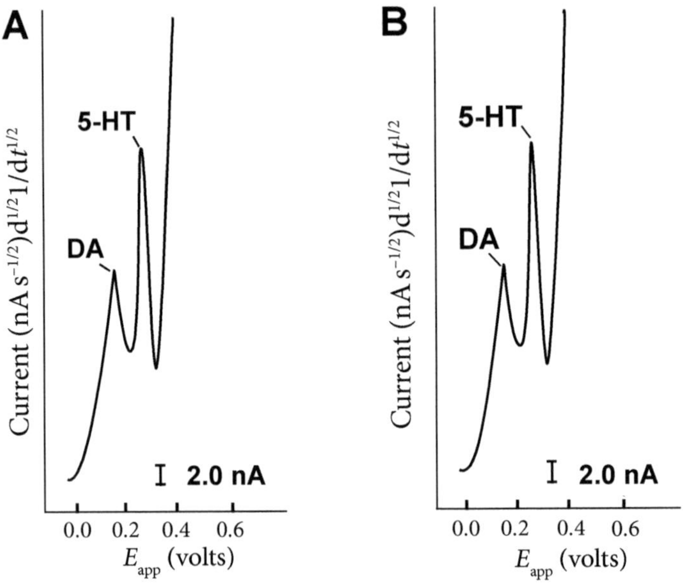

Figure 3.3 *In vivo* electrochemical images (NMI) of DA and 5-HT in vlNAcc before DHA administration (A) and after DHA administration (20–100 mg/kg ip) (B) (n = 5). The BRODERICK PROBE® laurate sensor was selected for the *in vivo* studies. Since little or no ascorbate was released from vlNAcc, the data show that little or no oxidative stress was produced in vlNAcc after DHA.

Representative NMI images, taken *in vivo* before and after DHA administration, are shown in Figures 3.3A and 3.3B, respectively. Selective images of DA and 5-HT release were recorded from the vlNAcc of brain in the male, anesthetized Sprague-Dawley murine animals. The brain is a lipid/protein medium that can enable enhanced detection of ascorbate, DA and 5-HT as is reflected in the *in vitro* PEA/BSA lipid/protein beaker environment.

Indeed, both before and after injectable DHA, baseline DA and 5-HT currents were approximately 5.6 nA and 9.7 nA, respectively. DHA was injected after baseline currents of DA and 5-HT had stabilized and currents had reached a steady-state. Thus, since DHA can cross the blood–brain barrier, little or no oxidative stress was seen because little or no ascorbate release was produced in the vlNAcc. The data suggest that ascorbate concentrations were lower than those of the monoamines at the recording specifications noted for these *in vivo* images: sensitivity [5.0 nA/V]; time constant [1 sec]; scan rate [10 mV/sec]. Charging current (-0.2±0.0 V) is not shown. Expectedly, *in vivo* images of the currents produced by endogenous DA and 5-HT in vlNAcc using the laurate sensor at a sensitivity of 5.0 nA/V, closely correlated with *in vitro* images of current produced by DA and 5-HT using the precalibrated pretreated PEA/BSA laurate sensor.

Also, expectedly, since ascorbate cannot cross the blood–brain barrier by facilitative transport, unless injectable ascorbate (5–35 mg/kg ip) did not significantly affect DA and 5-HT release in the vlNAcc (data not shown).

The technology described in this chapter provides the ability to study specific neurotransmitters in specific neuroanatomic substrates of living brain before and after drug treatment. This aspect of the technology provides a reliable internal control. Nonetheless, tests of saline (body weight in kg ip) effects on neurotransmitters in living brain were also performed. Saline did not significantly affect concentration of either monoamine (data not shown).

3.6.3 *In Vivo* Studies: Morphine

Representative NMI images, taken *in vivo* before (Figure 3.4A) and after administration of the first dose of morphine (10 mg/kg sc) (Figure 3.4B) and after the second dose of morphine (10 mg/kg sc repeated) (Figure 3.4C). Selective images of DA and 5-HT release were recorded from the vlNAcc of brain in the male, anesthetized Sprague-Dawley murine animals.

Figure 3.4A represents baseline data as this novel biotechnology (NMI) uses same animal control and baseline data represent 100% (control values). Baseline DA and 5-HT currents produced endogenously were approximately 5.6 nA and 9.7 nA, respectively. Morphine was injected after baseline currents of DA and 5-HT had stabilized and currents had reached a steady-state.

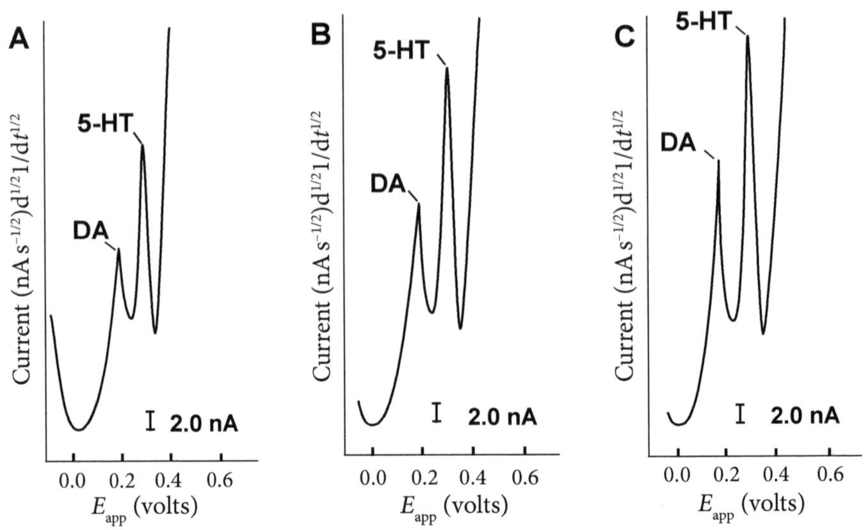

Figure 3.4 *In vivo* electrochemical images (NMI) of DA and 5-HT in vlNAcc before morphine administration (A) and after morphine administration (morphine (10 mg/kg, sc) (B) and a second morphine injection (10 mg/kg, sc) (C) (n = 5). Both doses of morphine significantly increased DA and 5-HT release in vlNAcc ($p < 0.001$) and did not significantly alter ascorbate release. Since little or no ascorbate was released from vlNAcc, little or no oxidative stress was produced in vlNAcc after morphine.

Therefore, Figures 3.4B and 3.4C show that there were significant differences in DA or 5-HT release as measured from baseline after the first dose and after the second dose of morphine, respectively; this is shown by current increases for the DA and 5-HT electrochemical signals. The data showed that the DA current increased from 5.6 nA from baseline to 6.7 nA after the first dose of morphine (Figure 3.4B) and to 8.0 nA after the second dose of morphine (Figure 3.4C) was administered. In addition, 5-HT current increased from 9.7 nA from baseline to 14.0 nA after the first dose of morphine (Figure 3.4B) and to 16/6 nA after the second dose of morphine (Figure 3.4C) was administered. The specifications for these *in vivo* recordings were: sensitivity [5.0 nA/V]; time constant [1 sec]; scan rate [10 mV/sec]. Charging current (–0.2±0.0 V) is not shown.

3.6.4 *In Vivo* Studies: Naloxone

Representative NMI images, taken *in vivo* after the u-opiate antagonist, naloxone (5 mg/kg ip) was administered are shown in Figure 3.5. Therefore, selective images of DA and 5-HT release were recorded from the vlNAcc within the same experiment in which morphine alterations on DA and 5-HT release were studied in *real time*. Figure 3.5 shows that naloxone dramatically decreased the current produced from the morphine-induced enhancement of DA and 5-HT electrochemical signals; the decreased current from electrochemical signals of both monoamines occurred selectively and immediately. Decreased currents, derived from naloxone administration, were compared with results on DA and 5-HT currents from the second morphine injection.

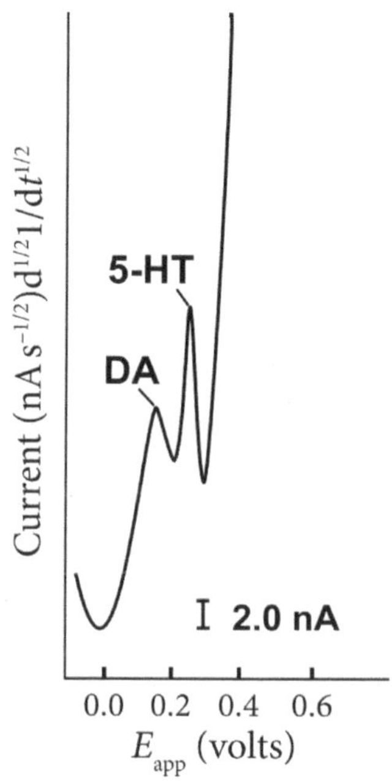

Figure 3.5 BRODERICK PROBE® *in vivo* electrochemical images (NMI) of DA and 5-HT in vlNAcc after naloxone administration (5 mg/kg ip) (n = 5). Naloxone was administered after the second morphine injection. Naloxone significantly decreased DA and 5-HT release in vlNAcc (p < 0.01) and did not significantly affect ascorbate release. Since little or no ascorbate was released from vlNAcc, little or no oxidative stress was produced after naloxone.

The data show that naloxone decreased the DA current from 8.0 nA to 3.1 nA thereby decreasing the DA current almost three-fold. In addition, the data show that the 5-HT current decreased from 16.6 nA to 5.5 nA after the administration of naloxone by three fold. The specifications for these *in vivo* recordings were: sensitivity [5.0 nA/V]; time constant [1 sec]; scan rate [10 mV/sec]. Charging current (−0.2±0.0 V) is not shown.

3.6.5 *In Vivo* Studies: Morphine-Naloxone-DA-5-HT

Figure 3.6A is a line graph and a time plot of the effects of morphine on DA release and 5-HT release in the vlNAcc at the two doses of morphine studied; the effect of the u-opiate antagonist, naloxone, on these morphine-induced monoamines was studied immediately after the morphine studies. As one can assess from Figure 3.6, an advantage of the biotechnology described in the present paper is the ability to monitor levels of neurotransmitters and neuromodulators over time, on line.

The line graph shown in Figure 3.6A shows that significant increases in DA and 5-HT release in the vlNAcc occurred over time after morphine (each point is significant over baseline-control values at $p < 0.05$, 95% C.L.) ($n = 5$) and effects were enhanced with repeated dose. Immediately after the repeated morphine injections, naloxone was administered and naloxone blocked morphine-induced DA and 5-HT release in vlNAcc. Decreased currents for electrochemical signals for DA and 5-HT were reduced by naloxone to below baseline by 32% for DA and approximately 40% for 5-HT at the final time point of the study.

ANOVA results for the study showed a significant difference in DA release between drug doses (first hour morphine, second hour morphine, naloxone) and relative to baseline (ANOVA, df = 18, $F = 109.8$, $p < 0.0001$). Tukey's *post hoc* test indicated that the first dose of morphine (10 mg/kg sc) significantly elevated DA over baseline ($q = 8.621$, $p < 0.001$). The second dose of morphine administered in the second 25 min, further increased DA significantly over baseline ($q = 16.9$, $p < 0.001$), and also relative to the initial first hour dose ($q = 8.781$, $p < 0.001$).

Figure 36B shows a histogram of the DA and 5-HT response to two morphine injections (10 mg/kg sc) and to a single injection of naloxone (5 mg/kg ip). Statistical significance is noted with asterisks (unpaired, two-tailed t-test, df = 7) over baseline. Two asterisks denote a significance of $p < 0.01$ and three asterisks denote a significance of $p < 0.001$).

Naloxone was found to significantly depress 5-HT relative to its baseline concentration ($q = 5.671$, $p < 0.01$) after administration of the second dose of morphine. Naloxone also decreased 5-HT compared to both hour 1 ($q = 13.43$, $p < 0.001$) and hour 2 ($q = 22.22$, $p < 0.001$) doses of morphine individually.

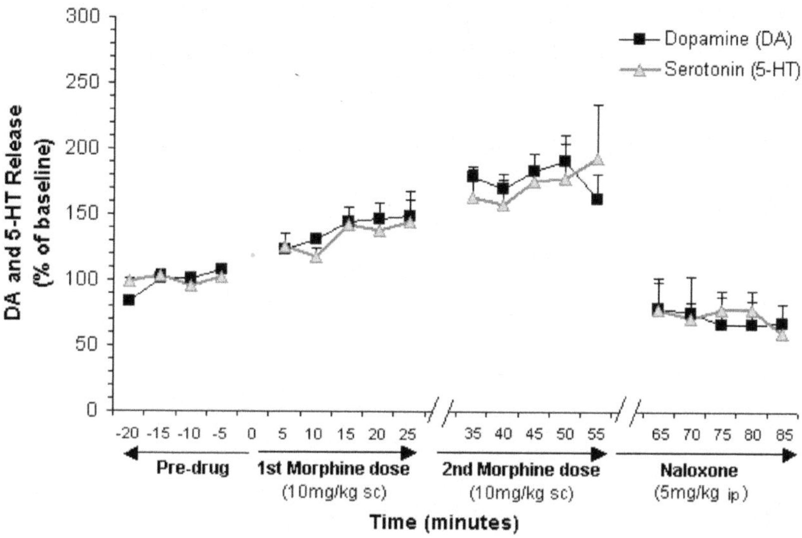

Figure 3.6A A line graph of the time course effects of cumulative morphine injections, followed by a single naloxone injection on DA and 5-HT release in vlNAcc. The x axis delineates treatment effects over time; the y axis delineates % change in monoamine release as compared with baseline (see text for statistics). Morphine increased DA release at both doses by 45% and 80 % and increased 5-HT release by 30% and 75%, respectively. Each point in time for both DA and 5-HT was significant over baseline at the 95% Confidence Limits (C.L.) ($p < 0.05$). BRODERICK PROBE® Electrochemical signals for DA and 5-HT that were increased by morphine, were reduced by naloxone to below baseline by 30% for DA and approximately 25% for 5-HT. The data are also presented in histogram form (Figure 3.6 B). Statistical significance over baseline is noted with asterisks (unpaired, two-tailed t test, df = 7). ** = $p < 0.01$, *** = $p < 0.001$).

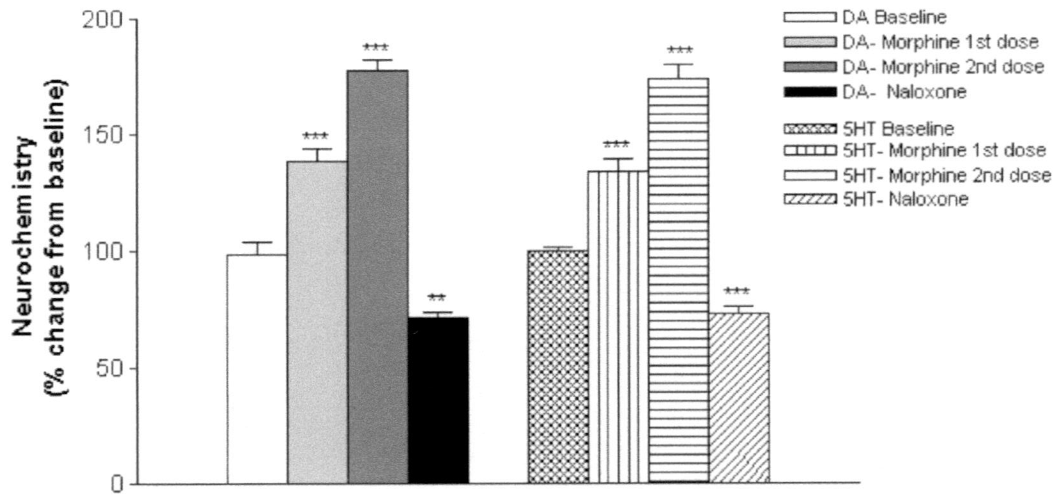

Figure 3.6B The Results from the BRODERICK PROBE®: Naloxone, administered (5.0 mg/kg ip) to animals after the second dose of morphine, was found to significantly decrease DA below baseline (q = 5.87, $p < 0.01$). Naloxone also decreased DA relative to both hour 1 (q = 15.37, $p < 0.001$) and hour 2 (q = 24.15, $p < 0.001$) doses of morphine. Unpaired, two-tailed t-test analysis supported these findings: morphine significantly elevated DA compared to pre-drug/endogenous release in both first hour (df = 7, $p < 0.0009$) and second (df = 7, $p < 0.0001$), and naloxone attenuated DA (df = 7, $p < 0.0015$) relative to baseline.

Unpaired, two-tailed *t*-test analysis supported these results. Morphine significantly elevated 5-HT compared to pre-drug/ endogenous release in both first hour (df = 7, $p < 0.0007$) and second hour doses (df = 7, $p < 0.0001$). Naloxone significantly attenuated 5-HT (df = 7, $p < 0.0003$) compared to baseline values obtained before the first injection of morphine.

- ANOVA results for the study showed a significant difference in 5-HT concentration following drug doses (first hour morphine, second hour morphine, naloxone) and compared to baseline (ANOVA, df = 18, $F = 90.7$, $p < 0.0001$). Tukey's *post hoc* test revealed that the first dose of morphine (10 mg/kg subcutaneous [sc]) after increased 5-HT significantly over baseline ($q = 6.993$, $p < 0.001$). The second dose of morphine in the second hour also elevated 5-HT significantly over baseline ($q = 15.27$, $p < 0.001$), and also relative to the initial first hour dose ($q = 8.783$, $p < 0.001$).

3.7 The Results Discussed

In order to assess the performance of new *N*-stearoylcerebroside and laurate BRODERICK PROBE® sensors, *in vitro* and *in vivo* studies were undertaken. *In vitro* studies with these sensors were performed in saline-phosphate buffer with sensors also pretreated with PEA/BSA, a simulation of the lipid/protein constituents of brain. Moreover, since the laurate sensor was found to have higher surfactant, i.e., electron-transfer properties than the *N*-stearoylcerebroside sensor *in vivo,* [10, 11], the laurate sensor was selected for the *in vivo* morphine-ascorbate oxidative stress studies presented here, albeit both sensors exhibit excellent operational stability, sensitivity and selectivity. Thus, additional studies were performed with the laurate sensor, *in vitro,* in saline-phosphate buffer, without pretreatment with PEA/BSA.

The results showed that in all NMI *in vitro* studies of these new sensors, three well-defined electrochemical images were recorded. The recordings showed that ascorbate was clearly differentiated from DA and 5-HT and DA from 5-HT, as increasingly higher concentrations of each neurochemical was added to the buffer. Neurochemicals were introduced simultaneously as well as separately into buffer. Repetitive studies of these sensors were undertaken. The increasing amplitude of the electrochemical images for each neurochemical occurred in the following order: ascorbate < DA < 5-HT. Concentrations of ascorbate were higher than those of the monoamines, that is, 3–300-fold higher than those of the monoamines. The data showed that these new *N*-stearoylcerebroside and laurate sensors exhibited excellent operational stability, sensitivity, and selectivity.

It is noteworthy that these are the first studies to be published in a journal format to quantify the performance of BRODERICK PROBE® *N*-stearoylcerebroside and laurate sensors, developed and manufactured in this laboratory. Previous studies on these sensors are published in national and international patents [10, 11]. Taken together, we see a multitude of BRODERICK PROBE® sensors. a library of formulae. Other laboratories have reported electrochemical microprobes which can separate ascorbate, DA and 5-HT signals *in vitro*; these microprobes are made of overoxidized poly(1,2-phenylenediamine) coated carbon fiber and DNA attached to carbon fiber *via* gold particles, respectively [24, 28]. However, to this authors' knowledge, there are no LIVE *in vivo* studies which have been done with these latter electrodes. DNA is happily electro-oxidizable as are its base pairs and it is exciting to think about a molecule such as pyrimidine, producing each of the four base pairs except guanine. The billions of applications of the BRODERICK PROBE® neurosensors, nanoprobes and biochips will keep the company, EAZYSENSE, going for several lifetimes. Actually, the whole world is invited.

For the *in vivo* studies, a morphine/ascorbate animal model was used. The rationale for the use of this animal model lies in the well-established biochemistry of monoamine oxidative mechanisms. When monoamines oxidize, reactive oxygen species, such as orthoquinones, are produced. Ascorbate in its role as a major cellular antioxidant provides neuroprotection against morphine-induced reactive oxygen species derived from the oxidation of monoamines. Thus, the very monoamines which morphine can induce to, e.g., relieve pain then actually further produce oxidative-stress metabolites which are neurotoxic and can by blocked by ascorbate. Thus, ascorbate is a reliable measure of oxidative stress. Indeed, ascorbate has been reported to be a *neuromodulator* in dopamine as well as in glutamate neurons [35].

Moreover, for the *in vivo* studies of ascorbate and the oxidized form of ascorbate, which is DHA, selective images of DA and 5-HT release were recorded from the vlNAcc of brain in the male, anesthetized Sprague-Dawley murine models. The brain is a lipid/protein medium and will enable enhanced detection of ascorbate, DA and 5-HT. The data showed that DHA (20–100 mg/kg ip), a molecule able to cross the blood–brain barrier, produced little or no ascorbate release, signifying that little or no oxidative stress occurred in the vlNAcc after DHA administration. It is important to mention though that another study used doses of DHA up to 500 mg/kg, given the caveat that these high doses were used to assess cerebral blood flow [25]. Finally, in the present study, the results showed that there were no significant differences in DA or 5-HT release as measured from baseline after DHA was administered at any time during injections of DHA.

The present data agree with a study in which sufentanil, a structural analog of fentanyl and a potent analgesic of the 4-anilidopiperidine class of opioids, increased extracellular DA in striatum but did not produce oxidative stress as shown by no change in the release of striatal extracellular ascorbate [36]. This laboratory has preliminary data which showed that fentanyl, *per se*, dramatically increased DA and 5-HT in the vlNAcc and did not change ascorbate release (unpublished data), thereby agreeing with the previous report that little or no oxidative stress was caused by the sufentanil and fentanyl opioids.

Other reports, in addition to [36] have also targeted nigrostriatal pathways in studies on morphine-ascorbate oxidative stress and in contrast to [36], ascorbate and xanthine oxidative stress metabolites were increased in response to morphine [19, 20]. The authors of these publications suggest that morphine increased DA and 5-HT and their oxidative stress metabolites only at striatal sites. The authors also suggest that a u-opioid receptor-mediated mechanism tonically controls extracellular ascorbate and glutamate concentrations at these specific neuroanatomic sites.

In this paper, studies on peripherally injectable ascorbate were performed. The findings were, expectedly, that injectable ascorbate did not significantly induce the release of ascorbate in vlNAcc, nor did injectable ascorbate affect DA and 5-HT release in the vlNAcc (data not shown). The results support other reports that ascorbate does not facilitate transport across the blood–brain barrier when ascorbate is present in its reduced form [3].

Other laboratories have performed studies using injectable ascorbate peripherally to assess morphine/ascorbate mechanisms [4, 34]. In the first study, ascorbate was injected peripherally to determine its effect on the morphine withdrawal syndrome and the authors report that their results were mixed. In the second study, peripherally injected ascorbate was used to study 3,4,dihydroxyphenylacetic acid (DOPAC) changes in striatum after morphine; the authors state that results were again inconsistent. These results were likely due to the inability of ascorbate in its reduced form, to use facilitative transport to cross the blood–brain barrier [3].

3.8 The Solution to Electrocatalysis: Abstract of Success: BRODERICK PROBE®

The *in vitro* and *in vivo* performance of new BRODERICK PROBE® *N*-stearoylcerebroside and laurate sensors, developed in the Broderick laboratory is presented. The *in vitro* and *in vivo* performance of these new sensors was evaluated using a patented, electrochemically based neuromolecular imaging (NMI) biotechnology. The sensors were placed in saline/phosphate buffer and were pretreated with a solution of phosphotidylethanolamine (PEA) and bovine serum albumin (BSA) which simulates the brain environment. Each new sensor was tested for operational stability, sensitivity and selectivity by pipetting graduated increases in concentrations of ascorbate, dopamine (DA) and serotonin (5-HT) into an electrochemical cell containing buffer. Multiple and repetitive images of electrochemical signals from ascorbate, DA and 5-HT were recorded. *In vitro* results showed that both new sensors produced three cathodic electrochemical, well-defined, and selective peaks for ascorbate, DA, and 5-HT. Dopamine and 5-HT were generally detected at nM concentrations, while ascorbate was generally detected at uM concentrations.

Concentrations of ascorbate were approximately 3–300 times greater than concentrations of DA or 5-HT. Further *in vitro* studies with the laurate sensor were performed without PEA/BSA pretreatment to serve as an additional control since the laurate sensor was selected to perform the *in vivo* studies. For the *in vivo* studies, a morphine/ascorbate animal model was used because this model has been linked to ascorbate, DA, 5-HT release and to monoamine oxidative metabolism at nigral sites [19, 20]. In the present paper, *in vivo* oxidative stress mechanisms were tested in a morphine/ascorbate animal model using NMI wherein laurate sensors were implanted at a mesolimbic site, the ventrolateral nucleus accumbens (vlNAcc), in anesthetized male, Sprague-Dawley strain of murine animals. In separate studies, animals were injected with (1) ascorbate, (5–35 mg/kg ip) or (2) dehydroascorbate (DHA) (20 to 100 mg/kg ip). Then, in another separate study, (3) morphine sulfate (10–20 mg/kg sc) was injected followed by a single injection of naloxone (5 mg/kg ip). Results showed that (a) neither ascorbate nor DHA injections produced ascorbate release in the vlNAcc, (b) morphine significantly increased DA and 5-HT release but did not alter ascorbate release in the vlNAcc, and (c) naloxone significantly reversed the increased DA and 5-HT release in the vlNAcc produced by morphine. Since ascorbate is released *in vivo* to compensate for oxidative stress, the data suggest that little or no change in oxidative stress occurred with morphine/ascorbate interactions in the vlNAcc. The data are in agreement with studies of sufentanil/ascorbate oxidative stress [36]. Ascorbate/monoamine oxidative stress mechanisms are important in the relief of pain, in reducing opiate addiction and in neuroprotection against stroke. Thus, new *N*-stearoylcerebroside and laurate sensors, which selectively detect and separate these critical neuromolecules, are clinically relevant.

3.9 To Conclude

In this work, data are reported that show that the pain reliever, morphine, selectively increased DA and 5-HT release in the vlNAcc in the male, anesthetized Sprague-Dawley murine animal species and this enhancement of the monoamines occurred after the administration of two cumulative doses of morphine sulfate (10 mg/kg sc and an additional 10 mg/kg sc). Moreover, naloxone, the u-opiate antagonist, blocked morphine-induced increases in monoamines at a dose of (5 mg/kg ip). The present data support previously published data which show that morphine increased extracellular DA in mesolimbic pathways specifically in NAcc [17, 18, 37]. The present data also receives support from other laboratories which showed that naloxone blocked the morphine-induced release of DA [23].

These data show for the first time, *in vivo*, *on line*, and *real-time* enhancement of selective DA and 5-HT release in vlNAcc produced by the opiate morphine and shows, *online*, the subsequent reversal of these morphine-induced monoamine effects by the u-opiate antagonist naloxone. Morphine produced significant increases of 40% and 80%, respectively, in DA release in vlNAcc at each dose of morphine administered. Morphine produced significant increases of 30% and 75%, respectively, in 5-HT release in vlNAcc at each dose of morphine administered. Current changes in NMI signaling of DA, after morphine was administered at both doses, were significantly increased as well. The DA current was increased from 5.6 nA to 6.7 nA to 16.6 nA, respectively; the 5-HT current was increased from 9.7 nA to 14 nA to 17 nA, respectively. Naloxone decreased the DA current almost three-fold, while naloxone decreased the 5-HT current by three-fold.

The paper presents an assessment of the operational stability, sensitivity and selectivity of two new patented sensors, i.e., the *N*-stearoylcerebroside and the laurate sensors. The data show that clearly defined and selective separation of the detection of ascorbate, DA and 5-HT was delineated *in vitro*. *In vivo* studies on morphine-ascorbate oxidative stress mechanisms were performed with the laurate sensor. The data show that little or no change in oxidative stress occurred with morphine/ascorbate interactions in the vlNAcc as shown by little or no increase in ascorbate release in the vlNAcc. Ascorbate/monoamine oxidative stress mechanisms are important in the relief of pain, in reducing opiate addiction and in neuroprotection against stroke. Thus, new *N*-stearoylcerebroside and laurate sensors, which selectively detect and separate these critical neuromolecules, are clinically relevant.

The paper sets up the Broderick Research Laboratory solutions to the electrocatalytic dilemma that plagued electrochemists/neuroscientists for many years. An important aspect of electrocatalysis is the fact that the phenomenon is not always a limitation. In fact, herein, Chapter 6, the benefits that electrocatalysis can bring to medicine, especially to the patient with diabetes, are emphasized.

It should be noted that the Broderick lab was the first to publish brain sensing as the expert in both fields of the brain and sensors. The BRODERICK PROBE® *was and is the first pre-medical and medical device to bring brain sensing into the MDPI sensing journals in 2006 (Basel, Switzerland and Beijing, China). At the same time, this neuroprobe was and is the first pre-medical and medical device to bring sensors into the neurology and psychiatry journals such as the Journal of Neural Transmission (formerly, Wien Austria, now, Berlin/Heidelberg, Germany, London, UK, New York, NY, USA).*

Thus, the Broderick lab has brought "brain sensing" to the world.

We move next to the drivers for the sensors to sense the brain. Broderick is thinking...

References

1. Abbott FV, Etienne P, Franklin KB, Morgan MJ, Sewitch MJ, Young SN (1992) Acute tryptophan depletion blocks morphine analgesia in the cold-pressor test in humans. *Psychopharmacol* (*Berl*) 108(1–2): 60–66.

2. Adams RN, Marsden CA (1982) New techniques in psychopharmacology. In: Iversen LL, Snyder SH, (eds.). *Handbook of Psychopharmacology*. New York: Plenum Press.

3. Agus *DB, Gambhir SS, Par*dridge WM, Spielholz C, Baselga J, Vera JC, Golde DW (1997) Vitamin C crosses the blood-brain barrier in the oxidized form through the glucose transporters. *J Clin Invest* 100: 2842–2848.

4. Alaei H, Esmaeili M, Nasimi A, Pourshanazari A (2005) Ascorbic acid decreases morphine self-administration and withdrawal symptoms in rats. Pathophysiology 12: 103–107.

5. Borman SA (1982) New electroanalytical pulse techniques. *Anal Chem 54*: (6) 700A (A698-&).

6. Broderick PA (1988) Distinguishing *in vitro* electrochemical signatures for norepinephrine and dopamine. *Neurosci Lett* 95: 275–280.

7. Broderick PA (1989a) Cathodic electrochemical current arrangement with telemetric application. US Patent #4, 883,057.

8. Broderick PA (1989b) Characterizing stearate probes in vitro for the electrochemical detection of dopamine and serotonin. *Brain Res* 495: 115–121.

9. Broderick PA (1991) In vivo voltammetric studies on release mechanisms for cocaine with γ-butyrolactone. *Pharmacol Biochem Behav* 40: 969–975.

10. Broderick PA (1995) Microelectrodes and their use in cathodic electrochemical current arrangement with telemetric application, US Patent #5,433,710.

11. Broderick PA (1999) Microelectrodes and their use in an electrochemical arrangement with telemetric application. US Patent #5,938,903.

12. Broderick PA, Pacia SV (2005) Imaging white matter signals in epilepsy patients: A unique sensor technology. In: Broderick PA, Rahni DN, Kolodny EH, (eds.). *Bioimaging in Neurodegeneration*, Totowa, NJ: Humana Press Inc., pp. 199–206.

13. Broderick PA, Pacia SV (2006) Identification, diagnosis, and treatment of neuropathologies, neurotoxicities, tumors and brain and spinal cord injuries using microelectrodes with microvoltammetry, US Patent #7,112,319.

14. Broderick PA, Pacia SV, Doyle WK, Devinsky O (2000) Monoamine neurotransmitters in resected hippocampal subparcellations from neocortical and mesial temporal lobe epilepsy patients: *in situ* microvoltammetric studies. *Brain Res* 878: 49–63.

15. Cherubini A, Ruggiero C, Polidori MC, Mecocci, P (2005) Potential markers of oxidative stress in stroke. *Free Radic Bio Med* 39(7): 841–852.

16. Coury LA, Huber EW, Heineman WP (1989) Applications of modified electrodes in the voltammetry determination of catecholamine neurotransmitters. *Biotechnology* 11: 1–37.

17. Di Chiara G, Imperato A (1988) Opposite effects of mu and kappa opiate agonists on dopamine release in the nucleus accumbens and

in the dorsal caudate of freely moving rats. *J Pharmacol Exp Ther* 224: 1067–1080.

18. Di Giannuario A, Pieretti S (2000) Nociceptin differentially affects morphine-induced dopamine release from the nucleus accumbens and nucleus caudate in rats. *Peptides* 21: 1125–1130.
19. Enrico P, Esposito G, Mura MA, Fresu L, De Natale G, Miele E, Desole MS, Miele M (1997) Effect of morphine on striatal dopamine metabolism and ascorbic and uric acid release in freely moving rats. *Brain Res* 745: 173–182.
20. Enrico P, Mura MA, Esposito G, Serra P, Migheli R, De Natale G, Desole MS, Miele M, Miele E (1998) Effect of naloxone on morphine-induced changes in striatal dopamine metabolism and glutamate, ascorbic acid and uric acid release in freely moving rats. *Brain Res* 797: 94–102.
21. Evangelou A, Kalfakakou V, Georgakas P, Koutras V, Vezyraki P, Iliopoulou L, Vadalouka A (2000) Ascorbic acid (vitamin C) effects on withdrawal syndrome of heroin abusers. *In Vivo* 14: 363–366.
22. Foucault R, Broderick PA, Rahni DN, Lombardi JR, Birke RL (2002) Neurotransmitter signatures: A correlation between Raman spectroscopy and the microvoltammetric BRODERICK PROBE®. NIH/NIGMS Symposium, New Orleans LA.
23. Herz A, Shippenberg TS (1989) Neurochemical aspects of addiction: Opioids and other drugs of abuse. In: Goldstein A, (ed.). *Molecular and Cellular Aspects of the Drug Addictions*. New York: Springer-Verlag, pp. 111–141.
24. Hocevar SB, Zivin M, Milutinovic A, Hawlina M, Hutton E, Ogorevc B (2006) Simultaneous in vivo measurement of dopamine, serotonin and ascorbate in the striatum of experimental rats using voltammetric microprobe. *Front Biosci* 11: 2782–2789.
25. Huang J, Agus DB, Winfree CJ, Kiss S, Mack WJ, McTaggart RA, Choudhri TF, Kim LJ, Mocco J, Pinsky DJ, Fox WD, Israel RJ, Boyd TA, Golde DW, Connolly ES Jr (2001) Dehydroascorbic acid, a blood-brain barrier transportable form of vitamin C, mediates potent cerebroprotection in experimental stroke. *Proc Natl Acad Sci USA* 98: 10989–10991.
26. Johnston PA, Chahl LA (1992) Chronic treatment with ascorbic acid inhibits the morphine withdrawal response in guinea-pigs. *Neurosci Lett* 135: 23–27.
27. Kissinger PT, Preddy CR, Shoup RE, Heineman WR (1996) Fundamental concepts of analytical electrochemistry. In: Kissinger

PT, Heineman WR, (eds.). *Laboratory Techniques in Electroanalytical Chemistry*, New York: Marcell Dekker Inc., pp. 11–50.

28. Lu L, Wang S, Lin X (2004) Attachment of DNA to the carbon fiber microelectrode via gold nanoparticles for simultaneous determination of dopamine and serotonin. *Anal Sci* 20(8): 1131–1135.

29. Meunier JC, Mollereau C, Toll L, Charles Suaudeau, Moisand C, Alvinerie P, Butour J-L, Guillemot J-C, Ferrara P, Monsarrat B, Mazarguil H, Vassart G, Parmentier M, Costentin J (1995) Isolation and structure of the endogenous agonist of opioid receptor-like ORLI receptor. *Nature* 377: 532–535.

30. Okuda AE, Tachibana S, Houtani T, Minami T, Masu Y, Nishi M, Takeshima H, Sugimoto T, Ito S (1996) Identification and characterization of an endogenous ligand for opioid receptor homologue ROR-C: Its involvement in allodynic response to innocuous stimulus. *Mol Brain Res* 43: 96–104.

31. Oldham K (1973) Semi-integral electroanalysis: Analog implementation. *Anal Chem* 45: 39–50.

32. Pacia SV, Doyle WK, Broderick PA (2001) Biogenic amines in resected human neocortex in patients with neocortical and mesial temporal lobe epilepsy: Identification with *in situ* microvoltammetry. *Brain Res* 899: 106–111.

33. Pellegrino LJ, Pellegrino AS, Cushman AJ (1979). *A Stereotaxic Atlas of the Rat Brain*. Plenum Press, New York.

34. Rajaei Z, Alaei H, Nasimi A, Amini H, Ahmadiani A (2005) Ascorbate reduces morphine-induced extracellular DOPAC levels in the nucleus accumbens: A microdialysis study in rats. *Brain Res* 16: 1053(1–2) 62–66.

35. Rebec GV, Pierce RG (1994) A vitamin as neuromodulator: Ascorbate release into the extracellular fluid of the brain regulates dopaminergic and glutamatergic transmission. *Prog Neurobiol* 43: 537–565.

36. Serra PA, Susini G, Rocchitta G, Migheli R, Dessanti G, Miele E, Desole MS, Miele M (2003) Effects of sufentanil on the release and metabolism of dopamine and ascorbic acid and glutamate release in the striatum of freely moving rats. *Neurosci Lett* 344: 9–12.

37. Spanagel R, Herz A, Shippenberg TS (1990) The effects of opioid peptides on dopamine release in the nucleus accumbens: An *in vivo* microdialysis study. *J Neurochem* 55: 1734–1740.

38. Wightman RM (1981) Voltammetric electrodes. *Anal Chem* 53: 1125A–1134A.

39. Wood PL (1993) Interrelationships of opioid, dopaminergic, cholinergic, and GABAergic pathways in the central nervous system. In: Herz A, (ed.). *Handbook of Experimental Pharmacology.*

Chapter 4

Broderick Is Thinking: Attention to Sensors and No Attention to Sensor Drivers?

Electric potential is the amount of electric potential energy per unit of charge that would be a property of a charged particle if the charged particle were to be placed within an electric field at some specified location. The concept of potential is location-dependent; it is independent of the amount of charge actually present on the particle possessing the electric potential. Therefore, electric potential difference is the difference in electric potential energy per unit of charge between two different locations within an electric field. Electrochemistry is the study of chemical processes that cause electrons to move. This movement of electrons is called electricity. Electricity can be generated by movements of electrons from one element to another in a reaction known as an oxidation/reduction ("redox") reaction.

4.1 The Critical Importance of Electric Potential, Electric Potential Energy, Potential Difference and Conductance

Innovatively, in the case of the BRODERICK PROBE® there is a way to loop the circuit back from low to high potential difference

Neuroimaging: Sensing Biochemistry in the Brain
Patricia A. Broderick

ISBN 978-981-4613-49-1 (Hardcover), 978-0-429-02974-5 (eBook)
www.jennystanford.com

and cycle it by way of initiating the reductive potential in the circuit. Then, the sensor would act optimally as an anion-cation exchange device for imaging new biomarker analytes. Thus, the Broderick reductive circuit is born.

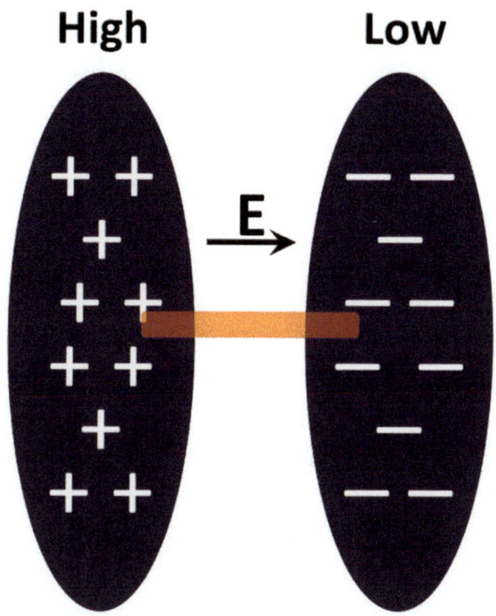

Figure 4.1 Shows the importance of the term, potential difference, the difference in energy potential between two points on a wired circuit. The difference is based on electron flow from high to low potential voltage in joules per coulomb. But, if there were a way to bring electrons from a low potential difference back to the high potential difference, electric potential energy could be cycled using an electronic sensing device, such as an electrode. The conductance would be continuous, therefore closing the wired circuit.

4.1.1 This Is Why She Did It

There was a reason why Broderick performed such a feat as to figure out electrical circuits to sense the brain by electron transfer and this rationale was for her to produce a real advancement in electrochemistry. Now, let us be mindful of the fact that this electron transfer will not occur in the noble gases because their ever so noble electron clouds are uniformly stable, albeit there is a compound of xenon and gold and gold is a noble atom! But, that is not necessarily important right now because Broderick is simply using this example to compare the utterly wonderful freedom that electrons have when they happily sit in "free radicals", a phrase that has taken on a "household" meaning. Everyone speaks of "free radicals". There is often an admonition to partake in substances that destroy "free radicals".

The fact is that free radicals need electrons. Thus, should one place that fact side by side with the fact that free radicals run rampant in such debilitating diseases as Alzheimer, Parkinson with Lewy Body Dementia and the epilepsies including autism, this BRODERICK PROBE® electron sink property cannot be overstated or underestimated. Taken together, the properties of the tiny nanoprobe may assume a higher level in discovery. You see, the mechanism for imaging biomarkers in such patients and murine models with neurodegenerative diseases described herein, derives from a circuit different from other more commonly used electrochemical circuits. In this inventive reductive circuit, for one example, redox reactions are detected in the *same quadrant of the classical electrochemical quadrangle*!

Make no mistake, circuits were and are part and parcel of the *KEY to success in the in vivo electrochemistry part of Neuromolecular Imaging*. Really, no one was able to see this sensor and sensor driver link to success. Broderick saw this, *a priori*. What may have happened? Why was this not seen by others? Perhaps, neuroscientists had targets that they had to reach to treat brain disorders but could not explain these targets analytically to the electrochemists. On the other hand, the electrochemists were in a quandary as to the exact parameters to catch, so to speak, the analyte that the neuroscientists wished to reach to achieve success in their endeavors within the areas of treatment and diagnostics.

But, Broderick's own brain neuronal circuits remained calm during this tumultuous time of debates among electrochemists and neuroscientists and electrochemists wanting to be neuro-scientists and *vice versa*. Patricia worked away in her 200 square foot lab in the *Albert Einstein College of Medicine* in the *Department of Psychiatry*. It is a pleasure to share here that this particular laboratory was especially constructed for the significant work of the BRODERICK PROBE® The designated space for Dr. Broderick was ordered to be renovated for her immediately by the esteemed *Chairperson of Psychiatry* at that time, who was Dr. Herman van Praag, MD, Ph.D. The author remembers working with Dr. van Praag with respectful pride and the person whom Dr. van Praag instructed to renovate the space that was built for her was Bernie in the machine shop.

Stepping up to the plate, so to speak, Broderick did not spin around in a circuitous route. She configured the circuit for the semi-derivative uncoupling of each neurotransmitter's image according to each selective signature, its redox potential. She knew what she had to do. Broderick, not a trained engineer, set out to configure a successful circuit for spectral analysis to image *in vivo* electrochemical events LIVE, to image molecularly with NMI and further, to set up the circuit for quantum mechanics, *Electroactive Photovoltaic Fiber Optics*.

4.1.2 Furthermore, This Is How She Did It

Patricia did it by connecting the reference relative zero cable to the negative terminal of the follower cathode operational amplifier (opamp) from the follower cathode, while sharing the operational amplifier with an extra ground, the auxiliary. In doing so, in one example of the work of these tiny nanoprobes, carbon/lipid/phosphatidyl polymers, the creation of a p+ and n− potential difference induces anionic analytes to become protonated. In effect, an anion/cation available structure is set up as electrons move freely wherever charge moves freely in the electron spaces within and around the atomic cloud. The resulting continuing "potential difference" between the electron source and the protonated sink reaches higher energy levels in joules *per* second as in a semiconductor, such as in the Broderick carbon allotropic semiconductor fullerene and protein fiber optic nanoprobes.

This circuit invention involves, in one example, inserting a BRODERICK PROBE® nanobiosensor in the brain, body or blood of a mammalian whether that be animal or human (not precluding the plant kingdom) and placing a miniature customized reference electrode and auxiliary electrode about seven mm away from the indicator. The reference electrode provides a relative zero voltage point. The auxiliary electrode maintains the current using electric potential control linked to the reference operational amplifier and thus, provides a ground as drift is eliminated. Voltage is applied between the indicator and the reference operational amplifiers. Excellent temporal and spatial stability is well known worldwide as a property of the BRODERICK PROBE® itself and this property is dependent on factors such as the stability of each operational amplifier as the tiny nanoprobe images signature electrochemical half waves online and in real time directly from the brain within seconds and sub-seconds even as the subject is walking.

It is exciting and it is, humbly spoken, the first time anyone has accomplished such a scientific and medical advance as the one spoken of herein. It is still unmatched. The mini neuro-probe is delicately inserted in discrete areas of the brain and selective profiles of each neurotransmitter and neurochemical has its very own signature at least one example of the novel inventive art. The sensing mechanism can and does select specific electrical potentials with respect to its opamp operational stability in concert with an opamp-driven reference within the internal circuit of the detector. In the quantum mechanic set of nanoprobes, near infrared enables photons to travel through fiber optics through the living skull without creating an aperture in the brain at all.

The signal emitted by any biological system is produced by the charge flowing, or current, between the reference and nanobiosensor indicator. The current is proportional to the size of the particular sensor or biosensor, the shape of the sensor and the mass of the neurotransmitter or neurochemical imaged. The electroactive signal is preferably processed in the inventive reductive circuit with three operational amplifiers, one for the biosensor indicator and is called a current measurer operational amplifier and one for the reference, a follower opamp. The auxiliary electrode in the inventive reductive circuit serves as an additional ground given the caveat that an additional ground is not mandatory.

The voltammetry circuit arrangement is semi-differentiated and can reliably be termed as semi-derivatized. The difference between the two voltammetry circuits, typically used in the field of electrochemistry, is minimal; the difference lies in a feedback loop. The semiderivative circuit is an electrical circuit that is comprised of a series of ladder networks of resistors (from about 2 kiloohms to about 2 megohms and capacitors (from about 5×10^{4} to 5×10^{-2} microfarads) which processes the current signal as the analog of the first one-half derivative of the undifferentiated signal. Either the actual resistors and/or capacitors are configured within the potentiostat such as the one preferable to this author or what is possible also, is to program algorithms mathematically sound that can mimic a ladder network of resistors and capacitors complementing each other in a relaxed staggered form.

Thus, Patricia did it, that is, she enabled a reduction circuit, as shown in Figure 4.2 by connecting the reference relative zero cable/lead to the negative terminal of the follower cathode operational amplifier (opamp) from the follower cathode from the IR_F Booster Recorder and further from the input operational amplifier; then, the cable was configured to the negative terminal of the cathodic current opamp in the internal part of the potentiostat. The auxiliary potential control opamp does not work on its own; it is controlled by the reference opamp. Although there is a connection from the nanobiosensor opamp to the auxiliary ground electrode potential opamp, this connection works *through* the reference opamp as is also seen in Figure 4.2.

4.2 A Circuit Diagram Illustrating the Broderick Patent Embodiment: The Reduction Circuit

The inventive art is explained further with five schematic drawings showing linear signally versus more sharply defined signal processing according to potentiometric designs such as semi-derivative. This is shown in Figure 4.3.

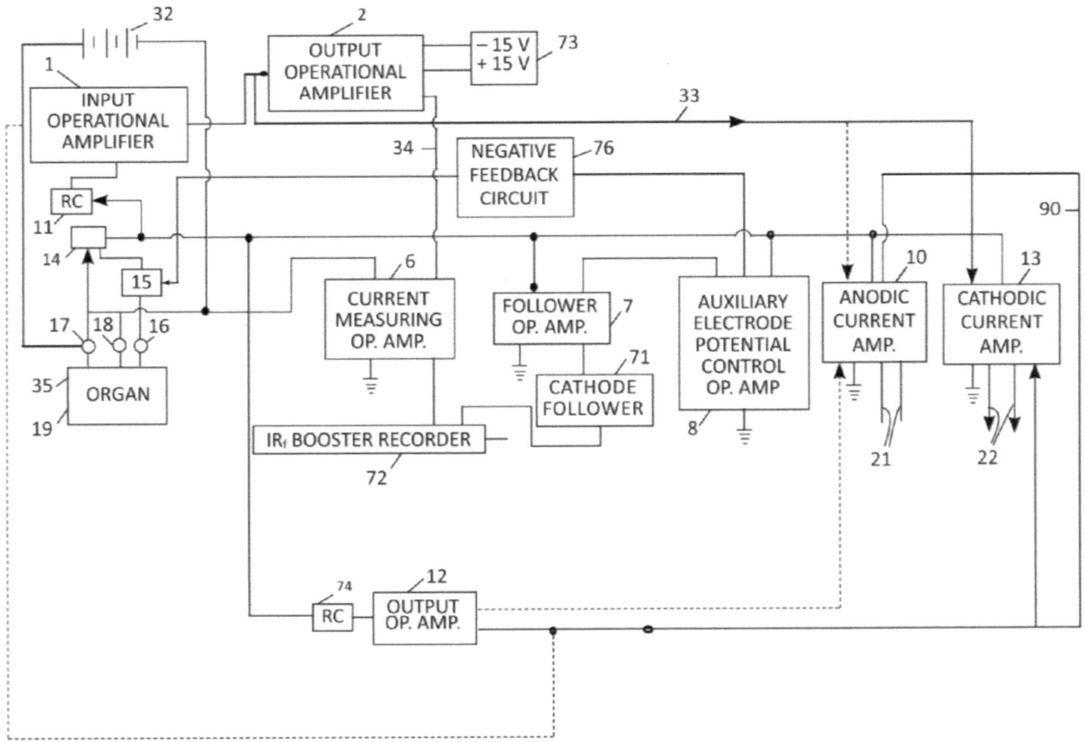

Figure 4.2 Circuit diagram illustrating an embodiment of a circuit arrangement according to the method of this invention. The diagram illustrates the circuitry for producing cathodic (reduction) currents to distinguish these from anodic (oxidation) currents, given the caveat that the BRODERICK PROBE® can be used as an electron source or sink, enabling anodic and cathodic currents to be distinguished according to this inventive circuit.

4.3 Linear Processed Signals versus Signally More Sharply Defined: Semiderivative Potentiometry

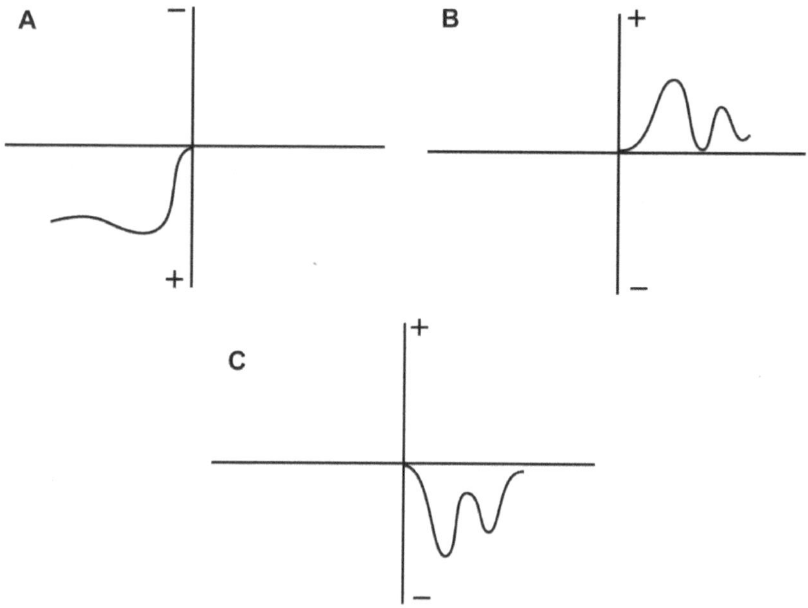

Figure 4.3A–C Representations of graphs showing signals which may be generated by linear and semiderivative scanning. (A) Graph drawn according to conventional electrochemical format wherein the axis indicating anodic (oxidation) current points downward and the axis indicating positive voltage points to the left using linear sensor scanning. (B) Semiderivative sensor signals drawn according to present day conventional format, wherein the axes indicating anodic (oxidation) current points upward and the axes indicating positive voltage points to the right. (C) Graphic shows semi derivative signal processing wherein the axes indicating anodic (oxidation) current points downward and negative potentials point to the right.

4.3.1 The Inventive Art Explained: Three Schematic Drawings—Figures 4.3A–C

Specifically, Figure 4.3A is a schematic diagram that depicts a graph drawn according to previously conventional electrochemical format wherein the *x* axis, indicating anodic (oxidation) current typically points downward and the axis indicating positive increasing voltage to the left. This reflects a classical voltammogram derived from an electrical circuit that is configured as linear. As voltage potential is scanned along the *x* axis, neurotransmitters and neurochemicals are imaged each according to a specific half wave potential.

Figures 4.3B,C are the focus of this treatise on a reduction circuit for electrochemistry. Figure 4.3B depicts semi-derivative voltammetry signals drawn according to conventional format, which is the one in use in the present-day vernacular, wherein the axes indicating anodic current points upward, that is, increasing volts in applied potential are shown in the upper right quadrant. In the inventive art, the reduction circuit allows oxidized and reduced species to be selectively imaged in the same electrochemical quadrant. Figure 4.3C shows another embodiment for the inventive reductive circuit.

4.3.2 The Inventive Art Explained: Schematic Drawings—Figures 4.3D and 4.3E

In Figure 4.3D, using the previously conventional electrochemical quadrangle, we see the method with which the electrochemists described analytes detected with the linear scanning circuitry. The analytes that are oxidized are drawn in the bottom left quadrant to denote anodic/oxidative current. Explaining electrochemistry in this way is the preferable format but neuroscientists found this way difficult to understand. Therefore, the scans were drawn in the upper right-hand quadrant of the electrochemistry quadrangle for easy access to scientists wishing to enter into the field of electrochemistry to monitor neurotransmitters that are oxidized. This method became simply, *par for the course.* Really, it was rare according to the author's knowledge, that neuroscientists understood the meaning of the electrochemical quadrangle.

In Figure 4.3E, using the previously conventional electrochemical quadrangle, the electrochemistry we see is the difference between the linear scanning circuit and the semiderivative circuit which is the first half derivative of the linear circuit and

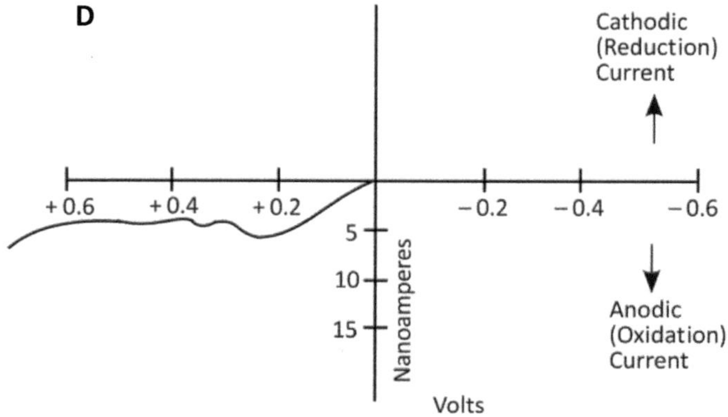

Figure 4.3D Diagram illustrating a linear scan voltammogram showing the production of biogenic chemicals using the conventional electrochemical notation. The resulting anodic voltammogram is shown then in the lower left hand quadrant of the electrochemical signal processing.

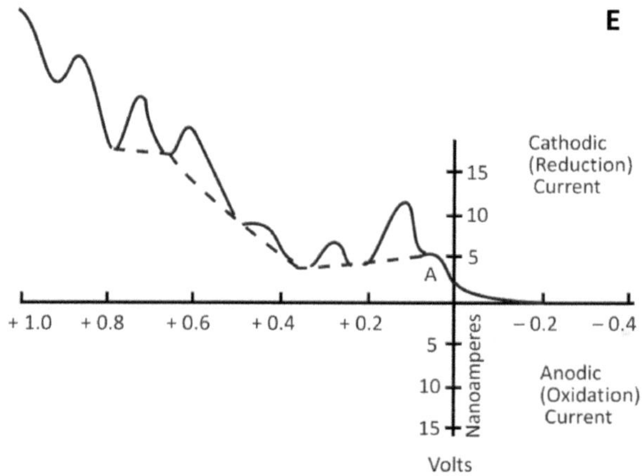

A = Charging current and non-faradaic
i.e., non-faradaic meaning non-physiological

Figure 4.3E The semi-derivative voltammogram, delineated in previously conventional electrochemical notation. The images drawn show the production of biogenic chemicals peaks at 150, 290, 520, 690, 790, 820, and 910 mV. This figure illustrates the usefulness of the method of this invention in medical diagnosis, as the peaks represent neurotransmitter and neurotransmitter-like peptides as detected by cathodic currents in accordance with the method of this inventive reduction circuit with semiderivative signal processing. These peaks, actually biomarkers for brain disorders were not enabled, to the author's knowledge, before the advent of the use of a reduction circuit using the BRODERICK PROBE® in its electron sink capacity.

one which provides much sharper curves and less interference from neighboring analytes. One readily sees that this is drawn in the upper left quadrant, which is the reduction quadrant, one configured for the BRODERICK PROBE® to act as an electron sink. Linear circuit results in waveforms in the lower left quadrant (Anodic) whereas semiderivative circuit using the reductive results in waveforms in the upper left quadrant (Cathodic).

4.4 Electron Transfer Kinetics with the BRODERICK PROBE®

Electroactive signals are drawn in what has now become present day conventional electrochemical format to show imaging of neurotransmitter profiles. The Broderick laboratory is video-tracking these profiles LIVE for the purpose of providing personalized medicine. The profiles, the signature of each neurotransmitter, embedded in the memory of the BRODERICK PROBE® polymer is imaged while the patient or the animal is in the natural state, in physiologically altered states of consciousness or while the subject is freely moving about. Moreover, the nanoprobe can image signals from the brain while the patient or animal is exhibiting fine movements of the limbs and the mouth. These fine movements are called stereotypies that are so typically associated with psychostimulant psychosis after cocaine or amphetamine, for example. Another example of the use of the nanoprobe is to image an epileptic seizure *in situ*. The nanotechnology of the BRODERICK PROBE® is amazing as clear electroactive signals are imaged from any part and discrete parts of the brain while the subject is walking!

It is extremely interesting that electrochemists began to use this type of quadrant format as shown in Figure 4.3E likely to help broaden the scope for neuroscientists and others to facilitate neuroscientists and others in reading redox currents. It may have come about for the neuroscientists and others who were without technical training in the art of electrochemistry or neuroengineering. This new format for electrochemical graphics did help neuroscientists to understand signal processing with electron transfer kinetics. The use of the quadrant in true electrochemical mechanics is rarely discussed among the neuroscientists or even among electrochemists.

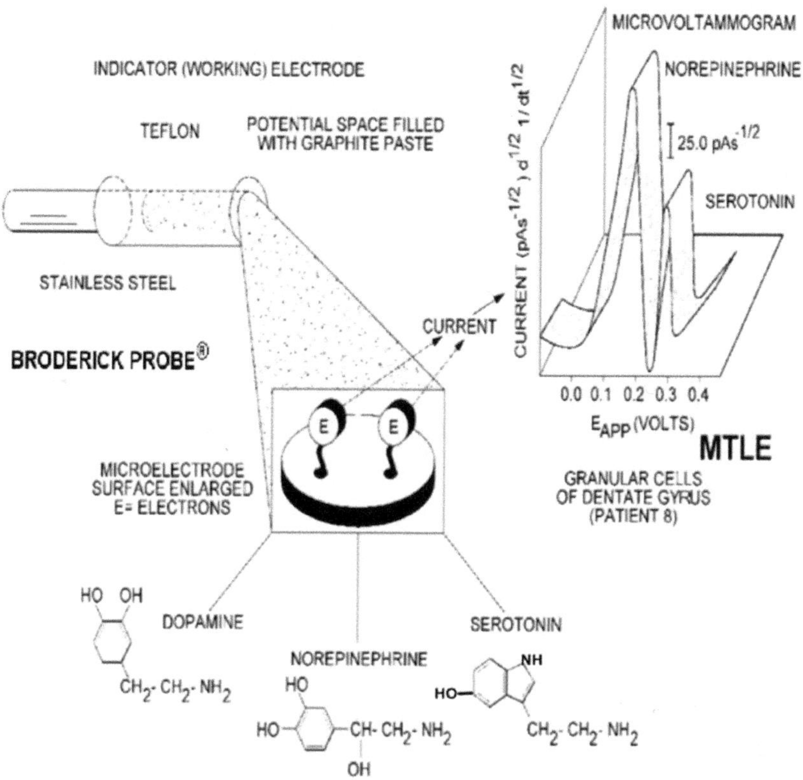

Figure 4.4 The mechanism of action of the BRODERICK PROBE® with NMI; this nanotechnology is based on electrochemical principles and brings new concepts of electron transfer, discovered in the Broderick laboratory and reported herein. **Upper left**: BRODERICK PROBE® which one cannot see with the naked eye. **Center**: Interfacial surface of the BRODERICK PROBE®. Notice the original NMI recordings at the upper right of the schematic diagram. Each neurotransmitter exhibits a signature along the *x* axis showing millivolts in the graphics. Neurotransmitters, dopamine (DA), norepinephrine (NE) and serotonin (5-HT) are selectively imaged. Potentials are based on the empirically derived half wave potentials for each neurotransmitter. The data are derived from granular cells of the dentate gyrus in a mesial temporal lobe epilepsy (MTLE) patient, *in situ*.

4.5 The Electrochemical Quadrangle: Cyclic Voltammetry

The use of the four quadrants of the classical electrochemical quadrangle also shows that this inventive reductive circuit is not cyclic voltammetry. Three cyclic voltammetry examples are given in Figures 4.5 to 4.7.

Thus, it is clear from Figures 4.5–4.7 that the inventive reductive circuit is not cyclic voltammetry.

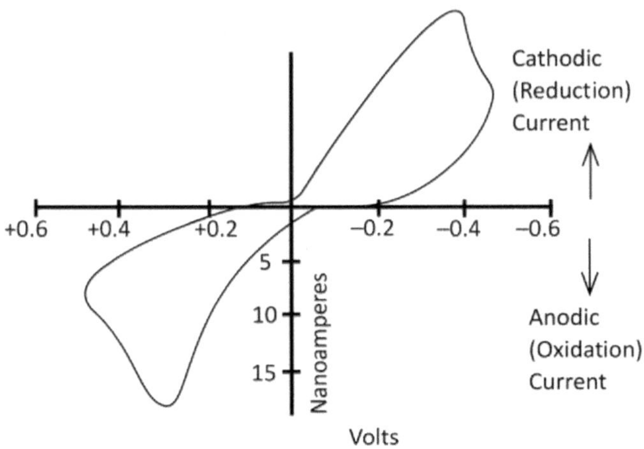

Figure 4.5 Representation of a typical linear scan cyclic voltammogram for a reversible redox reaction by electrochemical convention.

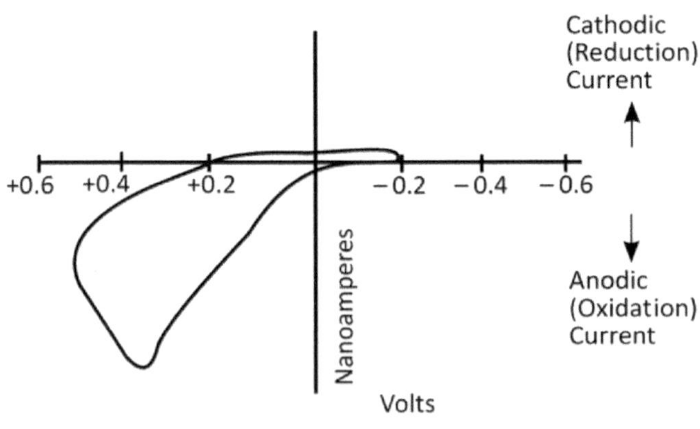

Figure 4.6 Representation of a typical linear scan cyclic voltammogram for an irreversible redox reaction producing a stable oxidized species only.

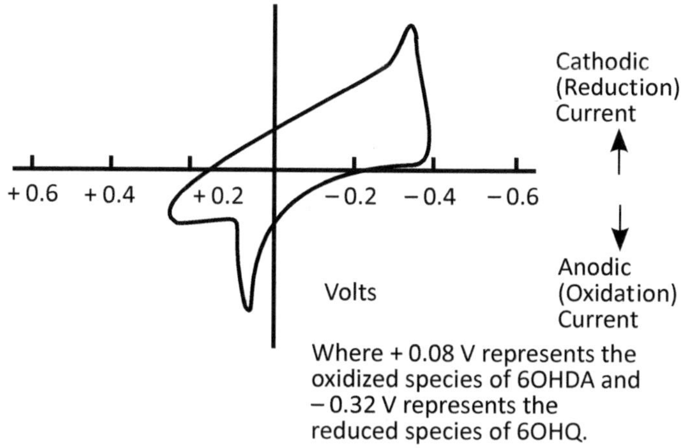

Figure 4.7 Representation of a typical linear scan voltammogram producing stable reduced and oxidized species. Conference Material from Reference [2].

4.6 Cathodic Processing is not a Function of Polarity

Inventive reductive circuitry does not appear to be due to a function of polarity. This empirical finding is shown in Figure 4.8. It is shown in these electrochemical graphics, using the test setting in the potentiostat, that whether applied potential in volts is in scan mode, upward or downward, the reductive quadrant voltammogram is not altered. Reduction circuitry is not dependent on polarity.

Thus, Broderick developed a working potentiostat with a primary reduction circuit. A much-needed sensor driver for the sensor field of advanced materials was developed, pioneered, and tested. Some of the key elements that were configured in the potentiostat circuit were critical parameters such as novel **sensitivity options** and novel **time constants**, advanced materials for nanobiosensors, circuits and cables specific for neuroanatomic substrates for animal work and separate operational stabilities for personal diagnostics and therapies for the patient with and without brain, body, and blood disorders. Delving into electrical circuits was and is a joy for Patricia as she did so while working in her lab and teaching Medical students at the CUNY Medical School. She interchanged capacitors and jumper cables with ease. Indeed, she was deeply engrossed in her work alone and peaceful in her laboratory.

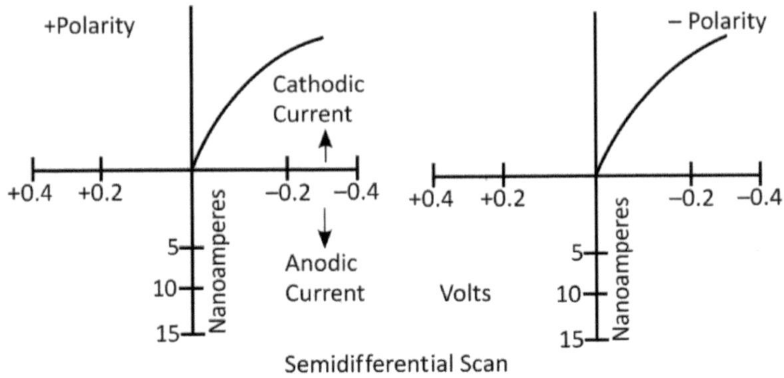

Figure 4.8 Diagram illustrating that semiderivative reductive processing of current is not a function of the directionality of E_{app} (applied potential). The data show that reductive processing does not appear to be a function of polarity.

Then, more good news arrived as Dr. John P. Morgan, MD, a fine Chairman and founder of the *Department of Pharmacology, in memoriam,* granted her a full-time assistant, Frank Phelan, BS. (Frank came to continue to work with Patricia from her fellowship at Cornell Medical School). Substantial grants from the *National Institute of Health* (NIH) were earned. Patricia achieved early tenure! Dr. Morgan gave Patricia leadership positions in the Lecture part of her Professorship. Patricia worked with Dr. Doris Clouet, PhD, at the NIH and Sr. Bridget and Patricia enjoyed the intellectual and professional friendship of Doris who is credited with paving the way for women in the world of science and medicine. Dr. Clouet and Sr. Bridget are respectfully spoken of in this monograph, *in memoriam.* Formidable colleagues and professional family friends during this time until this day and forward are distinguished psychiatrists and neuroscientists, Dr. Jack Jacoby and Dr. Joao Nunes.

Live imaging in the natural state of each animal and/or patient was made available for the first time. Online live imaging directly comparing individual natural cerebral state with dysfunctional cerebral states was and is made available for the first time using novel electrochemical techniques.

Close cause and effect relationships between brain and behavior provided novelty heretofore unseen. Cables were designed separately providing advanced material science. Work with animals was accomplished in the Broderick laboratory and reproduced from laboratory to laboratory in different Universities and companies internationally and in the Broderick laboratory with different teams of personnel, including medical and doctoral scientists. Each time, the original work was reliably reproduced.

Originality flourished. The work with all levels of professional life enabled personnel to higher levels of professional status, certification and academic degrees. The work accomplished in the patient was performed in epilepsy patients both during resection surgery intra operatively in *NYU Langone Medical Center Tisch Hospital* with approval from the *Institutional Review Board* as well as in post resection tissue taken from mesial and cortical temporal lobe epilepsy patients. The surgical suite was the place wherein LIVE imaging in temporal lobe epilepsy patients occurred intraoperatively using the BRODERICK PROBE® biosensor. Dr. Doyle, brain surgeon, *par excellence*, pronounced,

> "This day will be remembered in medical history as the first neurochemical surgery performed intraoperatively and successfully in the epilepsy patient."

Broderick did not stop at any time during her journey. For example, during her postdoctoral fellowship in the *Albert Einstein College of Medicine*, there were wires in the hallway in the process of discard. She said,

> "Let's computerize this BRODERICK PROBE® nanotechnology for more rapid dissemination of this brain nanotech for the world."

Then, with her assistant, Bill Lyons, Dr. Broderick stripped these wires and used the classical "interface bus device" to connect the scan rate, temporal, and spatial parameters within the potentiostat circuit settings with mother board settings to make the phenomenon of LIVE brain nanotech more accessible, facile and faster for the world. The brain/computer interface (BCI) and programming results are shown in Figures 4.9 and 4.10. The first computerized signal processing record of biogenic amine images from the linear circuit is shown In Figure 4.11.

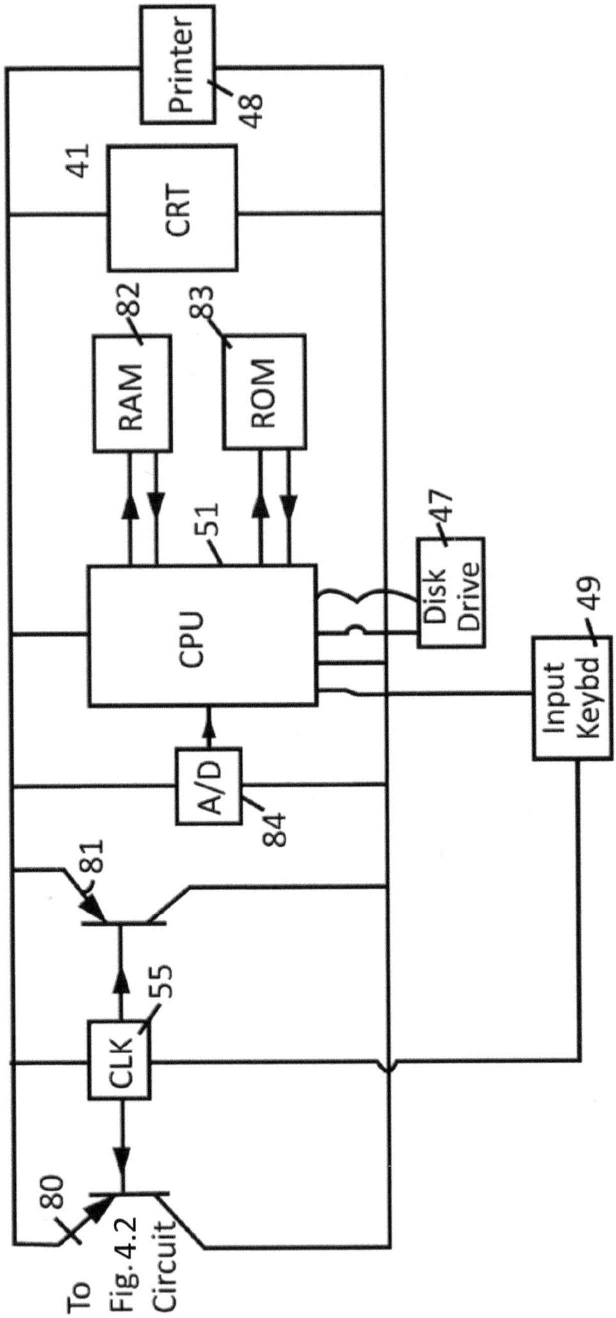

Figure 4.9 Schematic drawing illustrating a system for interfacing the detector with a computer.

4.7 The Brain Computer Interface: The BCI Circuit

The potentiostat/detector was successfully interfaced with the computer. The BCI circuit is shown in Figure 4.9. The computer program is shown in Figure 4.10.

4.8 The Computer Program

METHOD FOR CHANNEL O

1.	CHROMATOCHART V1.07	24.	CHART DELAY (1: 127) : 10
2.	RUN IDENTIFICATION/FILES OPTIONS	25.	CHART WIDTH. 100'S (1:4) : 4
3.	TITLE $ AUTO	26.	CHART Y SCALE (1:255) : 20
4.	DATE $ 9/25/85	27.	CHART OFFSET (–9999:9999) : –10
5.	TIME $ AM	28.	CHART HARDCOPY (Y:N) $ Y
6.	OPERATORS $	29.	
7.	CONDITIONS $	30.	INTEGRATION/REPORT OPTIONS
8.	DISK FILENAME.S,D $ AUTO, S6, D2	31.	SMOOTHING WIDTH (3:15) : 9
9.	SAVE DAT.FILE (Y:N) $ Y	32.	INIT. SLOPE THRESH. (1:32767) : 10000
10.	SAVE BAS.FILE (Y:N) $ Y	33.	MINIMUM BASELINE WIDTH (1:99) : 70
11.	SAVE PEK.FILE (Y:N) $ Y	34.	INITIAL PEAK WIDTH (1:99) : 10
12.	GRADIENT FILENAME.S,D $	35.	WIDTH % CHANGE (1:100) : 70
13.	STANDARD FILENAME, S,D $	36.	SHOULDER % WIDTH (1:100): 100
14.		37.	MIN. PEAK AREA (1:32767) : 100
15.	DATA ACQUISITION OPTIONS	38.	REVIEW BASELINE (Y:N) $ N
16.	ADALAB SLOT (0:7) : 2	39.	TIME UNITS IN REPORT (SEC:MIN) $ SEC
17.	RAM SLOT (0:7) : 0	40.	REPORT PRINTER SLOT #(0:7) : 1
18.	RAM ADDRESS, K (1:255) : 1	41.	CHART HARDCOPY (Y:N) $ Y
19.	RUN LENGTH, K (1:64) : 1	42.	CHART WIDTH, 100'S (1:4) : 2
20.	FIRST RUN # (1:127) : 13	43.	CHART X SCALE (1:20) : 1
21.	# REPEAT RUNS (0:127) : 0	44.	CHART Y SCALE (–255:255) : 20
22.	AVERAGE # POINTS (1:8) # 4	45.	CHART OFFSET (–9999:9999): –10
23.	SUM # POINTS (1:8) : 2		

EVENT CONTROL SCHEDULE

EVT	SECONDS	TYPE	BIT	ON/OFF	GOTO
1.	0	STATUS	1		
2.	1	SWITCH	2	OFF	
3.	121	SWITCH	1	OFF	
4.	136	STATUS	5		
5.	185	STATUS	1		
6.	242	SWITCH	1	ON	
7.	243	SWITCH	2	ON	
8.	244	SWITCH	3	OFF	
9.	245	SWITCH	3	ON	
10.	400	STATUS	1		
11.	401	STATUS	0		

Figure 4.10 One example of a computer program developed according to the inventive art in the Broderick patent portfolio.

4.9 The First Computerized Signal Processing in the Broderick Laboratory

Figure 4.11 The image shows the first computerized Broderick semi-derivative electroactive signal processing for the neurotransmitters, dopamine (DA) and serotonin (5-HT), major neurochemicals that play a significant role in neurodegenerative and psychiatric disorders. Moreover, the role in which these biogenic amines play dramatically, is their role in the natural physiological state. **Note:** the appearance of the first peak is DA and the second is 5-HT. Peak potentials are operationally stable. Computerized signals were matched with signals derived from the manual operation of the newly designed inventive detector.

In her *Albert Einstein College of Medicine* Research Laboratory, a precious site in the history of the BRODERICK PROBE® inventions and pioneering developments, Patricia brought about the marriage between the neuroscience of the brain and the science of sensors, the new "brain sensing tech". These "first of a kind" sensor and circuit materials are used safely in the human and in the animal brain. She taught herself surgical techniques, manufactured multiple types of sensors and she performed surgery in biological psychiatry to figure out "how does the brain work". Dr. Broderick wrote prolifically in the best scientific journals with her most prestigious mentors. It is well known that her scholarly proclivity reflected in original creativity and authorship, is indeed her custom. The cadre of imminent brain researchers with whom she has had the pleasure and the honor to work and publish during this period are mentioned in Chapter 2 of this monograph. It is testimony to the work and character of Dr. Broderick and that of her colleagues that her silver anniversary in Dominican religious life was shared with her brilliant chairman, Dr. van Praag and her eminent colleagues.

4.10 Operational Stability is a Property of the BRODERICK PROBE® in Long-Term Implants

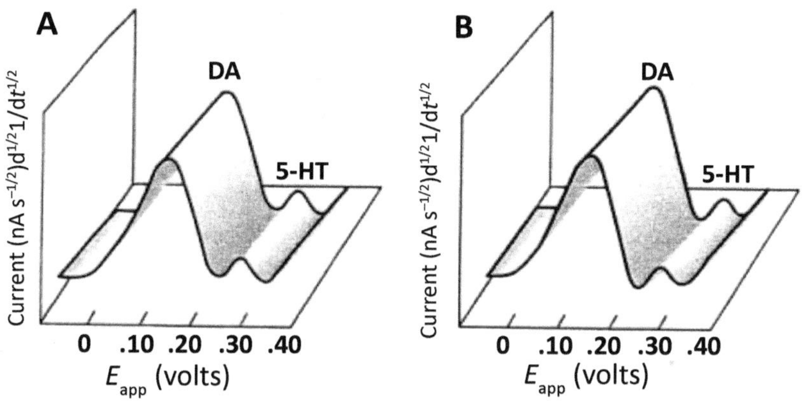

Figure 4.12 Images, recorded *in vivo* and in real time, from Nucleus Accumbens (NAcc), in freely moving and behaving, male, Fawn-Hooded laboratory animals (x axis is oxidation potential in millivolts and y axis is current derived from each neurochemical; (scale: 12.5 mm = 25 pAs$^{-1/2}$). (A) was recorded 2 weeks in freely moving animals implanted with BRODERICK PROBE® and (B) was recorded periodically on a controlled interval schedule for 7 months in the same freely moving animal (Wally). In the two images, the endogenous and selective release of neurotransmitters, dopamine (DA) and serotonin (5-HT) in NAcc in basal ganglia is videotracked while the animal is exhibiting normal motor activity in the open-field behavioral paradigm. Animal was not treated with drugs at any time. This work was performed in the CUNY Medical School and is shown in the *Broderick patent portfolio*.

...and from this novel and original empirical circuit discovery comes a myriad of opportunities to target analyte biomarkers *via* signal processing in the living and walking human and animal. The signal processing is performed in each subject in the natural physiological state and in altered states of consciousness as well as in brain and spinal cord dysfunction and/or trauma.

The proof of concept for this novel driver for sensors is delineated. The electrochemical electrocatalytic phenomenon is solved with online empirical and LIVE data just ahead in Chapter 5.

Chapter 5

Nanobionics: Electrocatalysis Solved—Proof of Concept

According to Ninjawords, nanobionics resides in the scientific art of engineering designs, in particular, electronic designs and further, those biologically based, indeed, perhaps, the imaging of the exact modus operandi, the mechanism of action, of a sound mind and body, that is a sound mind molecularly in concert with the body's molecular epitome. The BRODERICK PROBE® *is an electronic design that fits the molecular design of the neuromolecule and the electronic design of the neuron.*

Nanobionics can also be viewed or defined as the study of biomimetics or biomimicry in which we essentially imitate naturally biological systems that are in fact, electrical, and aligning these systems with electronic devices to investigate human natural and disease states that may not always directly be involved with cerebral complexities such as neurodegeneration. Therefore, within the context of the present monograph, an imaging textbook, the focus is on healthy as well as unhealthy brain and spinal cord. Thus, Patricia will utilize biomimetics to solve such neurochemical and electrochemical phenomena as "electrocatalytic reactions."

Neuroimaging: Sensing Biochemistry in the Brain
Patricia A. Broderick

ISBN 978-981-4613-49-1 (Hardcover), 978-0-429-02974-5 (eBook)
www.jennystanford.com

It may be an initial surprise to some that the biogenic, catecholamine, DA, and the biogenic, indoleamine, 5-HT, may indeed produce orthoquinones as byproducts when oxidized by the presence of air. As a veteran laboratory scientist, Patricia will share with you the interesting occurrence, we had, of placing dissolved DA and/or 5-HT[1] in air for as little as 10 min. The clear DA solution turns a black color, denoting that the orthoquinone has been produced and when studied voltammetrically, the orthoquinone is imaged at a later oxidation or half wave potential from that of the faradaic DA imaging signal. Interestingly again, the 5-HT solution turns from a clearly dissolved 5-HT solution to a solution that is now pink, denoting the transformation of 5-HT to the orthoquinone and indeed, imaging properties will, similarly to the DA orthoquinone, occur at a later peak or half wave potential. What stops these critical biogenic amines from oxidizing early? Well, *in vitro*, believe it or not, it is the insertion of the element, nitrogen, into the *in vitro* cellular system. *In vivo*, it is ascorbic acid that protects against free radical destruction.

How can this be? If ascorbic acid is protecting against free radicals, the free radicals produced by the biogenic amines (the catechols give the most heartache) should be protected. But, lo and behold, the orthoquinones turning the solutions black and pink become traitors, one might say, making the ascorbic acid reverse oxidize into catechols and indoles again. What a vicious cycle! The AA makes more DA, caused by the very "free radicals" that are supposed to protect. Thus, the infamous electrochemical electrocatalytic phenomenon began. The neuroscientist/electrochemist had to catch up.

The importance of nanobionics and the subsection, electro-catalysis, may well be that some electrocatalytic phenomena do hamper the quest for knowledge regarding neuromolecules in the brain, body and blood. In certain cases, then, evidence for

---

[1]We have found that the properties of the indoleamine, serotonin (5-HT) do not include electrocatalysis as does the interaction between DA and AA.

novel pharmacotherapeutics and new surgical devices which treat neurodegenerative and psychiatric disorders of the brain and spinal cord may be masked. This is what we are discussing in this chapter.

Therefore, let's discuss the discoveries in the Broderick lab to share the "know how" in solving electrocatalytic phenomena that have indeed puzzled and plagued neuroscientists and electrochemists for decades. The benefits of electrocatalysis are clear and are described in Chapter 6 herein. In this nanobionics, proof of concept chapter, let's look at the electroactive delineation of the neurotransmitter DA from AA, the latter being the critical cofactor (vitamin) in the biochemical dopaminergic pathway leading to the amino acid L-3,4-dihdroxyphenylalanine (L-DOPA) to the neurotransmitter DA.

Imaging each of these neuromolecules while in the process of separating these neuromolecules *at the same time*, while still these molecules are **unfolding** *on line* and *in vivo*, is shown in the following data. The art of nanobionics in the Broderick lab has continued on, peacefully and with unmatched determination, to shed considerable light on direct online live imaging of AA, distinct from the catecholamine, DA, as well as imaging AA distinct from the indoleamine, 5-HT by using the Broderick nanotechnology, *Neuromolecular Imaging* (NMI) and the BRODERICK PROBE®.

The following scans are drawn from raw data and show the sequence of events with exact concentrations of neuromolecules and laboratory research protocols needed in the mechanics of nanobionics to solve the DA/AA electrochemical "electrocatalytic reaction".

5.1 Introduction

The Equipment to Solve "Electrocatalysis": Manufactured on Site.

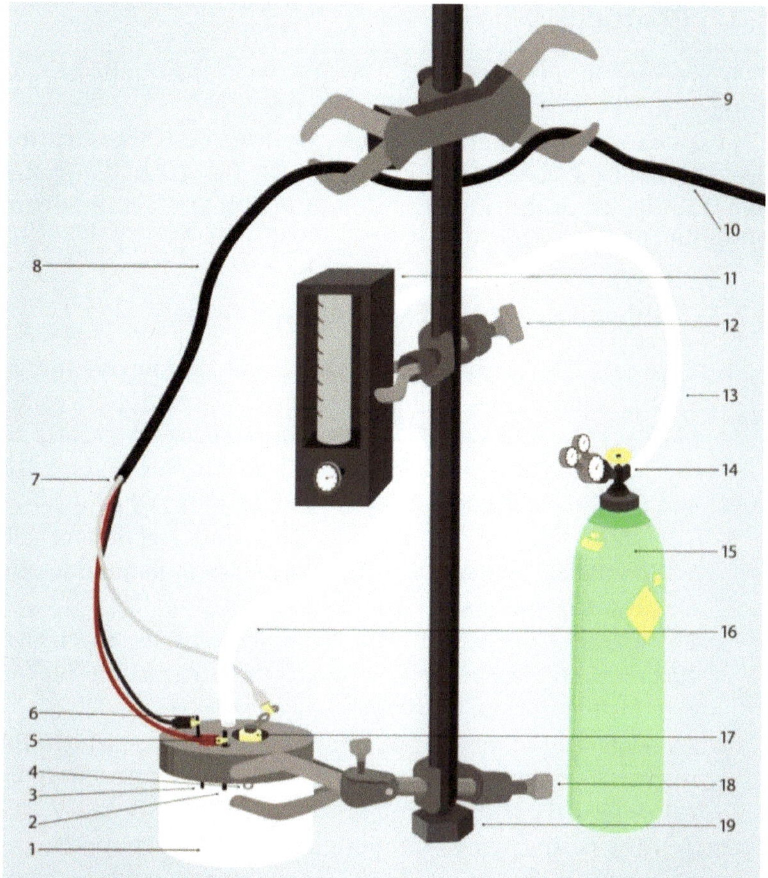

Figure 5.1 *The equipment used for in vitro, calibration studies for NMI and the* BRODERICK PROBE®. (1) Electrochemical cell manufactured on site. (2) Auxiliary-the tip is shown. (3) The meniscus of the buffer solution comprised of 0.9% physiological saline, phosphate buffer. (4) The carbon tip of the BRODERICK PROBE® nanobiosensor. (5) Alligator clip, a mechanical hold for the auxiliary electrode. (6) Alligator clip, a mechanical hold for the indicator nanobiosensor. (7) Electrical leads for indicator, reference and auxiliary electrodes. (8) Insulated Cable for three electrode assembly. (9) Clamp. (10) Insulated Cable. (11) Calibrated flowmeter. (12) Clamp. (13) Polypropylene tubing for transfer of Nitrogen (N_2) gas. (14) Gauge for calibration of N_2. (15) Gas tank for N_2 used to maintain buffer solution free of oxidized monoamine orthoquinones. (16) Polypropylene tubing. (17) Alligator clip, a mechanical hold for the Ag/AgCl reference microelectrode, manufactured on site and calibrated daily with commercial Ag/AgCl references. (18) Clamp. (19) Wrought iron laboratory stand.

5.2 Protocol

The protocol used to distinguish DA from AA and further distinguish DA and AA from 5-HT is as follows:

Experimental Design to Assess Operational Stability for the BRODERICK PROBE® to Distinguish DA from AA Using the Nanobionics Phosphotidylethanolamine (PEA)/Bovine Serum Albumin (BSA) Coating Model.

5.2.1 Objectives for the Experimental Design

1. To assess the relative ambient acute and chronic day-to-day operationally stable responses of the *N*-Stearoylcerebroside, (*S*-Type) BRODERICK PROBE® nanobiosensor in regard to the Neuromolecular Imaging (NMI) of biogenic amines DA and 5-HT and the known electrocatalysis of DA by AA.
2. To assess the operational selectivity and stability of the *S*-Type nanobiosensor for DA in particular in its interaction, electrocatalytically, with AA.
3. To identify optimized half wave potentials at which DA responds exclusively without the interaction of the electrocatalysis reaction by AA.
4. To identify optimized half wave potentials at which DA responds exclusively without electrocatalysis reaction by AA while 5-HT signaling is uninterrupted.
5. To assess the range of optimized half wave potentials at which minimum and maximum signaling of biogenic amines and 5-HT are imaged.
6. To assess the nanobionic features of the BRODERICK PROBE® as these features match peak capacitance of the *S*-Type nanobiosensor in relation to that of the resistance/capacitance potentiostat.
7. To provide actual raw data, proof of concept data that lend an explanatory note to the imaging of biogenic amines and AA for the controlled study of the imaging capabilities of the BRODERICK PROBE®.

Note: The *S*-Type BRODERICK PROBE® nanosensor is comprised as described in Chapter 3. In particular, it is comprised of the compound, *N*-Stearoylcerebroside. The plethora of nanobiosensors denoted by the internationally held trademark, BRODERICK PROBE® are referred to, in our laboratory, as alphabetic characters to maintain the security of each formula design.

5.2.2 Stock Solutions

DA 0.00576-gr in 10.00 mL Distilled Water (M = 0.000304M)

5-HT 0.00649-gr in 10.00 mL Distilled Water (M = 0.00305M)

AA 0.00594-gr in 10.00 mL Distilled Water (M = 0.011M)

Note: Stock solutions were prepared daily, just before the first assay was inaugurated. In other words, after establishing the control buffer baseline, these solutions were freshly prepared for the study of the first nanobiosensor. The same stock solution preparation was repeated for the daily study of each nanobiosensor thereafter.

5.2.3 Coating: PEA/BSA

A solution of phosphotidylethanolamine (PEA) 10% and bovine serum albumin (BSA) 1% was prepared freshly in distilled water.

Each BRODERICK PROBE® was treated as follows:

1. The carbon-based tip of the nanobiosensor was rinsed with saline and was positioned 0.5 mm above the saline phosphate buffer in the electrochemical cell, (cell was manufactured on site, see Figure 5.1)

   Note: Saline was allowed to drip from saline drop bottle over the tip gently for about 2 min to clean thoroughly from residual current.

2. The nanobiosensor was then placed in PEA 10%/BSA 1% for 30 min.
3. The nanobiosensor was then placed in 0.09% physiological saline for 30 to 45 min.
4. The nanobiosensor was placed in the electrochemical cell containing saline phosphate buffer and leads were attached as shown in Figure 5.1.
5. The electrochemical cell with nanobiosensor, reference and auxiliary leads attached, was purged with an initial laminar nitrogen flow for 2 min.
6. For each scan, the cell was purged with nitrogen for 1 min, followed by equilibrating the nanobiosensor in the CELL position on the potentiostat for 2 min, before the actual SCAN was prompted.

5.2.4 Nanobiosensors

Although many indicator nanobiosensors were used for proof of concept nanobionics data, those from nanobiosensor 18-6 are shown in sequence herein.

5.2.5 Procedure

After setting up the electrochemical cell and establishing a baseline for 30 mL, saline phosphate buffer, the following solution injections were made.

DA and 5-HT, respectively (ul): 2, 4, 5, 10.

(1 ul (uL) DA & 5-HT is equivalent to 10 nM *in vitro factoring further dilution*)

Then, 2, 35, 60 and 200 ul (in 100 ul aliquots) AA were added to the same solution as above. (1 uL AA is equivalent to 0.37 uM; hence, the concentration range is 0.37 uM to 0.37 nM). The data on the *y* axis (in centimeters) relate to current (nanometers); the signals are normalized against non-faradaic charge differences.

Scanning Protocol: Time Constant: 5 then 1; Chart Speed: 12.5 cm/min, then 5 or 1 at TC one and/or scan rate of 7.5 or 5 mV/sec; Gain: Generally at 2 nA/V.

A series of *in vitro* electrochemical images (NMI) of increasing aliquots of AA, DA and 5-HT in saline/phosphate buffer, derived from each *N*-Stearoylcerebroside nanobiosensor pretreated with PEA/BSA, are shown in Figs. 5.2–5.13.

The signature for each neurochemical, defined as the oxidation potential in mV, is shown on the *x* axis. Current is shown on the *y*-axis; concentration of each neurochemical is proportional to current according to the Cottrell Equation. Several and repetitive studies were performed. Three distinct and separate cathodic oxidation peaks were sequentially produced by the pretreated PEA/BSA, *N*-Stearoylcerebroside nanobiosensor throughout the protocol. The *N*-Stearoylcerebroside nanobiosensor exhibited superior operational stability, sensitivity and selectivity.

Specifications for each scan for gain (sensitivity), time constant, scan rate and chart speed are noted.

Biosensor	BRODERICK PROBE® "S"
Pretreatment	10% PEA/1% BSA
Indicator ID	YZ18-6
Gain	2 nA/V
Time constant	5 sec
Scan rate	10 mV/sec
Chart speed	12.5 cm/min
N_2 purge	9:46
Cell time	9:47
Scan time	9:49

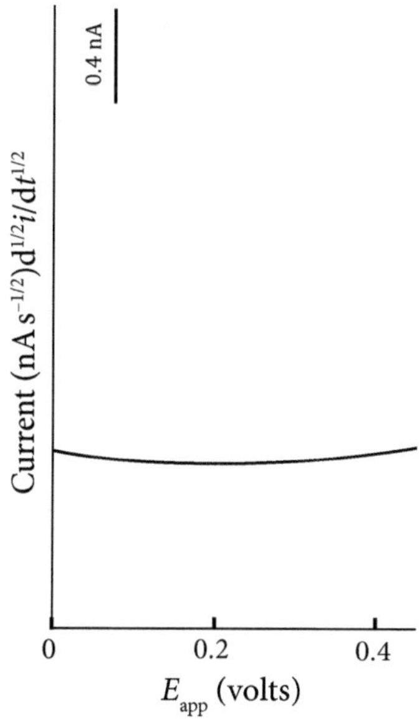

Figure 5.2 Results from buffer solution are shown. No neuromolecules have been added.

Biosensor	BRODERICK PROBE® "S"
Pretreatment	10% PEA/1% BSA
Indicator ID	YZ18-6
Gain	2 nA/V
Time constant	5 sec
Scan rate	10 mV/sec
Chart speed	12.5 cm/min
N_2 purge	9:51
Cell time	9:52
Scan time	9:54

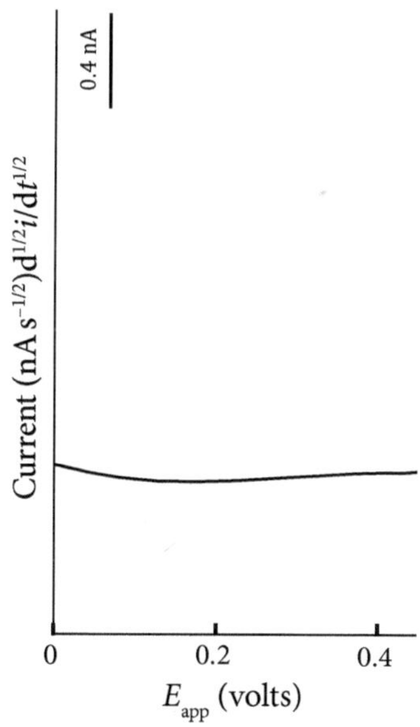

Figure 5.3 Another scan shows results from a repeated buffer solution. No neuromolecules have been added.

Biosensor	BRODERICK PROBE® "S"
Pretreatment	10% PEA/1% BSA
Indicator ID	YZ18-6
Gain	2 nA/V
Time constant	5 sec
Scan rate	10 mV/sec
Chart speed	12.5 cm/min
N_2 purge	10:02
Cell time	10:03
Scan time	10:05

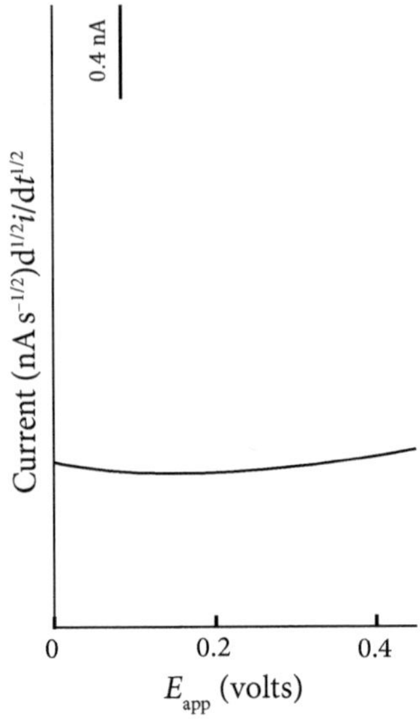

Figure 5.4 Another scan shows results from a repeated buffer solution. No neuromolecules have been added.

Biosensor	BRODERICK PROBE® "S"
Pretreatment	10% PEA/1% BSA
Indicator ID	YZ18-6
Gain	2 nA/V
Time constant	5 sec
Scan rate	10 mV/sec
Chart speed	12.5 cm/min
N_2 purge	10:07
Cell time	10:08
Scan time	10:10

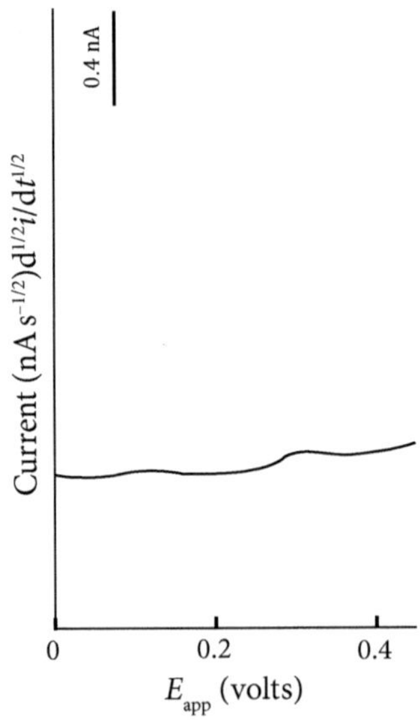

Figure 5.5 In this scan are the results from 2 ul of each neurochemical, AA, DA and 5-HT. This concentration is equivalent to 0.74 uM, 20 nM DA, and 20 nM 5-HT.

Biosensor	BRODERICK PROBE® "S"
Pretreatment	10% PEA/1% BSA
Indicator ID	YZ18-6
Gain	2 nA/V
Time constant	5 sec
Scan rate	10 mV/sec
Chart speed	12.5 cm/min
N_2 purge	10:13
Cell time	10:14
Scan time	10:16

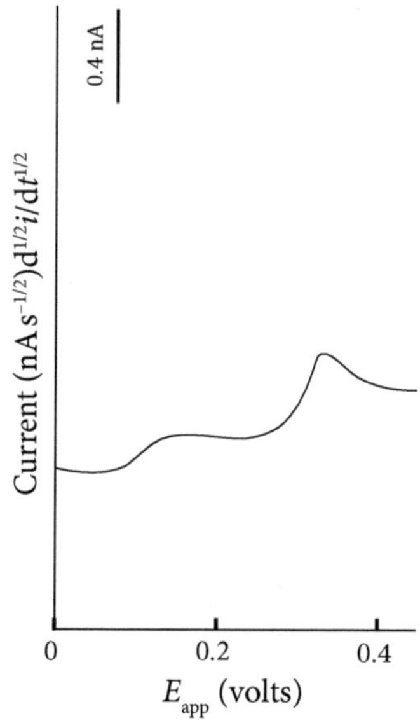

Figure 5.6 In this scan are the results derived from the same ul aliquots of DA and 5-HT with addition of 35 ul of AA equivalent to 11 uM.

Electrode	BRODERICK PROBE® "S"
Pretreatment	10% PEA/1% BSA
Indicator ID	YZ18-6
Gain	2 nA/V
Time constant	5 sec
Scan rate	10 mV/sec
Chart speed	12.5 cm/min
N_2 purge	10:20
Cell time	10:21
Scan time	10:23

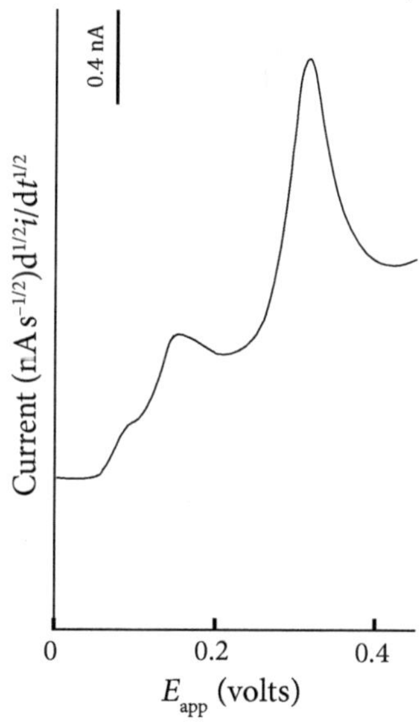

Figure 5.7 The results from this scan are derived from 4 ul of DA and 5-HT and 60 ul AA. These aliquots of neurochemicals are equivalent to 40 nM DA and 5-HT and 22 uM AA.

Biosensor	BRODERICK PROBE® "S"
Pretreatment	10% PEA/1% BSA
Indicator ID	YZ18-6
Gain	2 nA/V
Time constant	5 sec
Scan rate	10 mV/sec
Chart speed	12.5 cm/min
N_2 purge	10:25
Cell time	10:26
Scan time	10:28

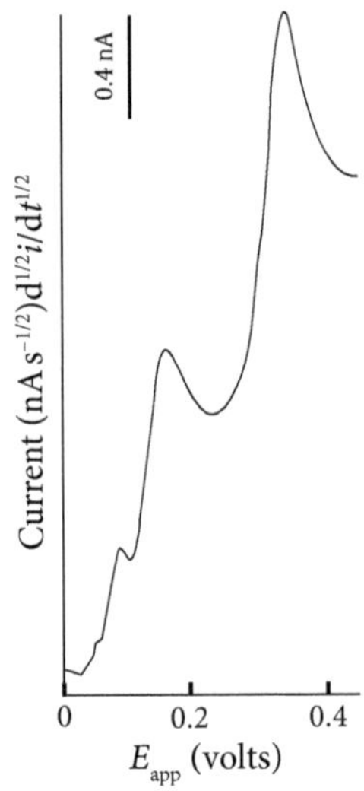

Figure 5.8 The results from this scan are derived from 5 ul of DA and 5-HT and 100 ul AA. These aliquots of neurochemicals are equivalent to 50 nM DA and 5-HT and 37 uM AA.

Biosensor	BRODERICK PROBE® "S"
Pretreatment	10% PEA/1% BSA
Indicator ID	YZ18-6
Gain	2 nA/V
Time constant	5 sec
Scan rate	10 mV/sec
Chart speed	12.5 cm/min
N_2 purge	10:31
Cell time	10:32
Scan time	10:34

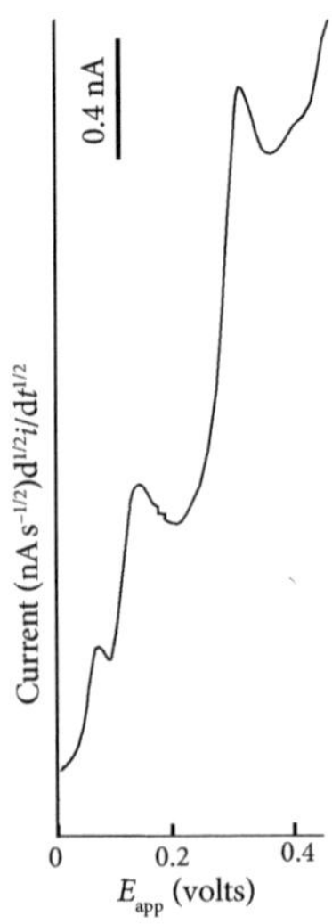

Figure 5.9 The results from this scan are derived from a repeated 5 ul of DA and 5-HT and 100 ul AA. These aliquots of neurochemicals are equivalent to 50 nM DA and 5-HT and 37 uM AA. Note that this nanobionic NMI Figure demonstrates the classical electrochemical column effect. Also, note that charging current, that is, non-faradaic current, is drawn a few mm less that seen in Figure 5.8.

Biosensor	BRODERICK PROBE® "S"
Pretreatment	10% PEA/1% BSA
Indicator ID	YZ18-6
Gain	2 nA/V
Time constant	1 sec
Scan rate	10 mV /sec
Chart speed	5.0 cm/min
N_2 purge	10:35
Cell time	10:36
Scan time	10:38

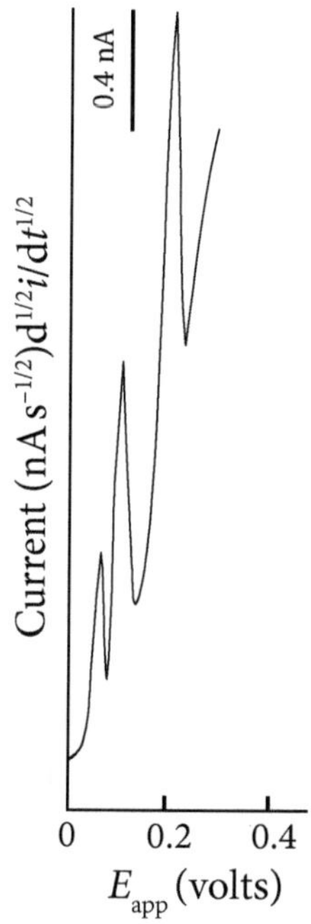

Figure 5.10 In this scan, there were no additional aliquots of neurochemicals; the specification for the use of chart speed in the field of electrochemistry is taught. Note that the chart speed is 5 cm/min producing even sharper peaks than previously specified for chart speed, while still working with the same nanobiosensor and while remaining in the potentiostat semiderivative electrical circuit modified for a reduction circuit. In this scan, the classical column effect, well known in electrochemical principles, is imaged. Thus, narrower peaks show the enhancement of the neuromolecular cathodic half wave peaks. Also, note that charging current, that is, non-faradaic current, is drawn a few mm less that seen in Figure 5.8. Another scan was taken, data are shown in Figure 5.13 wherein a scan rate of 7.5 mV/sec was used for teaching electrochemical principles. The results exhibiting sharp peaks remains clear but the image is about 4 mm less in nA and one to two mm less in mV.

Biosensor	BRODERICK PROBE® "S"
Pretreatment	10% PEA/1% BSA
Indicator ID	YZ18-6
Gain	2 nA/V
Time constant	5 sec
Scan rate	10 mV/sec
Chart speed	12.5 cm/min
N_2 purge	10:44
Cell time	10:45
Scan time	10:47

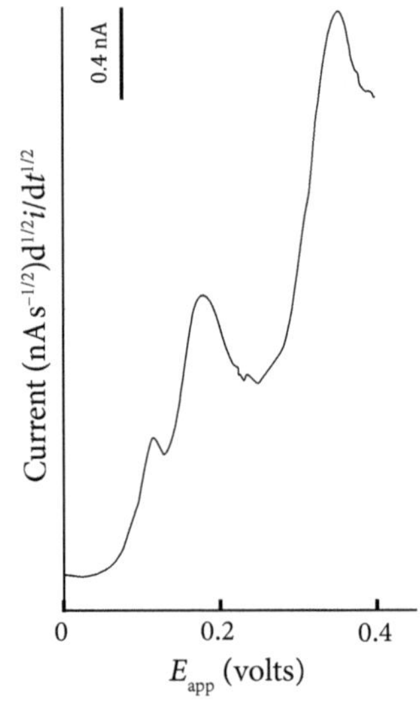

Figure 5.11 In the report of this scan, there were no additional aliquots of neuromolecules. Settings were returned to original specifications of 2 nA/V, 10 mV/sec, tune constant of 5 sec and chart speed of 12.5 cm/min. The data match the original. Thus, further validation of the data to original settings is shown.

Biosensor	BRODERICK PROBE® "S"
Pretreatment	10% PEA/1% BSA
Indicator ID	YZ18-6
Gain	2 nA/V
Time constant	5 sec
Scan rate	10 mV/sec
Chart speed	12.5 cm/min
N_2 purge	10:49
Cell time	10:50
Scan time	10:52

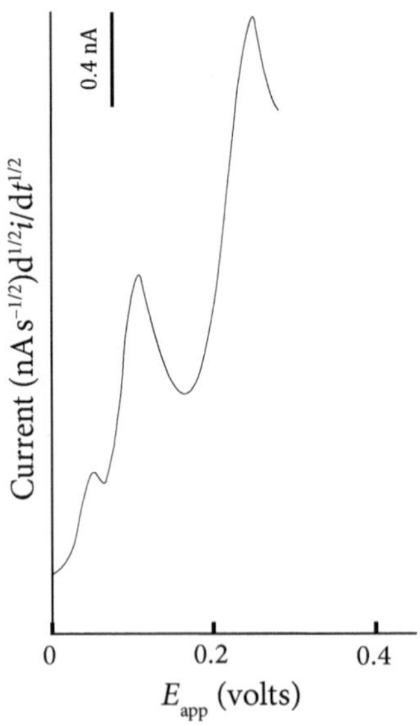

Figure 5.12 This scan shows the increase in DA and 5-HT signals after 10 ul (equivalent to 100 nM) of each neuromolecule was added.

Biosensor	BRODERICK PROBE® "S"
Pretreatment	10% PEA/1% BSA
Indicator ID	YZ18-6
Gain	2 nA/V
Time constant	1 sec
Scan rate	7.5 mV/sec
Chart speed	2.5 cm/min
N_2 purge	10:54
Cell time	10:55
Scan time	10:57

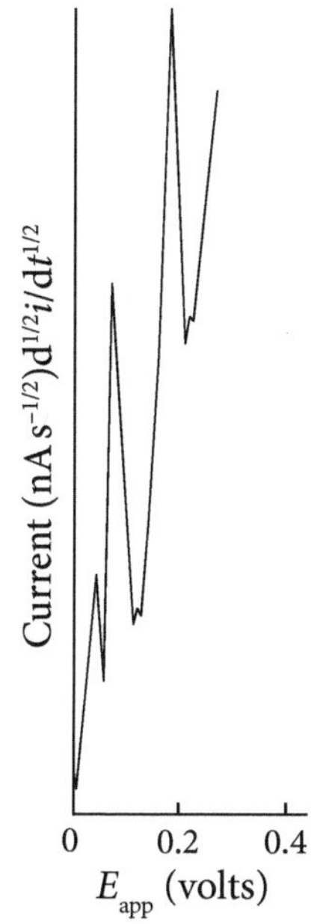

Figure 5.13 Settings are critical for sharp electroactive signals. Therefore, no further aliquots of neuromolecules were added to the solution in the electrochemical cell and indeed, three different settings were employed to teach neuromolecular imaging, electrochemical principles. The three settings were as follows: time constant, 1 sec; scan rate, 7.5 mV/sec; chart speed, 2.5 cm/min. The results further validate NMI and show the production of even sharper peaks than previously specified for chart speed as well as time constant and scan rate, while still working with the same nanobiosensor and while remaining in the potentiostat semiderivative electrical circuit modified for a reduction circuit.

5.3 Discussion

Demonstrated in this chapter is NMI and the BRODERICK PROBE® sensor technologies and their ability *via* electron transfer to encode voltage biomarkers for anodic and cathodic currents in the same oxidation electrochemical quadrant shown in Chapter 4. The carbon-based nanobiosensor uses its double layer capacitance, its many constituents, formulas and coatings to bidirectionally image reduction and oxidation analytes. The BRODERICK PROBE® as it relates to potentiometric-driven operational amplifiers for each constituent of the nanobiosensor assembly, the inventive art is described in detail in Chapter 4 herein. The nanobiosensor assembly is specific for discrete capacitance properties at a specific voltage and resistance.

This nanobionics biosensor is electronically designed to mimic pixels of distinct waveforms for each neurotransmitter and neurochemical specific for capacitance characteristics at specific applied energy potentials whether cathodic and/or anodic without the overuse of traditional cyclic voltammetric circuits although the nanobiosensor is amenable to such circuits. On the other hand, differential pulse and double differential pulse circuits are particularly suited to the BRODERICK PROBE® and significant experimentally derived biomarkers for neurotransmitters have been developed and pioneered by the Broderick Laboratory using these additional voltammetric circuits as well.

5.4 Conclusion

The BRODERICK PROBE® nanotechnology uses electron transfer to encode voltage biomarkers for nanobionic imaging of signature neurotransmitters identified in anodic and/or cathodic current kinetics. **Unfolding** of DA from AA is imaged on line and in real time and is reported here for the first time and to the author's knowledge, it has never been demonstrated as such before this demonstration in this break-through monograph or elsewhere. The scans are presented in electrochemical format in general use today. The inventive art further dramatically improves the art of electrochemistry with the inventive reduction circuit semiderivatized. These advancements are imminently important to the study of the brain and spinal cord in science and medicine.

Chapter 6

Lipid Biomimetics: A Nano-Microscope in the BRODERICK PROBE®

According to the Merriam-Webster Dictionary, a catalyst is a substance that enables a chemical reaction to proceed at a faster rate and/or under different chemical conditions than would occur in the usual case. An example of such a circumstance could be using a lower temperature enabling a faster reaction that would otherwise not be possible. An agent that provokes or speeds significant change or action in a chemical reaction is a catalyst.

Electrocatalysis can be defined as catalysis of electrochemical reactions that occur at the electrode-electrolyte interface. Realizing that the interconversion of electrical and chemical energy requires competent catalysts capable of expediting multielectron reactions at or even while diffusing near the sensor interface, a surface that's actually electrified, we need substances that can perform such a task. Metals such as platinum can perform electrocatalysis and are called heterogeneous as these types of catalysis involve active substrates with multiple and various catalytic sites, a priori. But, metals are not the only surfaces to provide a site for multiple catalytic reactions. In fact, redox mediation at molecular electrocatalysts are efficient at such a salient task albeit these types are associated with inert substrates.

Neuroimaging: Sensing Biochemistry in the Brain
Patricia A. Broderick

ISBN 978-981-4613-49-1 (Hardcover), 978-0-429-02974-5 (eBook)
www.jennystanford.com

In the case of the BRODERICK PROBE®, electrocatalysis is molecular and yet, multiple and controlled catalytic sites exist at the interfacial surface of the tiny carbon/lipid/phosphatidyl polymer; its substrate is not inert, comprising the carbon substrate with a heterogeneous component. The layer by layer synthesis, delineated in the Broderick patent portfolio [6, 8–12] shows the ease with which the BRODERICK PROBE® carbon polymer is synthesized and it ingeniously includes another homogeneous molecular component that relies on enzymatic influences via non-amino acids, actually coenzymes, such as ascorbic acid. The tiny probe has inherent catalytic property that is empirically determined as the lipid amplification number (LAN) [8, 9]. The carbon-based tiny BRODERICK PROBE® plays several roles as, for example, electron donor and acceptor. The nanoprobe allows for exquisite selectivity and is tunable. The nanoprobe is trainable; it has a memory. The tiny Broderick probe is a catalytic amplifier.

6.1 Introduction

In *Analytical Chemistry* [4] the future of catalytical chemistry as it has blossomed today, four decades later, is predicted. What is described is the difference between catalysis and amplification in addition to what is described as "catalytic amplification." The very elemental property of how to achieve catalysis and/or amplification of an analyte for its best possible production of either the analyte or its counterpart analyte, is outlined. In this way, Blaedel accurately describes today's analytical chemistry of molecular analytes. Indeed, the electrochemist, an expert in voltammetry, even alludes to our current and most exciting field of *nanotechnology* by talking about amplification and catalysis as *"amplification and catalysis processes may someday go to small molecules."* In *Accounts of Chemical Research, published by the American Chemical Society* [31], heterogenous electrocatalysis is shown to be expressively better controlled when graphite is conjugated with a metal, for example, platinum. Carbon polymorphism is described in Fujita et al. [26] and McCreery [40] delves deeply into the versatile polymorphic nature of the carbon interfacial surface and calls an interfacial surface that is made of carbon, "venerable."

Chemical and another form of chemical, electrochemical, amplification occurs usually before the final measurement step, that is, during the reaction process. Enzymes, antibodies, polymers, and proteins are some of the catalysts involved in chemical and electrochemical amplification. In this chapter, we are discussing the BRODERICK PROBE® electroactive polymer in the context of the precise steps that the polymer formulates its mechanics of amplification *via* for one example highly polar covalent bonds, highly electronegative carbon (C) oxygen (O), nitrogen (N) and fluorine (F) linking with carbons of the phosphotidyl groups, fatty acids, complex lipids and oil or oily substances.

Electrocatalytic amplification of an analyte by another analyte or catalyst is an accurate example of electrocatalytic amplification that occurs during and before the final measurement step. Using electrocatalytic amplification, the production of a new analyte from another can be magnified at least a hundred-fold or more. Such enormous production of an analyte that is specifically desired to perform a selected function is useful in science, medicine, and commercialization of a new end-product in any market whether that be industrial and/or medicine. In industry, one can produce signal amplification *via* catalytic conversion from one search engine to another by switching terminals or nonterminal ports using silicon photonics for securing networks, as does *Cisco Systems Inc.*, for example. In medicine, we use these catalytic phenomena for diagnostics and therapeutics for patients whether human or animal.

The specificity of the electrocatalytic amplification between the two molecules DA and AA seems to have caused this dilemma in the field of electrochemistry due to an unceasing desire to target the analyte DA in the central nervous system particularly because of its role in the mesocorticolimbic and nigrostriatal neuronal circuits. Why? The answer is well known to neuroscientists and physicians as DA is the key to elucidating the *modus operandi* underlying neurodegeneration, psychiatric affective disorders and sex related substance abuse dysfunction among others. Too, affective disorders, comorbid with lifelong addiction, are central to this cationic molecule, DA, and its relationship with the DA pathway cofactor, AA.

Thus, the importance of this cationic analyte, DA, that can also exist in its anionic form (a surprising property of DA to some, even well versed in science and medicine) cannot be emphasized enough, especially insofar as the diagnosis and treatment of movement disorders are concerned. The critical importance of this cationic analyte, DA, along with one of its cofactors such as AA, in the presence or absence of neurodegenerative brain diseases, such as Parkinson disease with Lewy body Alzheimer's disorders, ischemic stroke, hypoxic hypoxia, obsessive compulsive behaviors, and Lesch Nyhan stereotypies, cannot be underestimated. Indeed, in the ensuing chapters, the urgency to shed light on electrocatalysis of DA and AA in neurodegenerative dysfunction caused by epilepsy will be shown by detailing biochemical subtypes of temporal lobe epilepsy in the epilepsy patient.

Therefore, DA and AA taken together and/or alone, assume a critical importance in science and medicine. After all, as mentioned, AA blocks free radicals and thereby helps to alleviate brain disease by blocking the very molecules, the free radicals that are responsible for many brain and peripheral diseases. Thus, what the Broderick lab accomplished was and remains critical to the fields of neurology, psychiatry, brain science, and sensing nanotechnologies. The specific imaging of each molecule of DA as well as each molecule of AA online in real time, each separately and within subseconds as semiderivative images in real time ***during the very process of electrocatalytic amplification is*** reported here for the first time in this imaging textbook in Chapter 5. What is seen in Chapter 5 is the actual unfolding of AA from DA. This is a feat which may appear to an expert outside of the field of electrochemistry as seemingly effortless but clearly, the author will say, it is not.

Analytes that are primarily only cationic and primarily only anionic are difficult to separate in the same image while maintaining separate and selective waveforms at each half wave potential/oxidation potential. NMI, a nanotechnology technique, with the BRODERICK PROBE® series of nanobiosensors has imaged DA and AA in separate signals within the same recording *via* semiderivative analysis and a reduction circuit. The carbon/lipid/phosphotidyl polymer, one series of the BRODERICK PROBE®, was used as the basis for these studies. Thus, despite the fact that some missteps regarding the technical inability to detect DA separately from AA may have been made by some contributors from the neuroscientist/electrochemist group [43, 44], the solution to the dilemma of electrocatalysis between DA and AA, that has absorbed the minds of the neuroscientist and the electrochemist, is finally settling down.

The data in this chapter show further results of the operational sensing stability of the BRODERICK PROBE® multi-complex for imaging captious molecules. The BRODERICK PROBE® biomimetic polymers and the BRODERICK PROBE® biomimetic chips are added to the original Broderick polymer. The Broderick inventive sensors are placed in the ***simulated brain***.

6.2 Simulation Neuroscience

> "Reconstructing the brain recapitulates the history of neuroscience by evolving and accelerating its major steps, from early morphological descriptions of the nerve cell to later electrophysiological and biochemical studies of neural connectivity: synthesize and evolve available knowledge, methods and technologies into a new science, and take a quantum leap onto a path that, in the end, can lead to understanding the multiscale brain. Industrialization of neuroscience and the resulting large disconnected datasets are generating demand for integrative neuroscience; the scale of neuronal and connectivity maps is driving digital atlasing and reconstruction to piece together the multiple levels of brain organization, and the complexity of the interactions between molecules, neurons, microcircuits and brain regions is driving simulation neuroscience to understand the multiscale brain. To explore the origin and making of this paradigm shift, we reconstruct the deep historical paths leading to simulation neuroscience through the philosophical, experimental and theoretical phases of brain research, in particular, from the first observations of the nerve cell to modern attempts to digitally reconstruct and simulate the brain, by identifying the major scientific, technological and conceptual breakthroughs that have guided this passionate quest of humans to understand the brain and their own condition" [24].

6.2.1 The Simulated Brain in an Electrochemical Cell

The simulated brain is comprised of lipid and protein, primarily lipid, about 80%. A simulated brain was made in a beaker; it was comprised of the BRODERICK PROBE® microelectrode and the layered by layer coating of a lipid/protein complex, 10 % w/v PEA and 1% w/v BSA.

The purpose is to assess the operational stability for defining electroactive chemical *signals-to-noise* ratio for the *in vitro* and *in vivo* determination of neurotransmitters. NMI techniques were developed in the Broderick laboratory to gain insights into the mechanism of lipid amplification on interfacial sensor surfaces, in particular on the interfacial surface of the BRODERICK PROBE®. The science of advanced materials in the field of lipid media for sensor inventions is now enriched by studying a "***day-to-day and within days***," operationally stable response of the tiny probe to neurotransmitters, precursors and metabolites. The studies were performed by using a simulated brain, placed in an electrochemical cell, *manufactured on site* according to, compared with, calibrated pursuant with commercial requirements and pictured in Chapter 5 herein. Interestingly, the findings show that electroactive chemical signals are substantially amplified to correspond closer in magnitude, to *in vivo* signals of such neurotransmitters as measured by BRODERICK PROBE® nanobiosensors in specific neuroanatomic brain regions of the *mammalian human patient and animal model.*

6.2.2 Materials for Lipid Analysis: Fascinating Electroactive Biomimetic Polymers

Using the original lipid/phosphotidyl/complex with PEA/BSA added layer by layer in at least one example of these studies of lipid analyte signal amplification is a 12-carbon-atom chain saturated fatty acid, **Lauric acid.** The BRODERICK PROBE® was admixed and heated into what became, the carbon/phosphotidyl/fatty acid/laurate polymer complex; the polymer was then cooled at room temperature according to the Broderick, 1995 patented formulae [8]. Thus, the manufacture of Laurate nanobiosensors for these studies were prepared by admixing <1 mg aliquot of solid lauric acid stock, a mixture of ultra-pure carbon (1.5 g), oil (Nujol) containing alpha tocopherol (1.24 g), and lauric acid (100 mg). The redox properties for these nanobiosensors for biogenic amines and metabolites *in vitro*, are, in the order of detection from lowest to highest positive potential, ascorbate [0.08 V]; dopamine (DA) [0.13 V]; serotonin (5-HT) [0.29 V]; homovanillic acid (HVA) [0.43 V] and L-tryptophan (L-TP) [0.63 V]. Oxidation (half-wave) potentials for these biosensors, *in vivo*, are, in the order of detection from lowest to highest oxidation potential, ascorbate [0.09 V], DA [0.14 V], 5-HT [0.31 V]; HVA [0.44 V] and L-TP [0.64 V]. Each specified oxidation potential is within a range of ±0.015 V.

LAURIC ACID

A. Linear Formula

$CH_3(CH_2)_{10}COOH$

B. Structural Formula

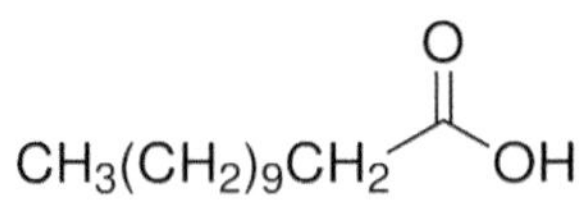

Figure 6.1 The linear (A) and structural (B) formulae for lauric acid.

In another example of compounds, that is, complex lipids, used in these studies about the mechanics of lipid analysis, is the part of the complex comprised of *N*-Stearoylcerebroside, its specifications are introduced in Chapter 3 herein.

The BRODERICK PROBE® *N*-Stearoylcerebroside is a ceramide comprised in one part of the 18-carbon saturated long chain fatty acid, stearic acid. The structural analysis of *N*-Stearoylcerebroside, one of the formularies of the BRODERICK PROBE® lends an explanatory note to the didactics of the formulae of the patent portfolio and in addition, shows that the cerebroside is also a ceramide; the two structures of C18-cerebroside and β-Galactosyl-C18-ceramide, are the same.

It is noteworthy to mention again that C18 is the linear formula of the saturated fatty acid, stearic acid. The chemical structure for this complex lipid is shown in Figure 6.2. The data were used to prepare the original carbon/phosphotidyl/ceramide complex. The chemical structure (Figure 6.2) and its properties are shown below for the original BRODERICK PROBE® *N*-Stearoylcerebroside, also called, the BRODERICK PROBE® β-Galactosyl-C18-ceramide biosensor.

N-STEAROYLCEREBROSIDE

LINEAR FORMULA

β-Galactosyl-C18-ceramide is $C_{42}H_{81}NO_8$

STRUCTURAL FORMULA

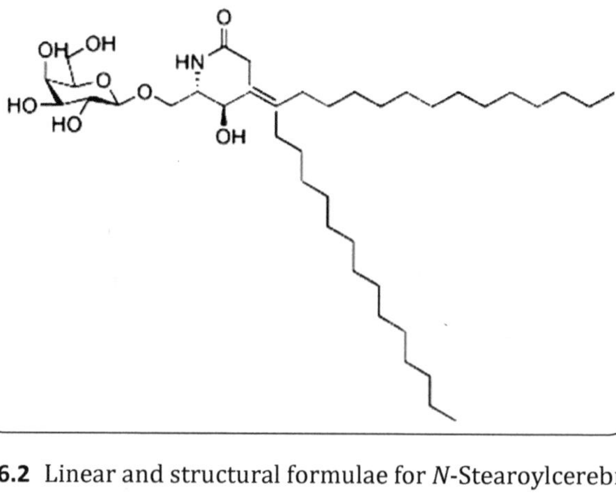

Figure 6.2 Linear and structural formulae for *N*-Stearoylcerebroside.

6.2.3 The Composition of the BRODERICK PROBE® β-Galactosyl-C18-Ceramide

Galactosyl is a glycosyl radical $C_6H_{11}O_5^-$, that is derived from galactose, a compound important in the etiology of the elusive Fabry's disease including juvenile arthritic aspects of the disease and its treatment.

Stearoyl is the radical $C_{17}H_{35}CO^-$ of stearic acid. Stearic acid is $C_{18}H_{36}O_2$.

6.3 Lipid Biomimetic Polymers from the BRODERICK PROBE® Nanoprobe

But, how does catalytic amplification in the nanoprobe occur? What operates the mechanics of this significant inventive finding? What happens in the lipid bilayer of the Broderick probe as it connects to the lipid bilayer of the brain to cause catalytic amplification?

The use of electronics is what is traditionally used to amplify an analyte signal that takes place after the final measurement step, that is, after the interfacial component of the nanosensor, in this case, after the Broderick carbon/lipid/phosphotidyl biomimetic polymer is exposed to either the layered lipid/protein environment or in another case, to the phosphatidyl neuronal cell membrane in neuroanatomic sites of mammalian brain.

In the simulated brain in the electrochemical cell, the interface of the nanosensor surface, a Broderick polymer is in contact with the meniscus of a phosphate solution. Cell deposition is triggered as a potential in voltage is applied to the nanosensor interface of the polymeric BRODERICK PROBE®. But, what is the reaction triggered by applying the potential to this carbon/lipid/phosphotidyl polymer probe? Broderick patents have given nomenclature for the amplification properties of the original Broderick probe complex. The lipid signaling amplification is named in the Broderick patents [8, 9] as *Lipid Amplification Number* (LAN) and the name was conceived in the eighties by the author after observing online, the controlled catalytic amplification inherent in the tiny probe.

The Broderick nanoprobe bonds through living neuronal networks as well as the network in which the nanoprobe is comprised.

Lipid biomimetics were selected as the messengers of amplification based on the phosphotidyl nature of neuronal membranes that is reported to be causal to beneficially amplify insulin for the diabetic patient [50] as well as the literature review that tells us that long chain fatty acids may be messengers for the distribution of lipidic ligands in the physiology of diabetes along with the carbohydrate, glucose [28].

6.3.1 Inside the Lipid Membrane

> "Insulin, for example, a hormone critical for the diabetic patient, a signal may reach a cell as one molecule of insulin. Then, inside the cell, the signal must be amplified so that the response is carried out multiple times rather than just a single molecular response. Thus, for insulin, molecular amplification is built into the system. Any molecule that catalyzes a reaction can do so multiple times producing more than one product molecule. Therefore, *each step* in the *signaling chain* has the potential for amplification. Since a signaling chain is several steps long, then amplification of the signal is in the hundred-fold multiplication. Membrane receptors can produce at least ten second messengers and each second messenger can generate the transcription of ten messenger ribose nucleic acid (mRNA) chains then the signal has been amplified 1000-fold" [50].

Although the pathophysiology of diabetes is characterized by elevated levels of glucose and long-chain fatty acids (LCFA), nuclear mechanisms linking glucose and LCFA metabolism are poorly understood. As the liver fatty acid binding protein (L-FABP) shuttles LCFA to the nucleus, where L-FABP directly interacts with peroxisome proliferator-activated receptor-α (PPARα), the effect of glucose on these processes was examined. In vitro studies showed that L-FABP strongly bound glucose and glucose-1-phosphate (K_d = 103 ± 19 nM and K_d = 20 ± 3 nM, respectively), resulting in altered L-FABP conformation, increased affinity for lipid ligands, and enhanced interaction with PPARα. In living cells, glucose stimulated cellular uptake and nuclear localization of a nonmetabolizable fluorescent fatty acid analog (BODIPY C-16), particularly in the presence of L-FABP. These data suggest for the first time a direct role of glucose in facilitating L-FABP-mediated uptake and distribution of lipidic ligands to the nucleus for regulation of PPARα transcriptional activity [28]. These data showing the role of glucose in synergizing fatty acid molecules in active transfer to lipidic ligands as these relate to PPAR, are supported by Huang et al. [29].

6.3.2 The Broderick Nanoprobe Makes Discoveries in the Simulated Brain

The first observation of the lipid biomimetic phenomenon took place by comparing pre-calibration data *in vitro* with post-calibration data *in vitro* after brain studies *in vivo*. This final step was called post-calibration and the first observation, one of hundreds of thousands of studies on this phenomenon, in our laboratory, was and still is, astounding. The first step of pre-calibration was the electronic application of potential, voltage, in millivolts using the semiderivative circuit of the Broderick detector described herein. The applied potential to the nano-biosensor yielded results in the 0.2 nA/V range depending on the molecular structures of the target. But, after the nanosensor had contacted the phosphotidyl neuronal cell membranes of the living brain and the same nanosensor was placed once again in contact with saline phosphate, the images yielded results that were excitingly in the 2.0 nA/V range. Therefore, in the simulated brain model, studies showed a significant amplification of the electroactive signals. The sensitivity of the nanosensor was dramatically higher, reflecting amplification of the signal as, one might say, using a simple analogy, as in a microscopic device, while realizing that the significance of the observation was of a higher magnitude than comparison with microscopy—using a pun here.

The sequence, *in vitro, in vivo* and then *in vitro* yielded dramatically higher sensitivity in terms of amperes/volt. The final step taken from the same BRODERICK PROBE® nanosensor, while the *in situ*, simulated nanoscale model was selected for study, yielded ten-fold or greater amplified responses of the BRODERICK PROBE® polymer to, for example, biogenic amines, catechol and/or indole, precursors, metabolites and non-amino acid catalysts like AA. When potential was applied to the current measuring operational amplifier during the final electric step potential in phosphate buffer we watched the action online and we were each and every time, amazed. When we took one layer off such as how graphene was ripped from carbon, strip by strip, the BRODERICK PROBE® LAN continued to prove controlled catalytic amplification. We are talking about hundreds of thousands of studies! Dr. Broderick with her numerous medical and pre-medical students whom she continually and carefully trained in the mechanics of the nanoprobe, discovered, delineated and defined the catalytic amplification. The author then figures out the nomenclature, LAN, subsequently, calibrating, publishing and patenting the same [8, 9].

But where and how was the amplified response causal to the Broderick nanoprobe?

6.3.3 The Role of Phosphorus: *In Vivo* Simulation Neuroscience

At least in part, this original carbon/lipid/phosphotidyl complex was incurring amplification *via* the brain. The first thought was: the key was the phospholipid in neuronal membranes in virtually each and every neuroanatomic substrate in the brain. Therefore, it had something to do with phosphorus! But phosphorus is not a conductor! This is dogma. But, has the dogma changed! Can the lipid messenger actually be phosphorus in the matrix of the Broderick neuro-sensor? Let's see! In fact, phosphorus, the element called "P" in the periodic table, has emerged electrochemically as electroactive! Indeed, there is amazing evidence that phosphorus is not redox insensitive! Phosphorus is redox sensitive. The evidence is shown below.

6.3.3.1 Evidence for the electroactive property of phosphorus

From: Delavaux-Nicot B, Guari Y, Douziech B, Mathieu R (1995) [19].

> "Condensation reactions of the phosphodihydrazide XP (Ph) $(NMeNH_2)_2$ (X = O **1a**, S **1b**) with ferrocene-1,1′-dicarboxaldehyde **2** afford in good yield the first examples of phosphorus ferrocenyl macrocycles $Fe[C_5H_4CH = NNMeP(X)PhNMeN = CHC_5H_4]_2$ Fe (X = O **3a**, S **3b**) and **3b** can be converted to the new compounds $Fe[C_5H_4CH_2NHNMeP(S)PhNMeNHCH_2C_5H_4]_2$ Fe **5**; Fe $\{C_5H_4CH = NNMeP(S)(Me)PhNMeN = CHC_5H_4\}_2Fe]$ $[CF_3SO_3]_2$ **6** by reaction with, respectively, LAH and CF_3SO_3Me; **5** represents a novel prototype of an anion receptor which electrochemically recognises the $H_2PO_4^-$, HSO_4^- and Cl^- anions" [19].

6.3.3.2 More evidence: Non-precious metals and metal free catalysts

Ferric Phytate was synthesized and carbonization to derive Fe–P Carbon!

From: Singh KP, Bae EJ, Yu J-S (2015) [46].

> "Phosphorus and nitrogen belong to the same group and bear somewhat similar chemical properties. Although Fe–N/C structure has been extensively studied as an electrocatalyst, no one has ever tried to evaluate the catalytic activity of Fe–P/C structure or to enhance the catalytic property of P-doped carbon catalyst in acidic medium.
>
> Until now, there have been several studies on transition metal phosphides, portraying the hydrogen evolution property of the catalysts. However, the catalytic property of metal–P catalyst for ORR remains untouched. Therefore, herein for the first time, we have tried to incorporate Fe–P structure in carbon matrix and measured its activity in both acidic and basic media. It is found that catalytic activity of already active P-doped carbon in basic medium becomes more pronounced, and in acidic medium, nonactive P-doped carbon becomes active with incorporation of Fe.
>
> Phytic acid (PA), which is considered to be principle storage form of phosphorus in plant tissues, is used as P and C source here. Ferric chloride ($FeCl_3$) is commonly used to precipitate out PA from plant samples, as PA binds with Fe irreversibly and forms insoluble complexes known as ferric phytate as shown in Scheme 1. To synthesize Fe–P bonded carbon, we have mixed $FeCl_3$ and PA followed by their polymerization at high temperature. The obtained polymer was then carbonized at various temperatures to obtain Fe-P-*x*, where *x* represents carbonization temperature of 800, 900, and 1000 °C. P-doped carbon (PA-900) is also prepared by carbonizing the PA polymer at 900 °C without Fe source" [46].

6.3.3.3 The evidence for the biogeochemical property of phosphorus: Redox between the proteomes and the *environtome*

Phosphorus is biogeochemical. It is a primary biogeographical element as it controls the traffic in the ecosystem, the environment.

We usually do not think of ourselves, even of part of ourselves, our proteins, as being transferred to the environment, the ecosystem that is presently called the *environtome*. But, indeed, we are transferred back and forth between and among the *environtome*. We, humans, plants and animals are transferred to the *environtome* and back again in real time and continually. We are the proteomic part of the proteome/*environtome* reality, always connected to the social proteome.[1] We, the biological proteome, are continually being transferred within the *environtome*, the environment, the ecosystem and phosphorus is the major biogeochemical element controlling the back and forth of the redox reactions between us and the environment: redox is continual and the element, phosphorus, comes forth as a dramatic and *critical* electroactive link between the proteome and the *environtome*.

Figure 6.3 shows the entwinement of the social proteome, biological proteome and the *environtome*.

Proteomics science and technology experts, ethicists and scholars examining independently the ethics-of-ethics, have autonomously worked together so as to develop robust accountability and more even and just distribution of power that is attendant to science, technology and innovation [41]. The veritable power of phosphorus as a messenger between the proteomes and the *environtome* comes forth as a *tour de force*. Phosphorus can be thought of as phosphates, a pentavalent oxidation state that includes the phosphate backbone of the nucleic acids, deoxyribose nucleic acid, DNA and ribose nucleic acid. RNA, and major metabolites such as adenosine triphosphate. (ATP). Redox processing of phosphates to phosphites and phosphonates and to even lower oxidation states are common. However, modern phosphorus cycling has increasingly identified reduced phosphorous compounds as playing a role as a biogeochemical element. These are the data that suggest that phosphorus is not redox-insensitive, but redox sensitive and

[1]Zhou M, Zhu S, Mo X, Guo Q, Li Y, Tian J, Liang C (February 14, 2022). Proteomic analysis dissects molecular mechanisms underlying plant responses to phosphorus deficiency. *Cells*. **11**(4):651. doi: 10.3390/cells11040651. PMID: 35203302; PMCID: PMC8870294.

reduced phosphorous compounds should be considered as part of the phosphorus biogeochemical cycle as well [42].

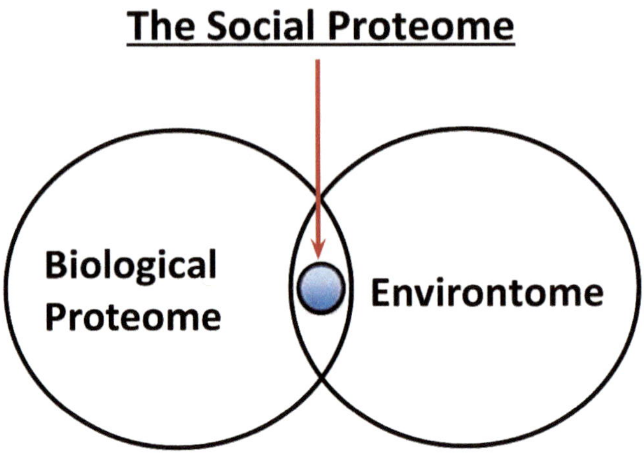

Figure 6.3 The *environtome* interacting with the biological proteome. The social proteome is a subset of the *environtome*, and situated at the intersection of the biological proteome and the *environtome*. The social proteome impacts on, and is impacted by the biological proteome [41]. Adapted with permission from Springer Verlag, (Wien, Austria), Berlin, Germany, London, UK and NY, USA.

It is well known that phosphatidic acids are anionic phospholipids important to cell signaling and in directly activating lipid-gated ion channels, they are lipid messengers. Phosphorus is the major element that transfers parts of us between the proteomes and the *environtome.* The sugar phosphate backbone brings us the molecules of life, DNA and RNA as it forms their structure. In fact, phosphate is so important that in double-stranded DNA, the molecular double-helix shape is formed by two linear sugar-phosphate backbones that take paths that are opposite and thus, the twist into the helix. The bonds are important to figuring out the exact molecule that is the lipid messenger for the BRODERICK PROBE®. Indeed, it is well known that the DNA backbone forms bonds with water since sugar-phosphate is negatively charged and also hydrophilic (loving water). These factors are amenable as lipid messengers and thus, the author concluded that the DNA-lipid BRODERICK PROBE® polymer images *via* phosphorus and DNA in the simulated brain.

It was time then to go back to more and more sensing scans for *in vitro, in vivo* and *in situ* sets of data for each nanosensor. What are other possibilities?

6.3.4 Polymerase Chain Reaction?

Is this BRODERICK PROBE® mechanism of action akin to a polymerase chain reaction (PCR) technique? Highly unlikely, Broderick thought to herself and taught her students the same because PCR makes multiples of a compound but cannot image or manufacture new chemical bonds or new molecules; the polymerase chain reaction is rather a copier machine, a xerox machine. In PCR, deoxyribose nucleic acid (DNA) is heated and split and copies are made of each side and multiplied accordingly.

"An organism's genome is stored inside DNA molecules, but analyzing this genetic information requires quite a large amount of DNA. In 1985, Kary Mullis invented the process known as polymerase chain reaction (PCR), in which a small amount of DNA can be copied in large quantities over a short period of time. By applying heat, the DNA molecule\'s two strands are separated and the DNA building blocks that have been added are bonded to each strand. With the help of the enzyme DNA polymerase, new DNA chains are formed and the process can then be repeated. PCR has been of major importance in both medical research and forensic science" (https://www.nobelprize.org/prizes/chemistry/1993/mullis/facts/).

New molecular entities are not made. So, the Broderick nanoprobe really is not PCR! The Broderick nanoprobe images electroactive signals and PCR does not. However, PCR has made a significant impact on public health during this pandemic[2] as a rapid testing procedure to assay the presence or absence of SARS-CoV-2.

The Nobel Prize in Chemistry was awarded to Kary Mullis, University of California, La Jolla, in 1993 for its discovery (https://

[2]Broderick, PA, ***Preferred Health Magazine*** *for the January 2021 Issue as Awardee,* ***Patient Preferred Physician****, The COVID Brain: A Cytokine Crisis: Major Depressive Disorder and Suicide.*

www.nobelprize.org/prizes/chemistry/1993/mullis/facts/). It is time for its successor to receive the award, the author says with a humble smile.

Dr. Broderick taught her medical and her pre-medical students that one likely conclusion for the LAN factor is chemical bonds. The BRODERICK PROBE® carbon-based polymer may covalently bond to phosphate in the 0.1 M phosphate buffer at the point of applied electric potential in millivolts *via* the detector's operational amplifier (op-amp) connected to this polymer complex. Covalent bonding among highly electronegative ions, carbon, oxygen, nitrogen and fluorine within the polymer likely was producing the amplified response at the interfacial surface of the BRODERICK PROBE®. A surfactant reaction at the nanosensor surface was definitely happening and this response is reported in the Broderick patent portfolio [6, 8, 9]. At least, surfactant mechanisms could be causal for amplification, catalysis or catalytic amplification since micelles are formed, the basis of which would support the concept of phospholipids as the lipid messenger.

Thus, the question continues to go back to where and how catalytic amplification occurs. Can catalysis be separated from amplification? Let's look at the Lipid Amplification Number (LAN) using controlled studies and the LAN operational stability of the BRODERICK PROBE® Lauric acid biosensor and compare the BRODERICK PROBE® Lauric Acid data with that of the BRODERICK PROBE® *N*-Stearoylcerebroside biosensor under the same protocol. Broderick thought that comparing a fatty acid complex with a phospholipid complex lipid would be helpful. It is well known that the triglycerides are classified as simple lipids because they are formed from two types of compounds: glycerol and fatty acids. In contrast, complex lipids contain at least one additional component, for example, a phosphate group (phospholipids) or a carbohydrate moiety (glycolipids). Dr. Broderick chose a phospholipid for the comparison.

Therefore, it is a downright pleasure to present to you the empirical results from our controlled catalytic amplification studies. *What follows are sensing operational stability studies of the Broderick nanoprobes day to day for three consecutive days,* ***day to day studies;*** *DA, 5-HT, DOPAC, and AA are the focus.*

6.4 Biomimetic Catalytic Amplification-Biomimetic Polymers: Proof of Concept

Empirical data of lipid amplification from the BRODERICK PROBE® *carbon/lipid/phosphotidyl polymer follows (Figures 6.4 to 6.10):*

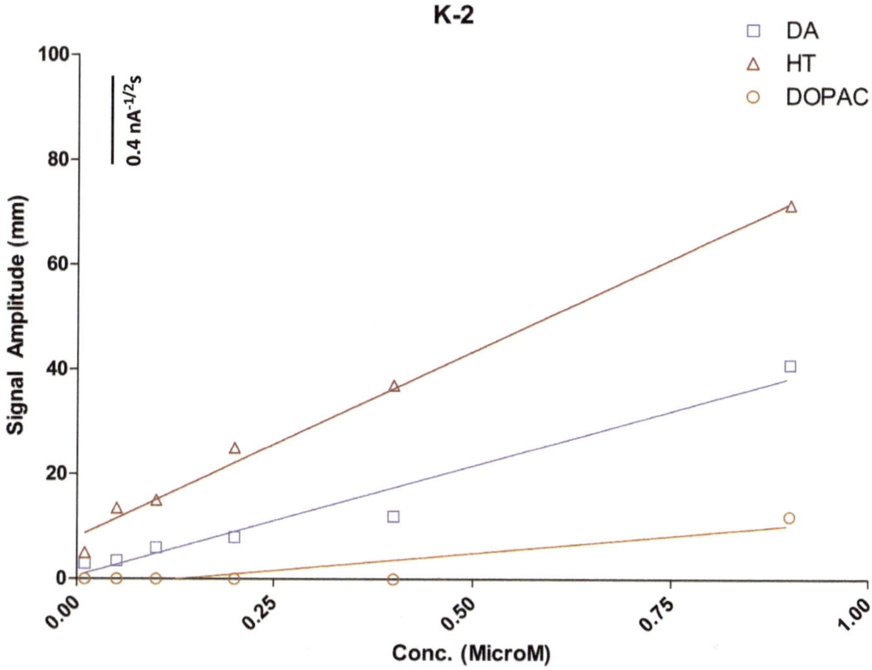

Figure 6.4 (K-2) A line graph portraying electroactive signal amplification for the synaptic release of functional amines and **DOPAC** when **K-2** is the coded *sensor* studied in the simulated brain. The sensor is comprised of a carbon/lipid/phosphatidyl **Laurate** polymer in the BRODERICK PROBE® reference library of novel biosensors and sensors. The sensor was phosphorylated further in phosphate buffer and layered by layer coated with the patented lipid/protein complex [12]. LAN results show that the sensor responds in the following order: indoleamine, serotonin (5HT) > catecholamine, dopamine (DA) > 3,4-dihydroxyphenylacetic acid (DOPAC). The x axis: concentration in micromolar units: the y axis: signal amplification-current in nA. Data are gleaned from three-day studies. Data shown in this line graph are taken from the third day of study. **Linear regression analysis: DA**: (F value_344.8; DFn.DFd-1.000,16.00; P < 0.000); 5-HT: (F value_1696.00; DFn.DFd 1.000,16.00; P < 0.0001); **DOPAC**: (F value_19,21 (DFn.DFd-1.000, 4.000; P < 0.0118).

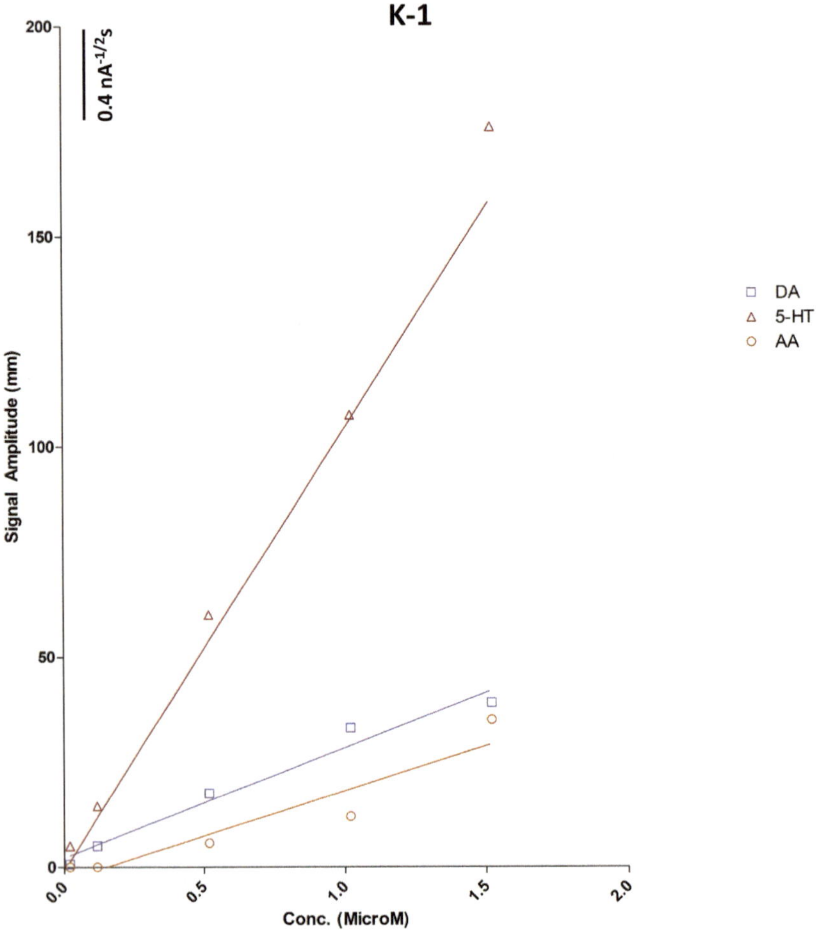

Figure 6.5 (K-1) A line graph portraying electroactive signal amplification for the synaptic release of functional amines and **AA** when **K-1** is the coded *sensor* studied in the simulated brain. The sensor is comprised of a carbon/lipid/phosphatidyl **Laurate** polymer in the BRODERICK PROBE® reference library of novel biosensors and sensors. The sensor was phosphorylated further in phosphate buffer and layered by layer coated with the patented lipid/protein complex [12]. LAN results show that the sensor responds in the following order: indoleamine, serotonin (5HT) > catecholamine, dopamine (DA) > ascorbic acid (AA). The x axis: concentration in micromolar units: the y axis: signal amplification-current in nA. Data are gleaned from three-day studies. Data shown in this line graph are taken from the third day of study. **Linear regression analysis: DA:** (F value 436.0; DFn.DFd 1.000,13.00; $P < 0.0001$); 5-HT: (F value_75.37; DFn.DFd-1.000,13.00; $P < 0.0001$); **AA**: (F value_99.35; DFn.DFd 1.000, 13.00; $P < 0.0001$).

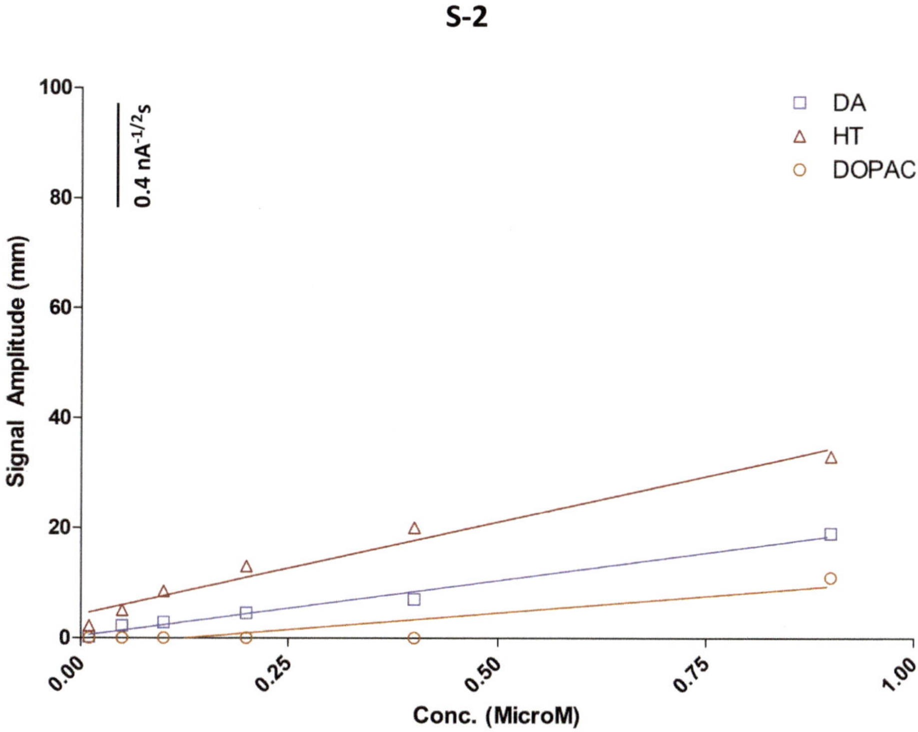

Figure 6.6 (S-2) A line graph portraying electroactive signal amplification for the synaptic release of functional amines and DOPAC when **S-2** is the coded *sensor* studied in the simulated brain. The sensor is comprised of a carbon/lipid/phosphatidyl ***N*-Stearoylcerebroside** polymer in the BRODERICK PROBE® reference library of novel biosensors and sensors. The sensor was phosphorylated further in phosphate buffer and layered by layer coated with the patented lipid/protein complex [12]. LAN results show that the sensor responds in the following order: indoleamine, serotonin (5HT) > catecholamine, dopamine (DA) > 3,4-dihydroxyphenylacetic acid (DOPAC). The *x* axis: concentration in micromolar units: the *y* axis: signal amplification—current in nA. Data are gleaned from three-day studies. Data shown in this line graph are taken from the third day of study. **Linear r**-current in nA. **Linear regression analysis: DA**: (*F* value_974.1; DFn.DFd-1.000,16.00; $P < 0.0001$) **5-HT**: (*F* value_500.7; DFn.DFd-1.000,16.00; $P < 0.000$): **DOPAC:** (*F* value_19,21 (DFn.DFd-1.000, 4.000; $P < 0.0118$).

S-1

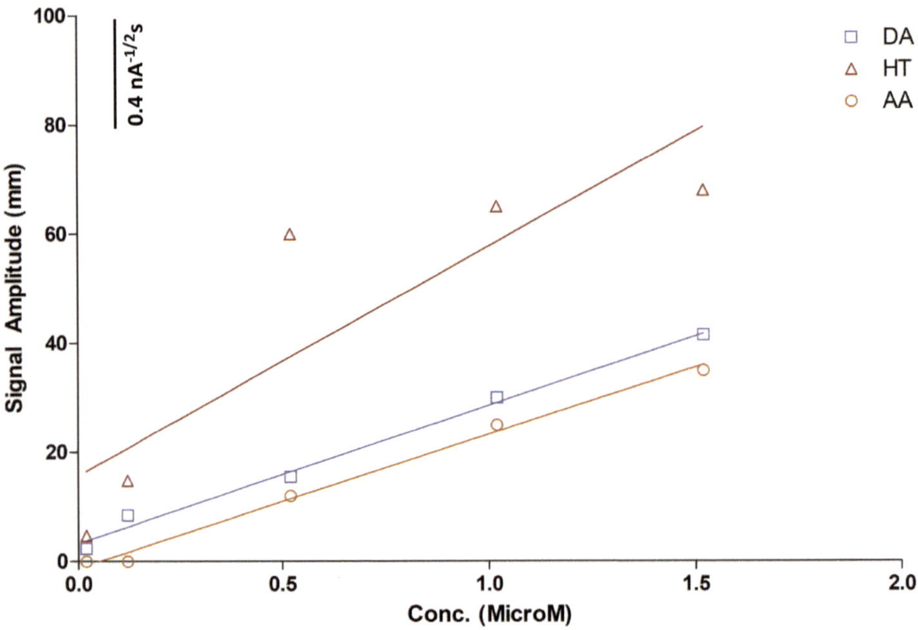

Figure 6.7 (S-1) A line graph portraying electroactive signal amplification for the synaptic release of functional amines and **AA** when **S-1** is the coded *sensor* studied in the simulated brain. The sensor is comprised of a carbon/lipid/phosphatidyl ***N*-Stearoylcerebroside** polymer in the BRODERICK PROBE® reference library of novel biosensors and sensors. The sensor was phosphorylated further in phosphate buffer and layered by layer, coated with the patented lipid/protein complex [12]. LAN results show that the sensor responds in the following order: indoleamine, serotonin (5HT) > catecholamine, dopamine (DA) > ascorbic acid (AA). The *x* axis: concentration in micromolar units: the *y* axis [2, 8]: signal amplification—current in nA. Data are gleaned from three-day studies. Data shown in this line graph are taken from the third day of study. **Linear regression analysis: DA**: (*F* value_138.1; DFn.DFd 1.000,13.00; $P < 0.0001$); **5-HT**: (*F* value_42.09; DFn.DFd 1.000,13.00; $P < 0.0001$); **AA** (*F* value_1755 (DFn.DFd 1.000, 13.000; $P < 0.0001$).

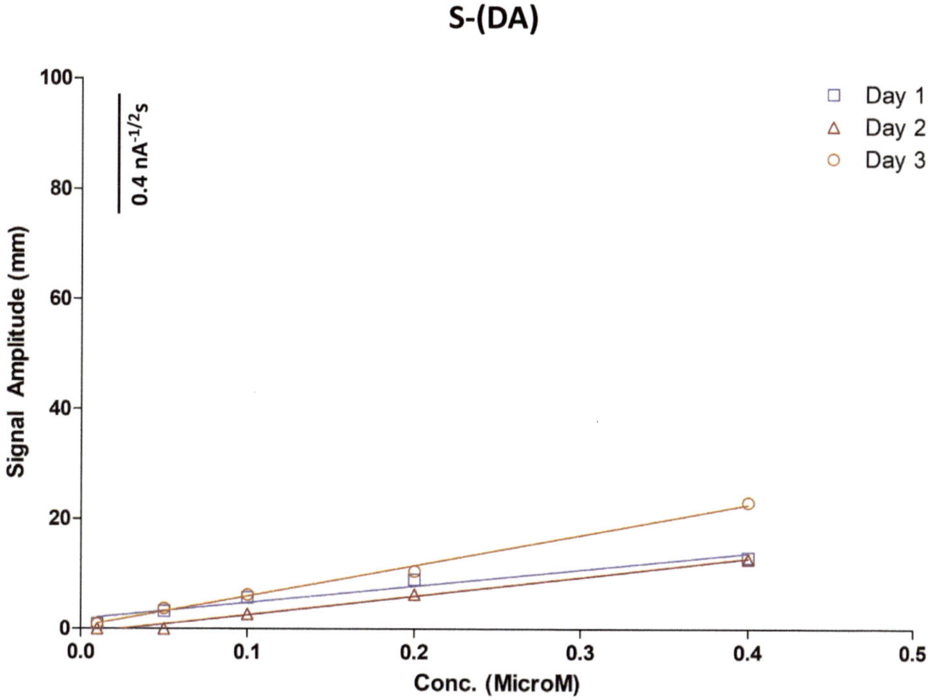

Figure 6.8 Day-to-day studies on the S code LAN for imaging the neurotransmitter **DA** by the BRODERICK PROBE® reference library of novel biosensors and sensors. This line graph shows evidence that the complex lipid, ***N*-Stearoylcerebroside** amplifies linearly according to concentration. The LAN for the catecholamine, DA amplifies more dramatically on the third day of study and amplification results are controlled with an operationally stable nanoprobe. The *x* axis: concentration in micromolar units: the *y* axis: signal amplification—current in nA. Data are gleaned from each of the three-day studies. Data shown in this line graph are taken from each day of the three days of study. **Linear regression analysis: Day 1**: (*F* value_119.2; DFn.DFd 1.000,8.000; $P < 0.0001$); **Day 2:** (*F* value_211.12; DFn.DFd 1.000,8.000; $P < 0.000$); **Day 3:** (*F* valuc_804.6; DFn.DFd 1.000, 8.000; $P < 0.0001$).

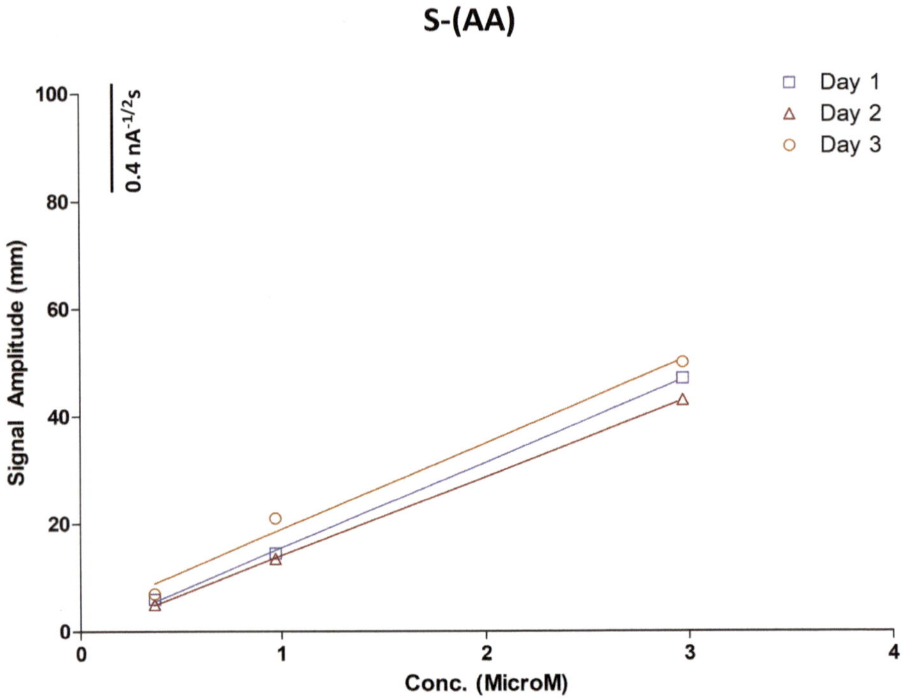

Figure 6.9 Day-to-day studies on the S code LAN for imaging the cofactor, **AA**, by the BRODERICK PROBE® reference library of novel biosensors and sensors. The line graph shows evidence that the complex lipid, ***N*-Stearoyl-cerebroside** amplifies **AA** proportionally to concentration. AA is imaged at higher detection limits than are the catecholamines and indoleamines. The LAN for the cofactor, AA, amplifies with dramatically catalytically controlled amplification; the results show an operationally stable nanoprobe. The *x* axis: concentration in micromolar units: the *y* axis: signal amplification—current in nA. Data are gleaned from each day of the three-day studies. Data shown in this line graph are taken from each day of the three days of study. **Linear regression analysis: Day 1**: (*F* value_1665; DFn.DFd 1.000,1.000; $P < 0.0156$); **Day 2**: (*F* value. 18050; DFn.DFd 1.000,1.000; $P < 0.0047$); **Day 3**: (*F* value_94,21 (DFn.DFn 1.000, 1.000; $P < 0.0654$).

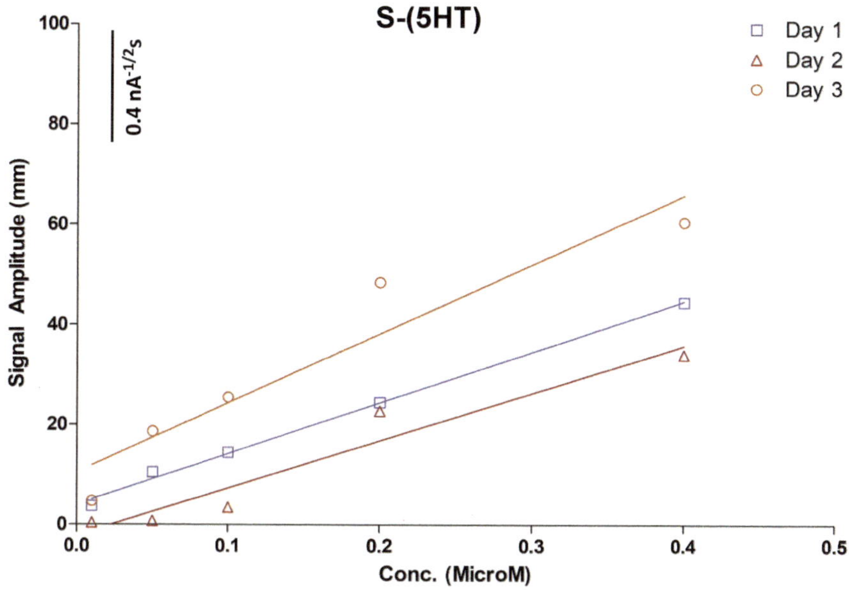

Figure 6.10 **Day-to-day** studies on the S code **LAN** for imaging the neurotransmitter, **5-HT** by the BRODERICK PROBE® reference library of novel biosensors and sensors. The line graph shows evidence that the complex lipid, ***N*-Stearoylcerebroside** undergoes signal amplification linearly. The LAN for the indoleamine, 5-HT increases proportionally with concentration during each day of study. The results show an operationally stable nanoprobe. Carbon has a particular propensity for adsorbing indoleamines an example of which is 5-HT [5, 32, 38]. The *x* axis: concentration in micromolar units: the *y* axis: signal amplification-current in nA. Data are gleaned from three-day studies. Data shown in this line graph are taken from each day of the three-day study. **Linear regression analysis: Day 1**: (*F* value_1607; DFn.DFd 1.000,8.000; $P < 0.0001$); **Day 2**: (*F* value_110.1; DFn.DFd 1.000, 8.000; $P < 0.0001$); **Day 3**: (*F* value_78.18 (DFn.DFd 1.000, 8.000; $P < 0.0001$).

6.4.1 The Merits of Electroactive BRODERICK PROBE® Biomimetic Sensing Polymers: Catalysis

- Sensors control catalytic amplification.
- Sensors are, in one example, PEA/BSA carbon-based polymers.
- BRODERICK PROBE® complex lipids catalyze differently from the BRODERICK PROBE® fatty acid complex.

- Catalysis is **Self-Catalysis** and the catecholamine, DA, is involved in the interaction with hydroquinone.
- DA-Quinone Self-Catalysis differs between BRODERICK PROBE® complex lipids and fatty acid complexes.
- DA-Quinone Self-Catalysis requires a physiological presence of AA.
- DA-Quinone Self-Catalysis may not need chemical bonds to AA.
- AA requires higher concentrations than does DA.
- DA Self-Catalysis is shown as electroactive orthoquinone is formed when solute in solution turns a black color.
- In DA Self-Catalysis, a renegade quinone remains adsorbed to DA during electron transfer kinetics.
- The renegade quinone forms a monolayer catalytically amenable to reverse DA redox chemisorption.
- New products are formed.
- Catalytic amplification in the BRODERICK PROBE® lipid bilayer materializes *via* chemical bonds.
- **LAN** is an acronym for *Lipid Amplification Number*-empirically derived by Broderick in her Laboratory.
- **LAN:** BRODERICK PROBE® ***N*-Stearoylcerebroside** polymer-less than BRODERICK PROBE®: **Laurate** polymer.
- DA-Quinone interactions are distinguished according to biomimetic catalytic polymer results.
- BRODERICK PROBE® biomimetic sensing polymers are catalytically controlled.
- BRODERICK PROBE® β-Galactosyl-C18-Ceramide is a carbohydrate sensor.
- Applying minimal voltage to the current measurer op-amp brings catalytic amplification *via* phosphorus.
- Phosphorus is electroactive.
- Phosphorus is a biogeochemical element.
- Phosphorus acts as messenger between the proteomes and the *environtome.*
- Phosphorus is a BRODERICK PROBE® biomimetic lipid messenger.

- A novel BRODERICK PROBE® phosphorus nanoprobe is introduced.
- A novel BRODERICK PROBE® DNA-Lipid nanoprobe is introduced.
- Phosphate is placed in the manufactured-on-site BRODERICK PROBE® electrochemical cell.
- Phosphate serves as a lipid messenger.
- Catalysis of AA and DA are controlled in the brain and the electrochemical cell by the BRODERICK PROBE®
- Sensors image AA and DA electroactive signals separately and selectively.

6.4.2 The Merits of Electroactive BRODERICK PROBE® Biomimetic Sensing Polymers: Assembly

- Sensors image serotonin 5-HT at a redox potential more positive than current redox potentials for DA or AA.
- Sensors image 5-HT at a more heightened amplification than AA and DA.
- Serotonin is an indoleamine.
- Indoles are known to adsorb to carbon with high adsorption magnitude.
- BRODERICK PROBE® adsorbs carbon/indole signals to a greater degree than DA or AA.
- Serotonin, the indoleamine, amplifies but appears not to be associated with the classical concept of "catalyze."
- Serotonin undergoes electron transfer to an electroactive orthoquinone when the solute in solution turns a pink color.
- Serotonin catalyzes to a quinone, likely an orthoquinone.
- Serotonin catalysis is not Self-Catalysis.
- Serotonin catalysis is not associated with AA.
- The 5-HT hydroquinone electron transfer kinetics differs from that of DA hydroquinone electron transfer kinetics.
- The 5-HT hydroquinone electron transfer kinetics appears not to be AA associated.

- Indoles and DNA are known to self-assemble at the water-lipid interface.
- Serotonin **Self Assembles**!
- Spontaneously formed lipid bilayers are assembled.
- Embedded in the newly formed lipid bilayers are amino acids, proteins and peptides.
- The mechanics for *serotonin* (5-HT) is amplification.
- The mechanics for serotonin amplification is primarily molecular Self-Assembly.
- The BRODERICK PROBE® separates catalytic amplification from amplification.
- A novel indole DNA BRODERICK PROBE® is introduced.
- Medical use of the sensor is clear.
- Industrial use of the sensor is clear.
- Commercial use of the sensor is clear.

6.4.2.1 *Modus operandi* for DA: Self-catalysis

McCreery jumped into the catechol and hydroquinone electrochemistry [22] of electrochemical electrocatalysis and published his results just four years after the famous and most essential meeting of neuroscientists and electrochemists, the conference held by the *New York Academy of Science* in Manhattan, NY, in 1983; the conference is discussed herein in Chapter 2. The conference was called *"Neurochemical Analysis of the Conscious Brain: Voltammetry and Push-Pull Perfusion"* and published in 1986. McCreery thought about these acutely intense discussions that took place and what the well-known electrochemist, Dr. McCreery, gives us is really important to bring us to a comfortable niche regarding what was missing during those tumultuous times.

The explanatory note is paraphrased but still in quotes; it is: "electrochemical electrocatalysis by quinones are not always unilateral in direction." In other words, it is best for us to step back and take a closer look at dogma. Sometimes it needs to be parsed more specifically. Indeed, sometimes, redox reactions are not catalytically *enhanced* at all, but in fact, there are DA-quinone *inhibitory* redox reactions, at least toward the redox heterogeneous glassy carbon catalytic interactions used by McCreery. The data show that a counter catalysis happens when a quinone monolayer is methylene blue, nitrophenyl and/or trifluoromethylphenyl. The electrochemical electrocatalysis that we were debating "back in the day" in fact, pointed to a duroquinone, anthraquinone and/or even dopamine (DA) itself as the latter monolayers will provide physisorbed and/or chemisorbed monolayers of several quinones that are increasingly upward catalytic toward DA oxidation and reduction.

The results of the BRODERICK PROBE® lipid biomimetic studies further bring us to another, albeit more precise look at dogma. The results from the Broderick Laboratory show that the degrees of catalysis are controlled by the BRODERICK PROBE® complex lipid, a ceramide bonded to Stearic acid/Stearate and the BRODERICK PROBE® fatty acid polymer complex comprised of Lauric Acid/Laurate [6, 8–12].

From: DuVall SH, McCreery RL (2000) [22].

> "Heterogeneous electron transfer kinetics for several catechols were examined on glassy carbon (GC) electrodes in aqueous solution. Electrode preparations yielded GC surfaces with low levels of oxides or adsorbed impurities, which exhibited strong adsorption of dopamine (DA) and related catechols. Conversely, modification of GC with an organic monolayer suppressed DA adsorption and, in many cases prevented electron transfer. By relating catechol adsorption to observed electron transfer, it was concluded that an adsorbed layer of catechol acts as an electrocatalyst for solution-phase redox components. Physisorbed or chemisorbed monolayers of several quinones, including duroquinone, anthraquinone and dopamine itself, are catalytic toward dopamine oxidation and reduction, but nitrophenyl, trifluoromethylphenyl, and methylene blue monolayers severely inhibit electron transfer. The magnitude of inhibition was affected by electrostatic attraction or repulsion between the surface and the redox system, but the major factor controlling electron-transfer kinetics is not electrostatic in origin. The most plausible mechanism is "self-catalysis" by an adsorbed quinone, which remained adsorbed during electron transfer to a redox couple in solution. The results are inconsistent with a redox mediation mechanism

involving a redox cross-reaction between adsorbed and solution quinone couples. An interaction between the adsorbed and solution quinone species during electron transfer appears to catalyze one or more of the steps in the "scheme of squares" mechanisms for hydroquinone/quinone redox systems. The results explain a variety of observations about catechol and hydroquinone electrochemistry, as well as provide more fundamental insights into quinone electron-transfer mechanisms." [22].

6.4.2.2 *Modus operandi* for 5-HT: Self-assembly

Carbon is the ***Base*** for DNA and in fact, carbon is the basis for life itself. The BRODERICK PROBE® is carbon based; the nanoprobe facilitates 5-HT to self-assemble. In another series of the BRODERICK PROBE® microelectrodes and sensors, we have non-invasive protein based photoluminescent nanosensors [10]. Another series of the nanoprobes are semiconductor carbon allotropes [12]. The Broderick nanoprobe holds a claim to the lipid molecule, phosphotidylethanolamine [12] without which, α-synuclein homeostasis is disrupted and the disease of Parkinson ensues [52] presumably leading to Alzheimer's and disorders of epilepsy including the autistic spectrum [11].

Self-assembly is known in the classical sense and the phenomenon is explained as the spontaneous and reversible organization of molecular units into ordered structures by, for example, non-covalent interactions. One of the first properties of a self-assembled system that this definition suggests is the exciting spontaneity of the self-assembly process. The interactions that are responsible for the formation of a self-assembled system act within themselves. This means that the nanostructure of the Broderick nanoprobe images the neurotransmitter, serotonin (5-HT) online, in real time as it assembles itself. The assembly is strongly bonded and is or becomes covalent in the process and the process involves molecules, thereby calling serotonin self-assembly as molecular self-assembly.

"Self-assembly can be classified as either static or dynamic. In static self-assembly, the ordered state forms as a system approaches equilibrium, reducing its free energy. However, in dynamic self-assembly, patterns of pre-existing components organized by specific local interactions are not commonly described as "self-assembled" by scientists in the associated disciplines. These structures are better described as "self-organized", although these terms are often used interchangeably." *https://www.chemeurope.com/en/encyclopedia/Self-assembly.html.*

6.4.2.3 The evidence for serotonin self-assembly: Indole self-assembly

From: De Marchi F, Cui D, Lipton-Duffin J, Santato C, MacLeod JM, Rosei F (2015) [20].

> "Model systems are critical to our understanding of self-assembly processes. As such, we have studied the surface self-assembly of a small and simple molecule, indole-2-carboxylic acid (I2CA). We combine density functional theory gas-phase (DFT) calculations with scanning tunneling microscopy to reveal details of I2CA assembly in two different solvents at the solution/solid interface, and on Au (111) in ultrahigh vacuum (UHV). In UHV and at the trichlorobenzene/highly oriented pyrolytic graphite (HOPG) interface, I2CA forms epitaxial lamellar structures based on cyclic OH···O carboxylic dimers. The structure formed at the heptanoic acid/HOPG interface is different and can be interpreted in a model where heptanoic acid molecules co-adsorb on the substrate with the I2CA, forming a bicomponent commensurate unit cell. DFT calculations of dimer energetics elucidate the basic building blocks of these structures, whereas calculations of periodic two-dimensional assemblies reveal the epitaxial effects introduced by the different substrates." [20].

6.4.2.4 The evidence for amino acid self-assembly

From: Makhlynets OV, Korendovych IV (2019) [37].

> "Enzymes require many, often hundreds, of amino acid residues arranged in a protein fold to promote catalysis. Now, self-assembly of a single amino acid — phenylalanine — in the presence of zinc is shown to form supramolecular structures that promote hydrolysis better than natural enzymes on a weight basis." [37].

6.4.2.5 The evidence for DNA-lipid self-assembly

From: Khalid S, Bond PJ, Holyoake J, Hawtin RW, Sansom MSP (2008) [33].

"DNA–lipid complexes are of biomedical importance as delivery vectors for gene therapy. To gain insight into the interactions of DNA with zwitterionic and cationic (dimyristoyltrimethylammonium propane (DMTAP)) lipids, we have used coarse-grained molecular dynamics simulations to study the self-assembly of DPPC and DPPC/DMTAP lipid bilayers in the presence of a DNA dodecamer. We observed the spontaneous formation of lipid bilayers from initial systems containing randomly placed lipids, water–counterions and DNA. In both the DPPC and DPPC/DMTAP simulations, the DNA molecule is located at the water–lipid headgroup interface, lying approximately parallel to the plane of the bilayer. We have also calculated the potential of mean force for transferring a DNA dodecamer through a DPPC/DMTAP bilayer. A high energetic barrier to DNA insertion into the hydrophobic core of the bilayer is observed. The DNA adopts a transmembrane orientation only in this region. Local bilayer deformation in the vicinity of the DNA molecule is observed, largely as a result of the DNA–DMTAP headgroup attraction" [33].

6.4.2.6 The evidence for lipid-protein self-assembly

From: An Z, Möhwald H, Li J (2006) [1].

"The preparation of nanotubes from human serum albumin (HSA) and mixtures of L-alpha-dimyristoylphosphatidic acid (DMPA)/HSA is described. The nanotubes were prepared via alternate adsorption of HSA of different/opposite charges (by variation of the pH) or by sequential adsorption of DMPA and HSA, respectively, onto the inner surfaces of porous anodic alumina templates. This simple layer-by-layer assembly results in a monodisperse size distribution and a uniform orientation. The nanotubes allow the specific incorporation of lipophilic components such as channels or receptors and may thus serve as probes or sensors for biological systems" [1].

6.5 Biomimetic Catalytic Amplification-Polymeric Biochips: Proof of Concept

6.5.1 Broderick Thought

How exactly do Fatty Acids Factor into the Lipid Biomimetics of the BRODERICK PROBE® Biochip? After all, certain facts about long-chain fatty acids that come from endogenous metabolism and nutrition are known. Indeed, long-chain fatty acids are known actors in the second messenger system as well as playing a role in the regulation of pathways that are responsible for cell signaling.

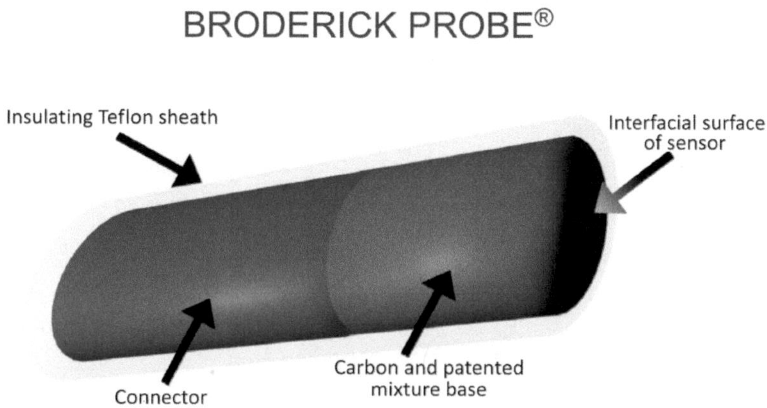

Figure 6.11 The BRODERICK PROBE® PEA/BSA biochip, the BRODERICK PROBE® PEA biochip and the BRODERICK PROBE® BSA biochip in 3D. **Credit:** Aisha Hashmi, CCNY student of Dr. Broderick in *Neurobiology Class, Medical 10000*, Fall, 2021.

Therefore, studies of the Broderick nanoprobes were performed using the original polymeric nanoprobes exposed for days in solutions of PEA–BSA combinations, one by one, for consecutive periods of 1, 24, and 48 h.

The "***within days***" data show further results of the operational sensing stability of the BRODERICK PROBE® multicomplex for imaging captious molecules. The BRODERICK PROBE® biomimetic chips are added to the original Broderick polymer. The inventive sensors were placed in the ***simulated brain***.

To study the role of fatty acids, the BRODERICK PROBE® biomimetic laurate chip was compared with the BRODERICK PROBE® biomimetic stearate chip. Both molecules are saturated fatty acids; the former is comprised of 12 carbons and the latter is composed of 18 carbons. Laurate is categorized as a medium chain fatty acid and stearate is categorized as a long chain fatty acid.

Therefore, it is a real pleasure to present to you the empirical results from our controlled catalytic amplification studies that further delineate how fatty acids play into the lipid biomimetics of the BRODERICK PROBE® nanoprobes.

Images of dopamine (DA) and serotonin (5-HT) responses using the BRODERICK PROBE® K-Laurate Biochip versus the BRODERICK PROBE® J-Stearate Biochip are shown in histogram form.

We studied the specific role of phosphate buffer detection, PEA/BSA, PEA alone, and BSA alone. These data are shown in histogram form in Figures 6.12 to 6.15.

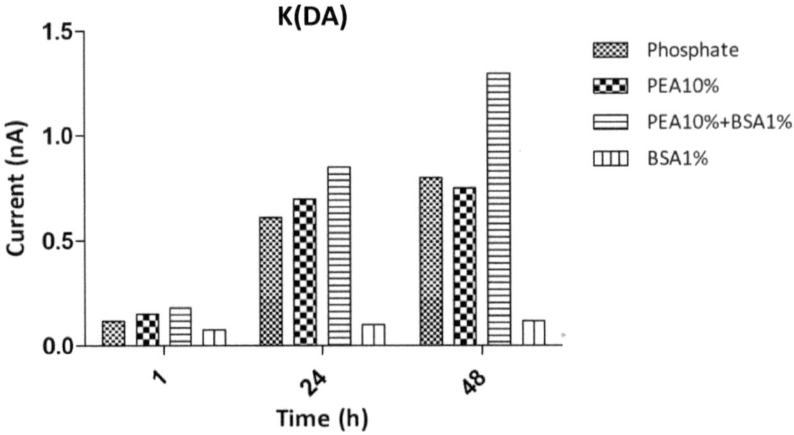

Figure 6.12 A bar graph (histogram) portraying electroactive signal amplification for the synaptic release of the functional amine, dopamine **(DA)** when **the K-Laurate Biochip** is the coded *sensor* studied in the simulated brain for *24–48 h* studies. The sensor is comprised of a carbon/lipid/phosphatidyl **Laurate** polymer in the BRODERICK PROBE® reference library of novel biosensors and sensors. The sensor was phosphorylated further in phosphate buffer and layered by layer coated with the patented lipid/protein complex [12]. The *x* axis: time in hours: the *y* axis: signal amplification-current in nA. LAN results show that the sensor responds in the following order: DA responded in linear fashion to each of the parameters with PEA/BSA leading the amplification response, followed by PEA, followed by Phosphorus. BSA was recognized the least by the K nanoprobe when sensing DA. There was not a significant difference in the PEA/BSA response when contrasting the 24 and 48 h data points. It is interesting that the lipid/protein response was the most dramatic response despite the lower response of the K nanoprobe to the albumin, BSA. The Lipid-Protein response is more potent than the lipid or the protein response alone.

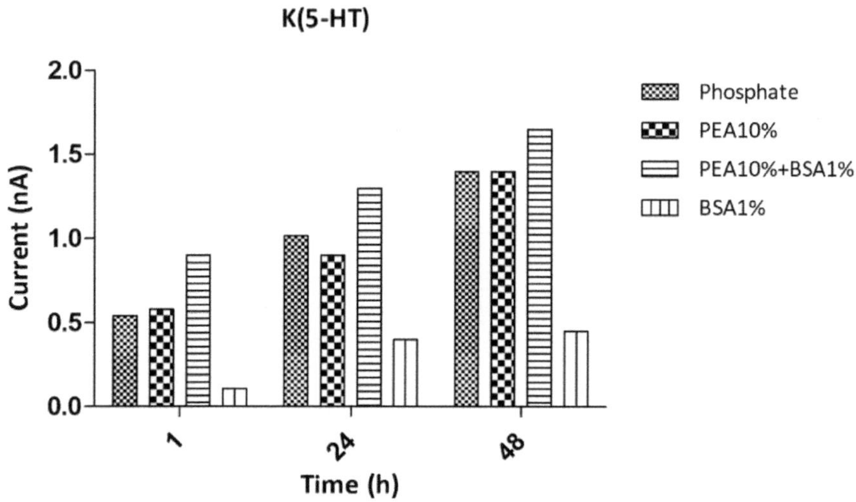

Figure 6.13 A bar graph (histogram) portraying electroactive signal amplification for the synaptic release of the functional amine, serotonin **(5-HT)** when **the K-Laurate Biochip** is the coded *sensor* studied in the simulated brain for *24–48 h* studies. The sensor is comprised of a carbon/lipid/phosphatidyl **Laurate** polymer in the BRODERICK PROBE® reference library of novel biosensors and sensors. The sensor was phosphorylated further in phosphate buffer and layered by layer coated with the patented lipid/protein complex [12]. The x axis: time in hours: the y axis: signal amplification-current in nA. LAN results show that the sensor responds in the following order: **5-HT** responded in linear fashion to each of the parameters with PEA/BSA leading the amplification response, followed by PEA, with notable competition by phosphorus. In fact, at the 24*h* data point, phosphate shows a heightened signal response over that of PEA. BSA recognized 5-HT the least by the K sensor but the response was still greater than that of DA. A difference between the phosphate and PEA response at the 48*h* data point was not statistically significant at the 95% CL. It is interesting that the response from the combination of lipid/protein was the most dramatic response despite the lower response of the K nanoprobe to the albumin protein, BSA. The Lipid-Protein response appears more potent than the that of the lipid alone or the protein response alone. The higher amplitude of current displayed by 5-HT shows the propensity of carbon to indoles such as 5-HT. Therefore, 5-HT signal responses are higher than electroactive signals for DA as expected.

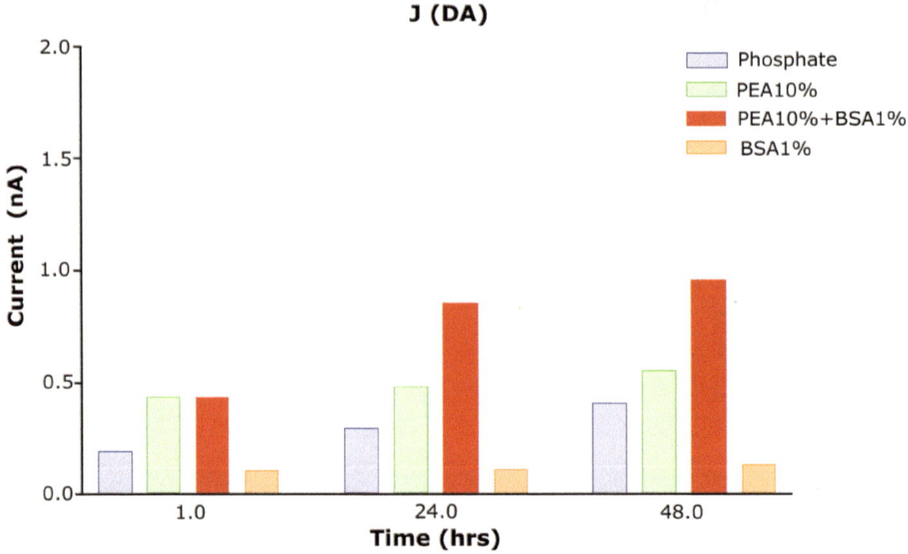

Figure 6.14 A bar graph (histogram) portraying electroactive signal amplification for the synaptic release of the functional amine, dopamine **(DA)** when **the J-Stearate Biochip** is the coded *sensor* studied in the simulated brain for *24–48 h* studies. The sensor is comprised of a carbon/lipid/phosphatidyl **Stearate** polymer in the BRODERICK PROBE® reference library of novel biosensors and sensors. The sensor was phosphorylated further in phosphate buffer and layered by layer coated with the patented lipid/protein complex [12]. The *x* axis: time in hours: the *y* axis: signal amplification-current in nA. LAN results show that the sensor responds in the following order: **DA** responded in linear fashion to each of the parameters with PEA/BSA leading the amplification response, followed by PEA, followed by phosphorus. BSA recognized DA the least by the K sensor. Albumin amplification increased each day with the J biochip. Long chain fatty acids, Stearate. incorporated in the tiny probe see the protein, albumin to a more heightened degree than does the middle chain fatty acid, Laurate, when imaging DA. It is interesting that the combination of lipid/protein response was the most dramatic response despite the lower response of the J nanoprobe to the albumin, BSA. The Lipid-Protein response is more potent than the lipid or the protein response alone.

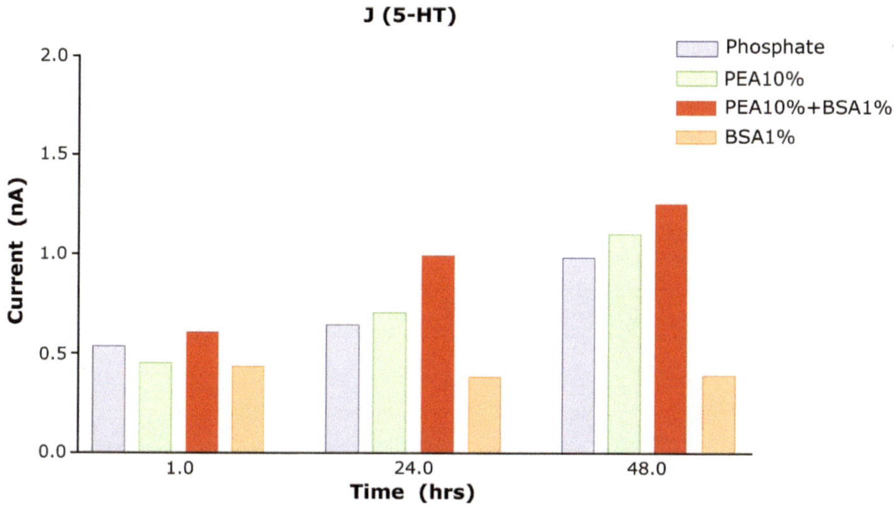

Figure 6.15 A bar graph (histogram) portraying electroactive signal amplification for the synaptic release of the functional amine, serotonin **(5-HT)** when **the J-Stearate Biochip** is the coded *sensor* studied in the simulated brain for *24–48 h* studies. The sensor is comprised of a carbon/lipid/phosphatidyl **Stearate** polymer in the BRODERICK PROBE® reference library of novel biosensors and sensors. The sensor was phosphorylated further in phosphate buffer and layered by layer coated with the patented lipid/protein complex [12]. The *x* axis: time in hours: the *y* axis: signal amplification-current in nA. LAN results show that the sensor responds in the following order: **5-HT** responded in linear fashion to each of the parameters with PEA/BSA leading the amplification response, followed by PEA, followed by phosphorus. BSA amplification by 5-HT was least when J coded sensor was used but the J sensor response to 5-HT was dramatically greater than that of DA and yet, slightly decreased from the 1*h* mark and then increased over time at the 48*h* mark. Long chain fatty acids, Stearate, incorporated in the tiny probe sees the protein, albumin to a more heightened degree than does the middle chain fatty acid, Laurate. It is interesting that the response signaled by lipid/protein was the most dramatic despite the comparatively lower response of the J nanoprobe to the albumin, BSA. Remember though, that the J nanoprobe electroactive signaling response to the albumin BSA is clearly better than that of the K sensor. The Lipid-Protein response is more potent than the lipid or the protein response alone. The higher amplitude of current displayed by 5-HT shows the propensity of carbon to indoles such as 5-HT. Therefore, 5-HT signal responses are higher than electroactive signals for DA as expected.

6.5.1.1 The empirical data for the electromechanics of fatty acids in the BRODERICK PROBE®

Below, the data are further provided for the reader in Table 6.1.

Table 6.1 Current responses for the K-laurate and J-stearate fatty acid Broderick Nanoprobe.

Table 6.1 Average relative response of DA and 5-HT in phosphate buffer using J and K electrodes

Duration (h)	K (DA) (nA)	K (5-HT) (nA)	J (DA) (nA)	J (5-HT) (nA)
1	0.12	0.54	0.19	0.53
24	0.61	1.02	0.29	0.65
48	0.80	1.4	0.40	0.98

Table 6.2 Average relative response of DA and 5-HT in albumin (1% BSA) + 10% PEA using J and K electrodes

Duration (h)	K (DA) (nA)	K (5-HT) (nA)	J (DA) (nA)	J (5-HT) (nA)
1	0.18	0.90	0.43	0.60
24	0.85	1.70	0.85	0.99
48	1.30	1.65	0.95	1.25

Table 6.3 Average relative response of DA and 5-HT in 10% PEA only using J and K electrodes

Duration (h)	K (DA) (nA)	K (5-HT) (nA)	J (DA) (nA)	J (5-HT) (nA)
1	0.15	0.58	0.43	0.46
24	0.70	0.90	0.45	0.70
48	0.72	1.40	0.55	1.10

Table 6.4 Average relative response of DA and 5-HT in 1% BSA only using J and K electrodes

Duration (h)	K (DA) (nA)	K (5-HT) (nA)	J (DA) (nA)	J (5-HT) (nA)
1	0.075	0.11	0.104	0.43
24	0.100	0.40	0.110	0.38
48	0.120	0.45	0.130	0.39

Note: K, laurate; J, stearate; DA, dopamine; 5-HT, serotonin; PEA, phosphotidylethanolamine; BSA, bovine serum albumin.

6.5.2 The Merits of Electroactive Biomimetic Sensors

"Biopolymers have the potential for use as a matrix for applications such as controlled release devices, environmentally sensitive membranes, mimic materials and energetic applications. Renewable resources (such as starch) can be utilized as polymer matrices for electroactive materials that are sensitive to their environment. Natural polymers are generally more environmentally-friendly and biocompatible than existing synthetic products. Thermoplastic starch is naturally insulative; however, the chemical, electrical, and mechanical properties of the biopolymer matrix can be tailored for specific functionality in a continuous process utilizing reactive extrusion. Conductance can be measured in the solid state by a direct-current resistance method. Ion-conducting materials, produced by doping thermoplastic starch and biopolymers with metal halides, have 5 orders of magnitude greater conductance than native materials. There is a correlation between polymer mobility and conductance. Plant or microbial biopolymers with ionic functional groups have shown promise for higher levels of conductance. The conductance approaches the level of synthetic polymer electrolytes [25]. The commercial use of biopolymers is clear.

6.5.2.1 Lipid-based biomimetic drug delivery platforms

Lipid-based biomimetics in drug and vaccine delivery are becoming more and more available [17]. An albumin coated trimethyl chitosan-based targeting delivery platform for photothermal/chemo-synergistic cancer therapy is introduced [30]. Lipid-based biomimetics comprised of the non-luminescent protein, albumin, is one that is marketed for tumor capture; it is a chemo-synergistic therapy for metastatic breast cancer [16] and lipid-based biomimetics for non-photoluminescent albumin is used to carry *temozolomide,* a chemotherapy for the treatment of anaplastic astrocytoma, and glioblastoma multiforme (GBM) (types of brain tumors) [45]. Novel biomimetic lipid carriers are used to capture *ponatinib* as a platform for multi-tyrosine kinase-targeted therapy for inhibiting tumorigenesis in patients presenting with cancer [54].

6.5.2.2 Defining the chains in the splendid fatty acids

- Short-chain fatty acids (SCFA) are fatty acids with aliphatic tails of five or fewer carbons (butyric acid).
- Medium-chain fatty acids (MCFA) are fatty acids with aliphatic tails of 6 to 12 carbons (lauric acid).
- Medium-chain triglycerides (MCT's) are fatty acids that form from medium chain triglycerides.
- Long-chain fatty acids (LCFA) are fatty acids with aliphatic tails of 13 to 21 carbons (stearic acid).
- Very long chain fatty acids (VLCFA) are fatty acids with aliphatic tails of 22 or more carbons (cerotic acid and the odd-chain fatty acid ceroplastic acid (C35).

6.5.2.3 Delving into the likes and the loves of fatty acids: Binding information

Zhu et al. [53] show how macromolecular crowded media affect fatty acids and BSA binding by using thermodynamic and spectroscopic methods. Molecular crowding enhanced the binding constant for saturated medium chain fatty acids but dramatically reduced the binding constant for unsaturated long chain fatty acids. The crowding media, which was polyethylene glycol, may well loosen the structure of the BSA protein and make more facile, fatty acid/protein binding. This data supports our data strongly as the BRODERICK PROBE® lipid biomimetics are bound in fatty acid/albumin compartments to aid in nutrition relating to medicine as well as fragrance and classical medicine including drug delivery nano-transports.

Bruce and Salter, 1996 [14] report on a series of stearic, oleic and palmitic fatty acids in their relationship to the incorporation of phospholipids for diet and cholesterol. The quote follows:

> "Unlike other saturated fatty acids, dietary stearic acid does not appear to raise plasma cholesterol. The reason for this remains to be established, although it appears that it must be related to inherent differences in the metabolism of the fatty acid. In the present study, we have looked at the metabolism of palmitic acid and stearic acid, in comparison with oleic acid, by cultured

> hamster hepatocytes. Stearic acid was taken up more slowly and was poorly incorporated into both cellular and secreted triacylglycerol. Despite this, stearic acid stimulated the synthesis and secretion of triacylglycerol to the same extent as the other fatty acids. Incorporation into cellular phospholipid was lower for oleic acid than for palmitic acid and stearic acid. Desaturation of stearic acid, to monounsaturated fatty acid, was found to be greater than that of palmitic acid. Oleic acid produced from stearic acid was incorporated into both triacylglycerol and phospholipid, representing 13% and 6% respectively of the total after a 4 h incubation. Significant proportions of all of the fatty acids were oxidized, primarily to form ketone bodies, but by 8 h more oleic acid had been oxidized compared with palmitic acid and stearic acid" [14].

Given the caveat that the six fatty acids studied are no longer classified or categorized as long chain fatty acids, the binding of six physiologically important fatty acids to defatted human plasma albumin was measured at 37 degrees in a calcium-free Krebs-Ringer phosphate buffer, pH 7.4 [2]. The data were analyzed in terms of multiple stepwise equilibria. With the saturated acids, the magnitude of the equilibrium (association) constants, Ki, increased as the chain length increased: laurate smaller than myristate smaller than palmitate smaller than stearate. The unsaturated 18-carbon, oleate, was bound more tightly than the saturated 18-carbon, stearate; by contrast, linoleate, another 18-carbon unsaturated fatty acid that produces hydroperoxides at different carbon positions, was bound less tightly than stearate. The equilibrium constants, K1 through K12, ranged from 2.4×10^{-6} – 3.5×10^{-3} m^{-1} for the 12-carbon saturated chain, laurate, to 2.6×10^{-8} – 3.5×10^{-5} m^{-1} for oleate.. Successive values of Ki decrease for each of the acids, indicating that major cooperative binding effects do not occur over the physiological range of fatty acid concentrations. The important conclusion from the studies show that "in no case could the Ki be segregated into distinct classes, suggesting that any grouping of albumin binding sites is somewhat arbitrary." The authors of this study show that binding energy, at least in this case, is not simply a statistical process dependent only on the fatty acid chain length and finally and interestingly, albumin preferentially contributes to the fatty acid binding process likely due to adapting to specific configurations of albumin as the fatty acid increases in length.

Also, very interestingly, analysis of chemical shifts suggested that the highest affinity sites for VLCFA are low affinity sites for long-chain fatty acids. In competition experiments with (13) C-labeled C22:0 (3 mol/mol of BSA) and unlabeled oleic acid, C22:0 bound to bovine serum albumin (BSA) in the presence of up to 4 mol of oleic acid/mol of BSA, but 1 mol was shifted into a different site. Our studies suggest that albumin has adequate binding capacity for the low plasma levels of VLCFA with 20 to 26 carbons, but the protein, albumin, may not be able to bind longer chain VLCFA [18].

Further to these studies, a scientist from the former group who have published extensively on fatty acid binding sites as these relate to the protein, albumin and its configurations, Hamilton in 1989 [27] has contributed *via* nuclear magnetic resonance (NMR) some vital information regarding the same. Therefore, Hamilton reports that a phospholipid, phosphotidylcholine, is involved in the interactive binding process. In fact, the report shows evidence that it is the albumin protein that binds to and then transfers fatty acids, such as laurate, for example, to the lipid bilayer membrane. The data are particularly decisive to our data as we track the *modus operandi* of the BRODERICK PROBE® lipid biomimetic properties of the BRODERICK PROBE® biomimetic polymers and biomimetic polymeric chips.

The empirical data that shows evidence for the role of the protein, albumin, binding to fatty acids as the protein carries its cargo in fatty acids to penetrate the lipid bilayer in physiological membranes, is striking. The protein inserts the fatty acid into the phospholipid of the lipid bilayer. The data fully and excitingly support our data reported here, in histogram form, that albumin plays a key role in lipid biomimetics. The data from the Broderick Laboratory show that the long chain saturated fatty acid, stearate in the lipid bilayer of the BRODERICK PROBE® sensing products transfers fatty acids *via* the protein, albumin, more efficaciously than does the shorter chain fatty acid, Laurate.

Thus, phosphorus is aided by a protein in the fatty acid mediation of lipid biomimetics with DNA as the ultimate key to these penultimate keys in the electromechanics of lipid biomimetics!

6.6 Medicinal and Cosmetic Nanoplatforms for Novel Sensing Biochips

One can carry around the *Broderick Biochips* in their pockets for cosmetics, skin emollients, fragrance and nutrition needs; the sensing biochips enable heuristic industrial, marketing, medicinal and anti-aging use. The BRODERICK PROBE® biochips are poised for commercial applications in the markets of food and nutrition, fragrance, perfumes, skin emollients and skin care to defy epithelial aging. The nanochips will work without an operational amplifier. How we look has a great deal to do with how we feel. The Journal, ***Psyche,*** is dedicated to topics directed at the brain's concept of one's beauty, fragrance and nutrition [13]. A sound mind leads to a sound body and brain reward rewards itself.

Some of these small but mighty chips are:

BRODERICK PROBE® albumin chip
BRODERICK PROBE® phosphotidylethanolamine chip
BRODERICK PROBE® phosphotidylethanolamine-albumin chip
BRODERICK PROBE® phosphate chip
BRODERICK PROBE® laurate chip
BRODERICK PROBE® stearate chip
BRODERICK PROBE® squalane chip
BRODERICK PROBE® squalene chip

Moreover, long chain fatty acids and triglycerides are also lucrative revenue and the press seeks them out. for example, certain coffees advertise that they contain middle chain triglycerides (MCT). The inventive neuro-probe fits right in here as saturated and unsaturated fatty acids are suitable commercial inventions in the Broderick patent portfolio.

From: McAllister MJ et al. (2020) [39].

"Ingestion of ketone supplements, caffeine, and medium-chain triglycerides (MCTs) may all be effective strategies to increase blood levels of the ketone body beta-hydroxybutyrate (D-BHB). However, acute ingestion of a bolus of lipids may increase oxidative stress (OS). The purpose of the study was to investigate the impact of adding varying amounts of MCTs to coffee on blood levels of D-BHB and markers of OS. Ten college-aged men ingested coffee with 0, 28, and 42 g of MCT in a randomized order. Blood samples were collected pre- as well as 2 and 4 h postprandial and analyzed for D-BHB, total cholesterol (TC), high-density lipoprotein cholesterol (HDL-c), glucose, triglycerides (TAG), insulin, and OS markers: advanced oxidation protein products (AOPP), glutathione (GSH), malondialdehyde (MDA), and hydrogen peroxide (H_2O_2). All three treatments resulted in a significant increase in D-BHB, HDL-c, and TC as well as a significant decrease in TAG, MDA, H_2O_2, and insulin. The 42 g treatment was associated with significantly higher levels of AOPP and MDA. Acute ingestion of coffee results in favorable changes to markers of cardiometabolic health that were not impacted by the addition of 28 g of MCT. However, 42 g of MCT caused significantly greater OS" [39].

As the previous Editor-in-Chief of the *Journal of Caffeine Research* (*JCR*), Patricia agrees with the McAllister et al. data. In fact, although Patricia has not ever had a cup of coffee, MCT added to coffee is keen! Patricia has published multiple works on caffeine and refers the reader to *JCR* and to Chapter 8 herein. There is a caveat. Additions of MCI's to more than scheduled doses of caffeine are not advised by McAllister nor this author. ***Dose is the turning point***.

6.7 The Merits of the Electroactive BRODERICK PROBE® Biomimetic Sensing Polymeric Chips

- BRODERICK PROBE® *sensing products make their way to the market. The philanthropist, entrepreneur and the industry itself, neutraceutical and pharmaceutical, sees a new way to the look at the tiny probes and biochips.*

- *The stearate and the laurate chains are part and parcel of our lives! The Broderick stearate chain images the protein, albumin better than does the laurate chain; the Broderick laurate chain images the biogenic amines better than does the stearate chain. Patricia published her stearate work early on; it is one of her earliest work in the body of her works [7].*
- *Of the approximate seventy million hits when one searches lipid bilayers/lipid biomimetics on the Internet, the Broderick biochips are the only "carry with you", biochips that can image neurotransmitters and essential molecules in the brain, body and blood with or without an operational amplifier. Just apply; mitochondria will assist in the mechanics.*
- *The mental health value and the lucrative financial value may well be unsurpassed.*
- *The* BRODERICK PROBE® *comprised of a fatty acid and phosphotidylethanolamine bonds with DNA at the water-lipid headgroup interface during the sequence of in vivo electric potential steps and the tiny nanoprobe in vivo spontaneously forms and self-assembles lipid bilayers online and in real time*

6.7.1 Key Drivers of the Squalene/Squalane Market

Squalene is originally obtained from shark liver oil [35, 36]. It is a natural Very Long Chain Carbon Fatty Acid, (VLCFA), a 30-carbon isoprenoid compound and squalene importantly plays a role as an intermediate metabolite in the synthesis of cholesterol! It is not susceptible to lipid peroxidation and provides skin protection and is used in skin products to defray the effects of ageing. Like laurate, squalene gives the skin a velvety feeling. In the Broderick library of sensor and nanosensor, squalene is shown to image biogenic amines [6, 8, 9] and patented images are in process of publication [13].

Fish oils are the source of nutrients important for health maintenance [48]. The most significant are essential fatty acids (EFA) of n-3 family, alkylglycerols and squalene. N-3 EFA are of great importance in atherosclerosis prevention. Alkylglycerols and squalene are modulators of immunity to infections and cancer. Human sebum also contains 13% squalene as one of its major constituents. **Squalane** is a saturated derivative of squalene and also found in these sources. Squalane is the preferred marketable form of the fatty acid. Fish oils are the craze now but Patricia is thinking how smart her mother was as Patricia was served *cod liver oil* every day!

6.7.2 Key Drivers of the Isoamyl Laurate Market

Isoamyl laurate is an ester formed by a reaction between ***isoamyl alcohol*** and ***lauric acid***. It is a colorless, clear, oily liquid with moderately strong waxy and peach-like fragrance. Isoamyl laurate is primarily used as a fragrance ingredient. It is also used as an emollient, since it contains lauric acid, which is beneficial for the skin. Generally, esters are fatty in nature and they provide moisturization to the skin. Isoamyl laurate forms a protective film on the skin or the hair, thereby protecting the skin from the environmental factors. Also, it does not let moisture evaporate from the skin and thereby, conditions the skin. It provides quick spread-ability and elegant, non-sticky, light and like, squalane, makes the skin feel velvety. It has good miscibility with other surfactants such as the Broderick nanoprobe.

Increasing use of isoamyl laurate in various skin care and hair care applications is projected to drive the global Isoamyl laurate market during the forecast period. Isoamyl laurate is a preferred choice for use in various skin care applications such as anti-aging creams and baby lotions, as it keeps the skin moisturized and protects it from air pollution. Thus, with the growth of the cosmetics industry, the demand for isoamyl laurate will increase significantly and exponentially [49].

Figure 6.16 shows a lauric acid sensogram from the Broderick patent portfolio.

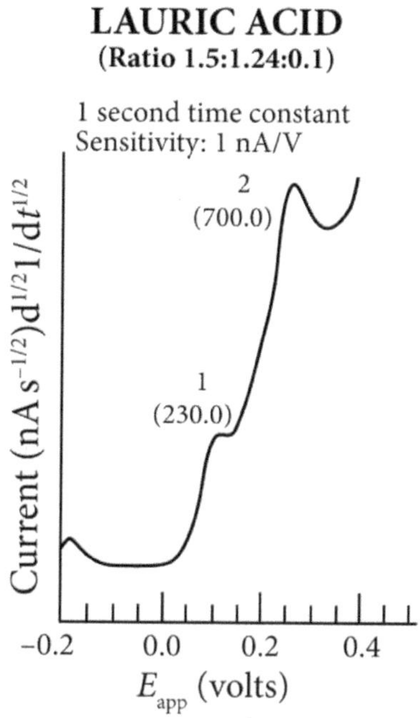

Figure 6.16 A Broderick sensogram of images online and in real time. Numbers in 1 and 2 denote the amplitude of the signal is peak height. The redox properties for these ***Broderick biochips*** for biogenic amines and metabolites *in vitro*, are, in the order of detection from lowest to highest positive potential, ascorbate [0.08 V]; dopamine (DA) [0.13 V]; serotonin (5-HT) [0.29 V]; homovanillic acid (HVA) [0.43 V] and L-tryptophan (L-TP) [0.63 V]. Oxidation (half-wave) potentials for these biosensors, *in vivo*, are, in the order of detection from lowest to highest oxidation potential, ascorbate [0.09 V], DA [0.14 V], 5-HT [0.31 V]; HVA [0.44 V] and L-TP [0.64 V]. Each specified oxidation potential is within a range of ±0.015 V. The biochip provides skin, hair and nutrition value; its value as a fragrance is impressively lucrative (cf. Figures in Chapter 3).

6.8 Proposal

Thus, we propose that phosphate is one of the lipid messenger catalytic amplifier, *in vitro* and *in vivo*. Phosphate is an anion. The lipid amplification, the LAN, may occur by covalent bonding in the carbon/lipid/phosphotidyl polymer and saline/phosphate solution, then in the phosphotidyl neuronal brain membranes and finally again in the saline/phosphate solution, activated by applied potential, in one method, by voltage, enabling a change in electric potential and electric potential energy enabling an energy change a new potential difference and facilitating movement of electrons and electron transfer. Highly polar and highly electronegative, C=O, C=N and C=F bonds in the carbon/lipid/phosphotidyl polymer allow the signal to amplify itself, within its own system. New molecular entities are transduced and signaling amplification occurs step by step, each step signaling further transduction and subsequent transformation, similarly to the brain's step by step signaling process Such is our hypothesis for lipid biomimetics with the BRODERICK PROBE®.

Understanding lipid bilayer membranes is a challenge; computer simulations help [23] and lipid-protein membranes can cause lipid deformation and lipids cause protein tilting [51].

But wait, the handy biochips tell more. DNA's sugar phosphate backbone binds to the protein, albumin, that transfers fatty acids selectively according to chain length to the phospholipid in the cell membrane. Broderick finds that a protein is the ultimate lipid messenger and its name is albumin.

6.9 Conclusion

Thus, it is concluded that phosphorus is likely one of the lipid messenger catalytic amplifiers or amplifiers and with DNA's phosphate sugar backbone, DNA is once again the master. Indeed, DNA makes the proteins and the enzyme catalysts anyway. One must smile now. Lipid biomimetics occurs first by contacting the interfacial surface of the BRODERICK PROBE® nanosensor with saline phosphate in solution at the meniscus. Then, as this polymeric nanosensor, with its multitude of formulae, amplifiers and circuits, targets the brain's lipid membrane, the phosphotidyl phosphate neuronal brain membranes, phosphorus is the messenger for lipid biomimetics through intracellular signaling via the second messenger system. Transduction continues with alteration in potential applied to the potentiostat continuously until stabilized at the post calibration step. The result is a new product, formed by catalytic amplification, useful products manufactured chemically and electronically, hundreds of folds; we called the final amplification number produced by Broderick polymer form of catalysis, Lipid Amplification Number (LAN).

References

1. An Z, Möhwald H, Li J (2006) pH controlled permeability of lipid/protein biomimetic microcapsules. *Biomacromolecules* 7(2): 580–585.
2. Ashbrook JD, Spector AA, Santos EC, Fletcher JE (1975) Long chain fatty acid binding to human plasma albumin. *J Biol Chem* 250(6): 2333–2338.
3. Bhilwade HN, Tatewaki N, Nishida H, Konishi T (2010) Squalene as novel food factor. *Curr Pharm Biotechnol* (8): 875–880.
4. Blaedel WJ, Bogulaski RC (1978) Chemical amplification: a review. *Anal Chem* (50) 8: 1026–1032.
5. Broderick PA (1987) Striatal neurochemistry of dynorphin-(1-13): *in vivo* electrochemical semidifferential analyses. *Neuropeptides* 10(4): 369–386.
6. Broderick PA (1989a) Cathodic Electrochemical Current Arrangement with Telemetric Application. U.S. *Patent No. 4,883,057. Issued.*
7. Broderick PA (1989b) Characterizing stearate probes *in vitro* for the electrochemical detection of dopamine and serotonin. *Brain Research* 495: 115–121.
8. Broderick PA (1995) Microelectrodes and their Use in Cathodic Electrochemical Current Arrangement with Telemetric Application. *U.S. Patent No. 5,433,710, Issued.*
9. Broderick PA (1999) Microelectrodes and their Use in an Electrochemical Arrangement with Telemetric Application. *U.S. Patent Serial No. 5,938,903, Issued.*
10. Broderick PA (2016) Noninvasive Electroactive Photonic Protein Sensor with Polymer Photovoltaic Optics for Memory Transduction Using Organic and Inorganic Elements as Platforms. *US Non-Provisional International Application No. PCT/US 16/68879.2020. US Application No.* 16/074,254, *Notice of Allowance, August 16th, 2021.*
11. Broderick PA (2020) Identification, Diagnosis, and Treatment of Tau-Associated Pathologies Using Electrodes. *US Provisional Application No.* 63/130,453. December 24th.

12. Broderick PA (2021) Identification, Diagnosis, and Treatment of Neuropathologies, Neurotoxicities, Tumors, and Brain and Spinal Cord Injuries using Electrodes with Microvoltammetry. *US Patent Serial No. 10,980,460, Issued.*

13. Broderick PA (2022) Prominent Papers in Psych—2021–2023 (PA Broderick, Ed., TE Schlaepfer, Ed. in Chief) *A Special Issue of Psych (ISSN 2624-8611). Basel. Switzerland, Beijing, China*: Molecular Diversity Preservation International (MDPI). *www.mdpi.com.*

14. Bruce JS, Salter AM (1996) Metabolic fate of oleic acid, palmitic acid and stearic acid in cultured hamster hepatocytes. *The Biochemical Journal* 316(Pt 3): 847–852.

15. Camin F, Bontempo L, Ziller L, Piangiolino C, Morchio G (2010) Stable isotope ratios of carbon and hydrogen to distinguish olive oil from shark squalene-squalane. *Rapid Commun Mass Spectrometry* 24(12): 1810–1816.

16. Cao H, Zou L, He B, Zeng L, Huang Y, Yu H, Zhang P, Yin Q, Zhang Z, Li Y (2017) Albumin biomimetic nanocorona improves tumor targeting and penetration for synergistic therapy of metastatic breast cancer. *Advanced Functional Materials* 27(11): 1605679.

17. Carmona-Ribeiro AM (2010) Lipid-based biomimetics in drug and vaccine delivery. In: Mukherjee A, editor. *Biomimetics Learning from Nature,* March.

18. Choi JK, Ho J, Curry S, Qin D, Bittman R, Hamilton JA (2002) Interactions of very long-chain saturated fatty acids with serum albumin. *J Lipid Res* 43(7): 1000–1010.

19. Delavaux-Nicot B, Guari Y, Douziech B, Mathieu R (1995) An efficient synthesis of the first electroactive phosphorus-containing bisferrocene macrocycles. *J. Chem. Soc. Chem Commun.* 5: 585–587.

20. De Marchi F, Cui D, Lipton-Duffin J, Santato C, MacLeod JM, Rosei F (2015) Self-assembly of indole-2-carboxylic acid at graphite and gold surfaces. *J Chem Phys* 142: 101923.

21. Díaz-Saldívar P, Huidobro-Toro JP (2019) ATP-loaded biomimetic nanoparticles as controlled release system for extracellular drugs in cancer applications. *Int J Nanomedicine* 5(14): 2433–2447.

22. DuVall SH, McCreery RL (2000) Self-catalysis by catechol and quinones during heterogeneous electron transfer at carbon electrodes. *J Am Chem Soc* 122(28)6: 759–6764.

23. Enkavi G, Javanainen M, Kulig W, Róg T, Vattulainen I (2019) Multiscale simulations of biological membranes: The Challenge

to Understand Biological Phenomena in a Living Substance. *Chem Rev* 119 (9): 5607–5774.

24. Fan X, Markram H (2019) A brief history of simulation neuroscience. *Front. Neuroinform.* 13 (32): https://doi.org10.3389/fninf.2019.00032.
25. Finkenstadt VL, Willett JL (2005) Preparation and characterization of electroactive biopolymers. *Macromolecular Symposia* 227(1): 367–372.
26. Fujita M, Umeda T, Yoshida M (1995) Polymorphism of carbon forms: Polyhedral morphology and electronic structures. *Phys Rev B* 51: 13778–13780.
27. Hamilton JA (1989) Medium-chain fatty acid binding to albumin and transfer to phospholipid bilayers. *PNAS* 86 (8): 2663–2667.
28. Hostetler HA, Balanarasimha M, Huang H, Kelzer MS, Kaliappan A, Kier AB, Schroeder F (2010) Glucose regulates fatty acid binding protein interaction with lipids and peroxisome proliferator-activated receptor α. *J Lipid Res.* 51(11): 3103–3116.
29. Huang H, McIntosh AL, Martin GG, Petrescu AD, Landrock KK, Landrock D, Kier AB, Schroeder F (2013) Inhibitors of Fatty Acid Synthesis Induce PPAR α-Regulated Fatty Acid β-Oxidative Genes: Synergistic Roles of L-FABP and Glucose. *PPAR Res* 2013: 865604.
30. Hu H, Qi Q, Dong Z, Yu X, Mo Y, Luo J, Wang Y, Du S, Lu Y (2020) Albumin coated trimethyl chitosan-based targeting delivery platform for photothermal/chemo-synergistic cancer therapy. *Carbohydrate Polymers* 241: 116335.
31. Jackson MN, Surendranath Y (2019) Molecular control of heterogeneous electrocatalysis through graphite conjugation. *Acc. Chem. Res.* 52: 12, 3432–3441. *https://doi.org/10.1021/acs.accounts.9b00439.*
32. Kaushik NK, Kaushik N, Attri P, Kumar N, Kim CH, Verma AK, Choi EH (2013) Biomedical importance of indoles. *Molecules* 18(6): 6620–6662.
33. Khalid S, Bond PJ, Holyoake J, Hawtin RW, Sansom MSP (2008) DNA and Lipid Bilayers: Self-Assembly and Insertion. *J. R. Soc. Interface.* 5: 241–250. *https://doi.org/10.1098/rsif.2008.0239.focus.*
34. Kim S-K, Karadeniz F (2012) Biological importance and applications of squalene and squalane. *Review Adv Food Nutr Res* 65: 223–233.
35. Lewkowicz N, Lewkowicz P, Kurnatowska A, Tchórzewski H (2006) Biological action and clinical application of shark liver oil. *Pol Merkur Lekarski* 20(119): 598–601.

36. Lewkowicz P, Banasik M, Głowacka E, Lewkowicz N, Tchórzewski H (2005) Effect of high doses of shark liver oil supplementation on T cell polarization and peripheral blood polymorphonuclear cell function. *Pol Merkur Lekarski* 18(108): 686–692.

37. Makhlynets OV, Korendovych IV (2019) A single amino acid enzyme. *Nat Catal* 2: 949–950.

38. Ma Q, Zhang X, Qu Y (2018) Biodegradation and biotransformation of indole: Advances and Perspectives. *Front Microbiol* 1(9): 2625.

39. McAllister MJ, Waldman HS, Rentería LI, Gonzalez AE, Butawan MB, Bloomer RJ (2020) Acute coffee ingestion with and without medium-chain triglycerides decreases blood oxidative stress markers and increases ketone levels. *Can J Physiol. Pharmacol.* 98(4): 194–200.

40. McCreery R (2008) Advanced carbon electrode materials for molecular electrochemistry. *Chem Rev* 108: 2646–2687.

41. Özdemir V, Dove ES, Gürsoy UK, Şardaş S, Yýldýrým A, Yýlmaz ŞG, Ömer Barlas I, Güngör K, Mete A, Srivastava S. (2017) Personalized medicine beyond genomics: alternative futures in big data-proteomics, environtome and the social proteome. Broderick PA, Jacoby JH. Editors, *Special Issue,* In: *J Neural Transm* (Vienna). Jan;124(1): 25–32. doi: 10.1007/s00702-015-1489-y. Epub 2015 Dec, 8. PMID: 26645377.

42. Pasek M (2019) A role for phosphorus redox in emerging and modern biochemistry. *Curr Opin Chem Biol* 49: 53–58.

43. Perez XA, Bressler AJ, Andrews AM (2007) Determining Serotonin and Dopamine Uptake Rates in Synaptosomes Using High-Speed Chronoamperometry. In: Michael AC, Borland LM, editors. In: *Electrochemical Methods for Neuroscience.* Chapter 7, Boca Raton, FL: CRC Press/Taylor & Francis.

44. Rebec GV (2007) From Interferant Anion to Neuromodulator: Ascorbate Oxidizes its Way to Respectability. Michael AC, Borland LM. Editors, In: *Electrochemical Methods for Neuroscience.* Chapter 9, Boca Raton (FL): CRC Press/Taylor & Francis.

45. Rubio-Camacho M, Encinar JA, Martínez-Tomé MJ, Esquembre R, Mateo CR (2020) The interaction of temozolomide with blood components suggests the potential use of human serum albumin as a biomimetic carrier for the drug. *Biomolecules* 10(7): 1015–1037. https://doi.org/10.3390/biom10071015.

46. Singh KP, Bae EJ, Yu J-S (2015) Fe–P: A new class of electroactive catalyst for oxygen reduction reaction. *Journal of the American Chemical Society* 137 (9): 3165–3168.

47. Song J, Kim Y-S, Lee DH, Lee SH, Park HJ, Lee D, Kim H (2019) Neuroprotective effects of oleic acid in rodent models of cerebral ischaemia. *Sci Rep* 9: 10732.

48. Szostak WB, Szostak-Wegierek D (2006) [Health properties of shark oil]. *Przeglad Lekarski* 63(4): 223–226.

49. Transparencymarketresearch.com Isoamyl Laurate Market – Global Industry Analysis, Size, Share, Growth, Trends, and Forecast, 2019 - 2027. https://www.transparencymarketresearch.com/isoamyl-laurate-market.html Nov 2020.

50. Turk J, Gross RW, Ramanadham S (1993) Amplification of insulin secretion by lipid messengers. *Diabetes* 42(3): 367–374.

51. Venturoli M, Berend Smit B, Sperottoz MM (2005) Simulation studies of protein-Induced bilayer deformations, and lipid-induced protein tilting, on a mesoscopic model for lipid bilayers with embedded proteins. *Biophysical Journal* 88: 1778–1798.

52. Wang S, Zhang S, Liou L-C, Ren Q, Zhang Z, Caldwell GA, Caldwell K, Witt SN (2014) Phosphatidylethanolamine deficiency disrupts α-synuclein homeostasis in yeast and worm models of Parkinson disease. PNAS 111(38) E3976-E3985. https://doi.org/10.1073/pnas.1411694111.

53. Zhu T-T, Zhang Y, Luo X-A, Wang S-Z Jia M-Q, Chen Z-X (2018) Difference in binding of long- and medium-chain fatty acids with serum albumin: The role of macromolecular crowding effect. *Journal of Agricultural and Food Chemistry* 66 (5): 1242–1250.

54. Zinger A, Baudo G, Naoi T, Giordano F, Lenna S, Massaro M, Ewing A, Kim HR, Tasciotti E, T. Yustein JT, Taraballi F (2020) Reproducible and characterized method for ponatinib encapsulation into biomimetic lipid nanoparticles as a platform for multi-tyrosine kinase-targeted therapy. *ACS Applied Bio Materials*. 3(10): 6737–6745. doi: 10.3390/nano10112172.

Chapter 7

Sex-Dependent Schizophrenia from Cocaine Addiction: Brain Reward from the Auditory Circuit

Further seminal discoveries in scientific and medical research found by the BRODERICK PROBE® *series of sensors and biosensors enable the assessment of pre-clinical and clinical manifestations of brain disorders as the sensors have controlled electrocatalysis with superior operational stability. This is particularly important in DA/AA interactions.; the unfolding of DA from AA is shown herein. In Chapter 6, the author, Patricia, renders the proof of concept empirical data to crystallize the exact modus operandi for each layer by layer of the nanoprobe as well as she discerns the LAN for the* BRODERICK PROBE® biomimetic *polymers and* BRODERICK PROBE® *polymeric biochips.*

Moreover, then, it is also particularly important for us to enable controlled electrocatalysis to study brain disorders that involve neuroanatomic brain substrates that predominantly undergo signally via other dual catecholamines, such as dopamine (DA) and norepinephrine (NE). NMI with the BRODERICK PROBE® *biosensor has proven to be critical in the selective separation of such multifaceted complexities in neurotransmitter interactions. Importantly, these creative NMI contributions to the solution of neurotransmission complexities were soon followed by the seminal empirical report that the neurotransmitter, DA, was found in a brain stem nucleus, the pons. This was surprising since the pons was always believed to be highly innervated, even exclusively, with NE and a*

Neuroimaging: Sensing Biochemistry in the Brain
Patricia A. Broderick

ISBN 978-981-4613-49-1 (Hardcover), 978-0-429-02974-5 (eBook)
www.jennystanford.com

further surprise ensued when DA was found in another brain stem nucleus, so closely connected neuronally to the pons, that is, the locus coeruleus, well known as noradrenergically innervated in high concentrations. With all of these seminal studies, access to delineate brain disorders are made possible as never before. Thus, this chapter discusses, the topic of dopaminergic/noradrenergic sex-dependent schizophrenia induced by the psychostimulant, cocaine. Preclinically. the studies were performed by making use of startle stimuli and responses were monitored in the auditory circuit, localizing, a priori, in the brain stem, pons and then tracking to the midbrain, Ventral Tegmental Area, the brain reward/addiction pathway.

7.1 Seminal Discoveries in Neuromolecular Imaging (NMI): Images of Dopamine and Norepinephrine Are Separated by 50 mV

Neuroanatomic brain substrates involve multiple interactive neurotransmitter complexities. This laboratory immediately tackled the task of separating electroactive signaling for the catecholamines, DA and norepinephrine (NE), neurochemicals that dually innervate the brain in its physiological natural, temporally synchronous state and/or its physiologically disordered, asynchronous state. Thus, this successful separation of the catecholamines, the first of these seminal discoveries, is shown in Figure 7.1.

Figures 7.1 (A–D) show the earliest of these four discoveries.

Figure 7.1(A) shows neuromolecular imaging (NMI), performed by the BRODERICK PROBE® nanobiosensor, selectively images dopamine (DA) and serotonin (5-HT) *in vitro* in separate electroactive signals. Images for the catecholamines, DA and NE, and the indoleamine, 5-HT show a distinct signature difference and signals are operationally stable in the presence of and without interference from metabolites, 3-4-dihydroxyphenylacetic acid (DOPAC) and 5-hydroxyindoleacetic acid (5-HIAA). The operationally stable nanobiosensor provides selective differentiation of DA and 5-HT while in the presence of other possible interfering chemicals, ascorbic acid (AA) and uric acid (UA) [7].

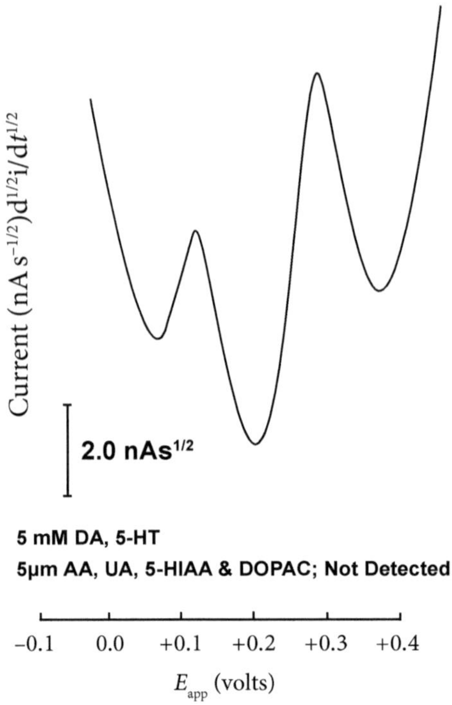

Figure 7.1(A) NMI: the electrochemical signal for the catecholamine neurotransmitter, dopamine (DA), is shown selectively separated from the electrochemical signal for the indoleamine, neurotransmitter, 5-HT. The data shows the response of the carbon/lipid/phosphotidyl stearate biosensor to aliquots of 5 uM DA; 5 uM 5-HT. The neuromodulator, Ascorbic acid (AA) at 5 uM, the biomolecule, Uric Acid (UA) at 5 uM, in addition to the metabolite of 5-HT, 5-hydroxyindoleacetic acid (5-HIAA) at 5 uM as well as the metabolite of DA, 3,4-dihydroxyphenylacetic acid (DOPAC) at 5 uM were aliquoted into the saline phosphate buffer solution. At equivalent concentrations to the monoamines, the latter electroactive molecules were below detection limits.

Figures 7.1(B) (C) and (D) show that DA and NE are separated by a range of 50 millivolts. Increasing concentrations of NE and 5-HT in saline phosphate buffer in the electrochemical cell (manufactured on site and calibrated in accordance with commercial apparatus) show correlated sharp increases in the designated signals for NE and 5-HT according to experimentally

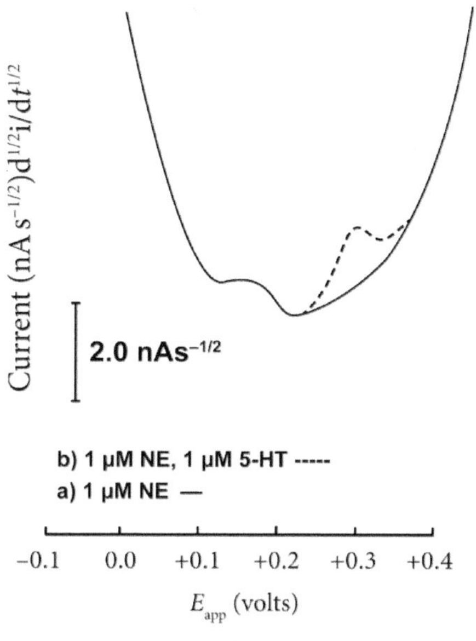

Figure 7.1(B) NMI: the electrochemical signal for the catecholamine neurotransmitter, norepinephrine (NE) is shown selectively separated from the indoleamine, neurotransmitter, 5-HT. NMI was performed with concentrations listed as follows: (a) 1 uM NE; (b) 1 uM NE; 1 uM 5-HT.

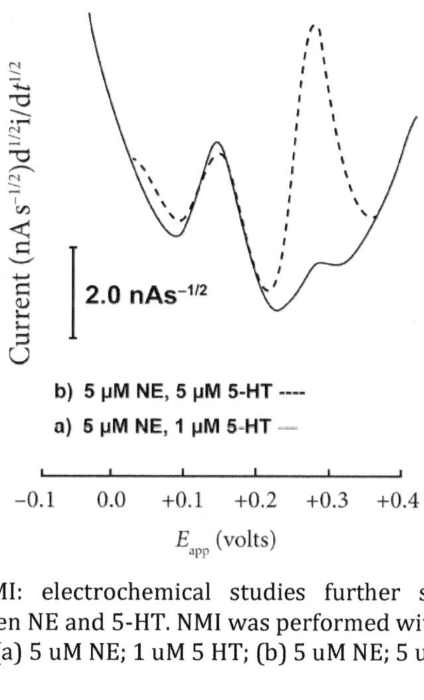

Figure 7.1(C) NMI: electrochemical studies further showing a clear distinction between NE and 5-HT. NMI was performed with concentrations listed as follows: (a) 5 uM NE; 1 uM 5 HT; (b) 5 uM NE; 5 uM 5-HT.

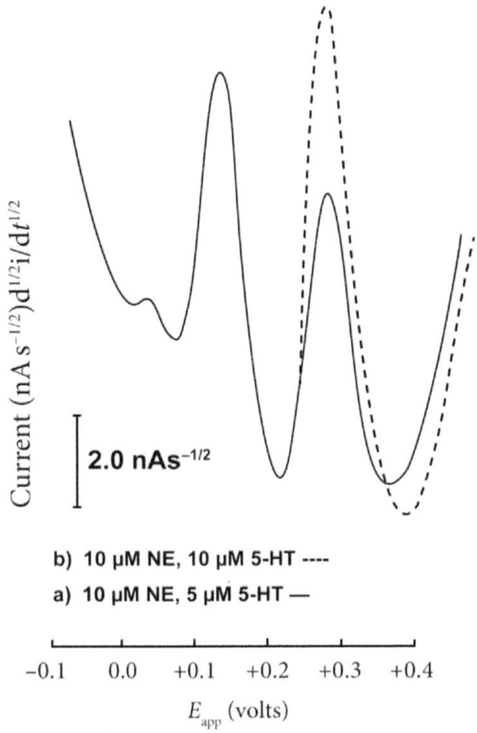

Figure 7.1(D) NMI: further electrochemical studies, showing a clear distinction between NE and 5-HT. NMI was performed with the following concentrations: (a) 10 uM NE and 5 uM 5 HT; (b) 10 uM NE and 10 uM 5-HT.

derived half wave/oxidation potentials specific for each neurotransmitter/neurochemical. Therefore, the catecholamines, DA and NE are clearly differentiated from each other and further, from electroactive signaling for the indoleamine, 5-HT [5]. It is concluded that there is no interference from chemicals closely allied molecules with half wave/oxidation potentials for biogenic amines, as demonstrated by thousands of *in vitro, in situ* and *in vivo* studies accomplished in the Broderick Laboratory.

The electrical circuit, the driver, uses semidifferential/ semiderivative treatment of microvoltammetric NMI data in conjunction with a graphite stearate nanobiosensor in this example. The data are relevant to the interpretation of electrochemical signals in neuroanatomical substrates, richly innervated by dopaminergic, noradrenergic, serotonergic neuronal circuitry comprised of multifaceted neuroactive intermolecular anion/cation exchange processes. Current response is concentration dependent according to the Cottrell Equation. The data are raw data, as seen in real time with the inventive art of polymeric carbon/lipid/phosphotidyl matrices as one of the novel series of BRODERICK PROBE® sensors. This nanobiotechnology is known for its operational stability *in vivo, in vitro* and *in situ and is also* known for its excellent temporal and spatial resolution [6, 8–13].

7.2 NMI: Images of Dopamine and Ascorbic Acid Separated by 50 mV

The second discovery of these four earliest seminal discoveries enabled the selective separation of dopamine (DA) from ascorbic acid (AA) *via* NMI. These precise signatures of complex neurochemistry in the brain of patients and animals are published in the fifth chapter of this monograph. The actual unfolding of the molecular image for DA from AA in real time is astounding as one sees this happening in fact, in real time and on line. Much has been written about this neuromolecular interaction but no one and/or no other has seen this remarkable phenomenon. Three distinct and separate cathodic oxidation peaks are imaged as shown in the Broderick patent portfolio [6] using a semiderivative electrical reduction circuit for use in a potentiostat designed in our laboratory very early on to perform cation/anion exchange (Chapter 4 of this monograph).

7.3 The Third and Fourth Discoveries: Neuronal Dopamine Tracts in the Highly Noradrenergic Pons and Locus Coeruleus in the Brain Stem

The last of the seminal discoveries enabled the selective separation of DA from NE in brain nuclei, such as the pons. Dopamine was seen in the highly innervated noradrenergic neuroanatomic substrate, the pedunculopontine nucleus of the pons in the brain stem. Surprisingly, then, DA was found in another brain stem nucleus, the locus coeruleus, so closely related to the pons; the locus coeruleus is comprised of NE in high concentrations.

Dopaminergic innervation of the locus coeruleus is reported by Ornstein et al. 1987 [57] by using biochemical radioautographic evidence, also reported in the form of tissue localization of DA [47], also discussed by studying Dopamine Transporter (DAT) members of both the D1 and D2 family of receptors [71]. Hypothalamic A11 and A13 and midbrain ventral tegmental area (VTA) A10 cell groups [45] have been identified as sources of this innervation (Figure 7.2).

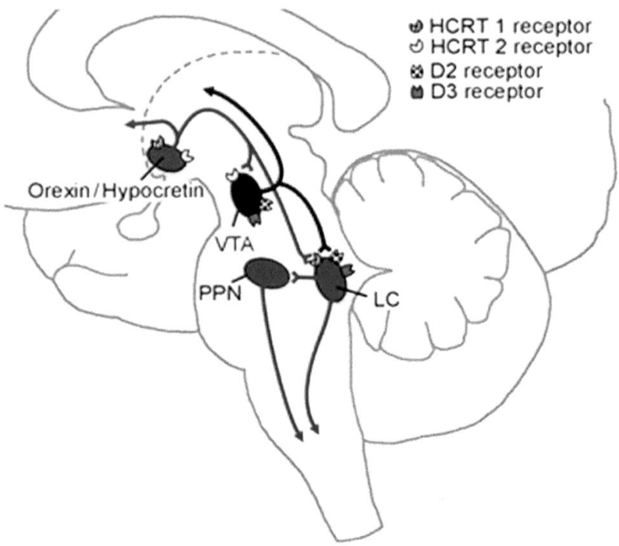

Figure 7.2 Although a model of REM-sleep generation, this advancement shows that it incorporates novel midbrain DA neurons and dopaminoreceptive pontine structures. HCRT 1, hypocretin 1; HCRT 2, hypocretin 2; LC, locus coeruleus; PPN, pedunculopontine nucleus; VTA, ventral tegmental area [19].

Keating and Rye [34] commented on the discovery by Crochet and Sakai [19] regarding DA in noradrenergic nuclei in the brain stem with the following warning:

Because DA neurons express D-2 and D-3 autoreceptors, these same indirect pathways to pontine neurons influencing REM-sleep may also account for the effects of focal manipulations of mesocorticolimbic DA circuitry upon the quality, quantity, and timing of REM-sleep cited above. Distinguishing the individual contributions of midbrain DA neurons and pontine dopaminoceptive structures to normal and pathological states of wake and sleep (e.g., in narcolepsy/cataplexy and PD), will require a significant amount of additional anatomical, physiological and pharmacological data. The findings of Crochet and Sakai point us in the right direction, but investigators in the field have a lot of catch-up to play before DA circuitry can be properly incorporated into current heuristic models of sleep and REM sleep regulation [34].

7.3.1 The Pons in the Brain Stem

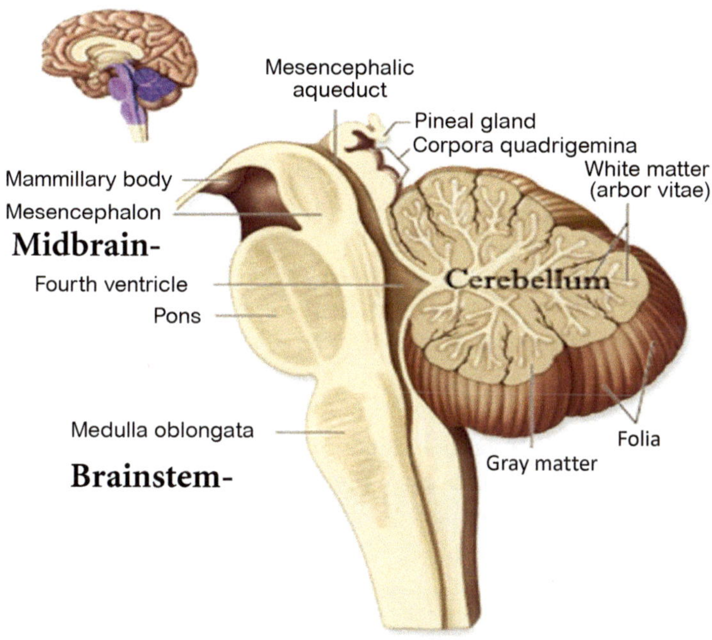

Figure 7.3 Anatomic detail of the pons in its relationship to surrounding structures in the brain stem. *Adapted with permission from Academic Press* [33].

7.3.2 The Pons in the Brain Stem: Neuroanatomy and Neurochemistry

The brain stem can be subdivided and consists of the medulla (myelencephalon), pons (metencephalon) and midbrain (mesencephalon), the reticular formation, and monoaminergic neurotransmitter systems as well as the cranial nerves and associated nuclei.

Straddling the dorsal surface of the pons is the cerebellum and spanning the length of the brain stem are the nuclei for cranial nerves numbered nerves 3 to 12.

The pons represents the most rostral portion of the "hindbrain" and it is separated from the midbrain via the superior pontine sulcus, and from the medulla by the inferior pontine sulcus. The two components of the vestibulocochlear nerve (VIII), and the facial (VII), abducens (VI), which also lies in the floor of the fourth ventricle-and the trigeminal nerve (V) are the cranial nerve nuclei associated with the pons.

Neurons of the reticular formation receive specific sensory input from the skin, muscles, joints, and vestibular system which they then integrate. That is, the reticular formation as a whole is not only concerned with arousal, but the integration and coordination of behavior in response to arousal, such as through the integration of trunk, limb, arm, head and eye movements. Hence, the reticular formation exerts both generalized and specific activating influences and is thus concerned with sensory-motor integration (mediated by the excitatory and inhibitory neurotransmitters, glutamate and GABA) and modulatory functions (mediated by norepinephrine and serotonin).

In part, the activating influences of the reticular formation are subserved by clusters of chemically active neurons which secrete and distribute norepinephrine (NE), serotonin, (5-HT), and acetylcholine (ACh). Specifically, clusters of NE neurons (designated A1 through A7), are scattered throughout the lateral, ventral, and anterior reticular formation and tegmentum.

NE neurons project throughout the brain and spinal cord are concerned with emotional arousal, promote neuroplasticity, and also serve a neural protective function when subject to considerable and prolonged emotional stress. Specifically, the more anteriorally located cell cluster, the locus coeruleus (A4 and A6) projects to the spinal cord, the brain stem, cerebellum, and cerebral cortex. Other NE cell clusters project throughout the brain stem, hypothalamus, limbic system and limbic striatum (A1–A3), as well as to the spinal cord and lower brain stem (A5 and A7).

5-HT neurons are also concerned with emotional arousal, as well as with sensory filtering. 5-HT neurons are clustered throughout the brain stem, forming the raphe nuclei in the midline, dorsal and median portions of the reticular formation (B1–B9). 5-HT neurons project throughout the forebrain.

Brain stem ACh neurons are clustered in the pons, forming the pedunculopontine nucleus and the laterodorsal tegmental nucleus. These neurons project throughout the brain stem, diencephalon, and limbic forebrain. Additional clusters of ACh manufacturing and secreting neurons are located in the septum and substantia innominata ("extended amygdala") [33].

7.3.3 The Acoustic Nerve for the Startle Stimulus Is in the Pons

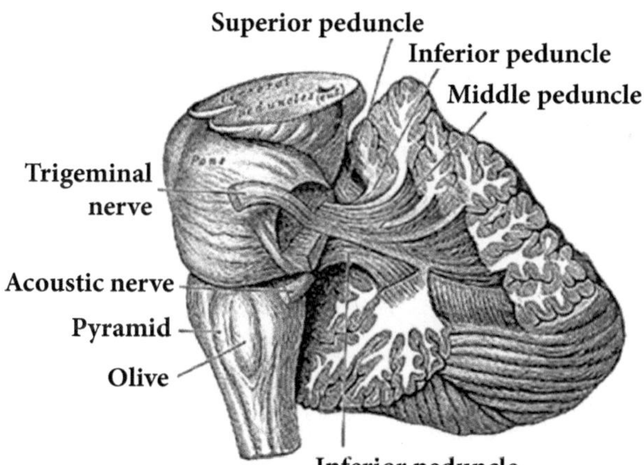

Figure 7.4 Depicts a microscopic cross section of the pons in the human brain that shows the position of the acoustic nerve. *Adapted with permission: Cristian Rivera Foundation: This Foundation works toward a cure for inoperable brain pontine gliomas in pediatrics. https://www.cristianriverafoundation.org/.*

7.4 The Pons and the Acoustic Startle Response

The brain stem is of the utmost importance since lesion studies show that the inferior and superior colliculi, the pons and the caudal pontine reticular nucleus are critically important in the execution of the startle response. Leitner and Cohen studied the role of the inferior colliculus in the startle reflex as early as the late 1980s [42]. Neuronal networks, whether studied empirically or postulated theoretically, lend an explanatory note to the mechanism of action for acoustic startle and well rely on the anatomic brain structure called the pons.

7.4.1 Brain Circuitry Mechanisms for the Auditory Startle Stimulus and Response

Rudomin in 2002 reported that sensorimotor gating often occurs via presynaptic inhibition of the activated sensory neurons [61, 65, 68]. There are three proposed neuronal networks to describe the action of neurotransmitters within the auditory stimulus/startle brain circuits.

Neural circuits for the Acoustic Startle to Auditory Stimuli and the neurotransmitters related to the circuitry are postulated and reported by Koch et al. [35–39], Fendt et al. [23], and Leumann [43]. Thus, the neural network underlying the mechanism of action of PPI is reported to be the nigrostriatal pathway, mesocorticolimbic circuitry linking the limbic cortex, striatum, pallidum, and pontine reticular formation by monoamines, and indoleamines [5, 7]. Another proposal is the link between the mesocorticolimbic circuit components of ventral tegmentum through to the brain stem pons [35–39]. Another is the proposed circuit which describes links, connections between the colliculi and the ventral tegmentum [23]. The final circuit described, links the auditory stimulus with the dorsal cochlear nucleus with dorsolateral tegmental nucleus by glutamate and links the auditory stimulus to the pons and motor neurons by acetylcholine, the amino acid neurotransmitters [43].

Figure 7.5 delineates the circuit postulated by Koch et al. [35–39].

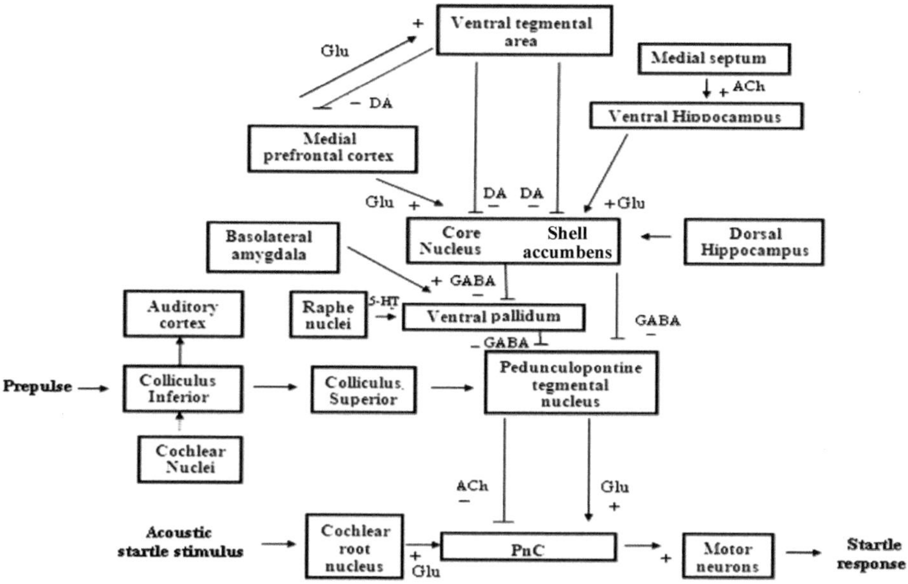

Figure 7.5 Shows connections postulated by Koch et al., wherein the Acoustic Startle Stimulus signal processes through the cochlear root nucleus to the pons nucleus to motor neurons to produce the startle response. The pons nucleus receives input by glutamate from the pedunculopontine tegmental nucleus from the shell of the nucleus accumbens and further from the Ventral Tegmental Area. The glutamate pathway is the excitatory pathway whereas the Acetylcholine pathway plays the inhibitory action on the startle response [35–39]. *Abbreviations*: Glu, Glutamate; CRH, Corticotropin-releasing hormone; NA, Noradrenaline; DA, Dopamine; SP, Substance P; OM, Somatostatin; nC, Caudal pontine reticular nucleus. Here one sees the neuronal links between ventral tegmental area and the PnC leading into a link between the mesocorticolimbic brain reward circuitry with the pons brain stem circuitry for the auditory acoustic stimulus for startle.

7.4.2 The Pons Performs Sensorimotor Gating

Since the brain directs behavior and vice versa, the behavior "startle," a test of learning avoidance with an element of reflexive action became a behavioral target of discovery for us because the startle test can be used to decipher brain disorders characterized by the inability to dampen unwanted streams of unceasingly and often contradictory sensory information. Thus, we searched the auditory system in the brain for the pons, a channeling endogenous brain device that reduces stress, noise, for example, to lessen unavoidable shock, even fright from shock. This discovery, then, has enabled the possible acoustic startle disease aspects of the pons to be elucidated and treated clinically. The pons performs sensorimotor gating! Pre-Pulse inhibition (PPI) is the normal suppression of the startle reflex when an intense stimulus is preceded by a weak non-startling pre-stimulus. PPI is widely used as a model for sensorimotor gating processes and has been shown to be impaired in various neuropsychiatric disorders, including schizophrenia; here, it is used as a tool to study schizoaffective disorders from cocaine drug use and abuse.

7.5 Cocaine

Cocaine acts in the mesocorticolimbic dopamine neuronal pathway to produce euphoria, brain reward (*joie de vivre*). Moreover, the reinforcing brain reward properties of cocaine are due to increased synaptic release of dopamine and serotonin selectively and in separate oxidation potentials and with operational stability in mesocorticolimbic circuitry, as signaled by electron transfer with patented Neuromolecular Imaging (NMI). Also, with NMI, dopamine and serotonin antagonists e.g., atypical and typical antipsychotic agents were shown to block cocaine-induced release of monoamines while blocking psychostimulant behavior online [6, 8–13].

7.5.1 Schizophrenia/Cocaine Induced Psychoses

A study by Chapman *et al.* [17] brought about a view of schizophrenia which related schizophrenia to the process of cognition in the cerebral areas of attention and perception. Thus, the notion of schizophrenia as a "*breakdown of the ego*" [2] was revised. Schizophrenia is sometimes defined as a basic disorder of cognition [25, 48, 49], likely leading to sensory overload. To remark, "*too many things are coming into my mind*"; "*the thoughts feel bigger*"; "*the person I am talking to looks bigger.*" Therefore, the inability to gate sensory stimuli and channel the flow of sensory information to avoid sensory overload may well describe the schizophrenic/cocaine brain.

The hypothesis that correlates schizophrenic and cocaine psychoses is rooted in their neurochemical and behavioral similarities. Indeed, reports of cocaine-induced psychotic patients are happening at an alarming rate [4, 27, 41, 46, 50, 52, 53, 55, 62, 63, 66, 69, 70, 76, 78]. According to the mental health clinician's handbook, *Diagnostic and Statistical Manual of Mental Disorders,* the broad category of cocaine-related disorders can be subdivided into cocaine use and cocaine disorders e.g. delusions (hanging on to beliefs that are clearly nonexistent) and hallucinations (hearing and seeing things) [18]. A clinical case of cocaine psychosis presenting with hallucinations and delusions, is reported [56]; in this report, Nunes and Broderick publish a true narrative about a patient in the emergency department who had binged crack cocaine. Patient heard "voices in his head who were telling him to kill himself and the patient continued to believe there was a plot against him." These auditory hallucinations can be defined as "strange perceptions of sounds" [80].

Thus, cocaine psychosis is a major psychopathology [67] and hyperfunction of dopaminergic systems is critical to drug-induced psychosis [44]. Too, what complicates the situation further are data which show that about 50% of the patients who suffer from schizophrenia are also substance abusers at some time during their illness [15, 54]. It is now known that substances with psychomimetic properties such as cocaine are used and abused worldwide and consumption of such substances can and will induce psychotic reactions including primary psychotic disease [1, 3, 16, 21, 24, 51, 60, 64, 73, 74, 75, 79]. The data are supported by an innovative combination of preclinical an clinical empirical data using NMI in psychiatry [56].

7.5.2 The Acoustic Startle Response: What Does It Mean?

The acoustic startle response (ASR) is a reflex reaction to a loud sound or stimuli and is believed to have evolved as a defensive or learning adaptation mechanism among mammals. In murine animals, it can be elicited within milliseconds as a full body contraction of skeletal muscles. Many scientific investigators have explored both the pre-clinical and clinical aspects of the startle pathway to elucidate the underlying neuroanatomical mechanisms and pathways which give rise to learning, adaptation, anxiety as well as providing the *modus operandi* for the effects of cocaine psychosis and various fearful stimuli which can influence survival and even state of consciousness of an animal or in fact, human society. Such is the case in the twentieth year of the twentieth century with the advent of the SARS/Covid-19 (SARS-CoV-2) betacoronavirus that has dramatically changed life as we presently know it and this virus protein molecule appears to have done so on a global scale [14, 20, 22, 26, 28–30, 32, 40, 58, 59, 77].

These previous studies have been insightful in defining the one common set of nuclei that signal to the caudal pontine reticular formation (PnC) to express conditioned fear in acoustic startle. The set of nuclei are in the limbic brain, also known as the "feeling" brain and indeed, the set of nuclei we want to talk about is the "amygdala" and likely its extension, the substantia innominata, reticular nuclei that relate to one of four precise pathways that are in charge of signaling the conditioned fear stimulus to the caudal pontine reticular formation (PnC).

Fendt et al. 1996 [22] differentiated these pathways and different inputs to the PnC by testing the hypothesis that somatostatinergic projections from the amygdala or CG (central gray) to the PnC mediates fear-potentiated startle. This study consisted of three parts;

First, in double-labeling experiments, it was shown that the CG is the main source of somatostatinergic input to the PnC.

Second, electrophysiology with murine animals, the somatostatin agonist sandostatin was iontophoretically applied on PnC neurons with extracellular recordings of PnC activity. Sandostatin alone had no effect on tone-evoked or spontaneous activity, however, sandostatin significantly attenuated the increase of neuronal activity following the "excitatory" glutamate.

Third, in a behavioral study with murine animals, sandostatin was injected into the PnC. The first round of injections did not affect the baseline ASR but in the second round, murine animals, those that had undergone sustained fear conditioning followed by sandostatin, the data showed that sandostatin did block fear potentiation. Sandostatin may have had the effect of blocking fear potentiation due to a reduction of glutamate in evoked PnC neurons.

7.5.3 Hypothesis

> *"N-Methyl-D-Aspartate (NMDA) receptor-mediated effects on PnC neuronal activity in the ASR pathway can be modulated by somatostatin! The authors hypothesize that since there are many pathways from the central amygdala to the PnC, it is possible that the amygdala stimulates the PnC by a direct NMDA pathway, simultaneously inhibiting the PnC by an indirect somatostatinergic pathway from the CG. It is important to note that the circuits in their entirety must be intact and work together as an interactive network in order for the ASR to function"* [22].

7.5.4 An Acoustic Startle Response: Empirical Studies

This laboratory has studied the effect of cocaine using an ASR paradigm. We further advanced the field of cocaine and startle stimuli by assessing differences in sex/gender responses to substances of abuse further based in startle [14]. Therefore, the following publication from our laboratory is repurposed and is reprinted with permission from MDPI Journal Publications, specifically from *Brain Sciences*. MDPI Journals are produced by collaboration between Basel, Switzerland and Beijing, China.

7.5.5 Acoustic Startle Response: Sex Differences—The Data

Adult male and female Sprague-Dawley animals (2 months of age, weight matched for sex) were purchased from Charles River Laboratories, USA. The animals were housed in the Animal Care Facility under the auspices of the City College Institutional Animal Care and Use Committee (IACUC) in compliance with National Institute of Health (NIH) guidelines. Food (Purina Rat Chow) and water was available *ad libitum*.

Males were housed separately from females, three animals (male or female) per chamber. Studies were performed in groups of six, equal numbers of females and males during two seasons (Fall/Spring), with the exception of an additional group consisting of six females, isolated from males during the experimental period.

The auditory stimulus, acoustic startle response, termed pre-pulse inhibition (PPI), consisted of four trials, comprising three tests: habituation, injection and post-injection. The habituation period, which was the first test, was preceded by a 5 min acclimation period and each of the three tests was preceded by an acute injection of cocaine or physiological saline (animal weight in kilograms). The cocaine dosages administered were 10.0 mg/kg, 20.0 mg/kg and 40.0 mg/kg (dissolved in distilled water) and injected intraperitoneally (i.p.).

The acoustic startle chamber is an SR-LAB™ model (San Diego Instruments, San Diego, CA, USA) automated to record the startle response of small animals. The chamber consisted of a cylindrical enclosure on a plexiglass base. Background noise, pre-pulse stimuli and startle stimuli were controlled by the SR-LAB™ software. Startle responses were transduced by a piezoelectric accelerometer mounted on a seismic sensor plate platform, located below the cylinder in which the animal is placed.

Startle response to auditory stimuli in humans according to sex difference was studied by Swerdlow et al. [72]. Startle response via muscle movements was studied by pre-pulse inhibition (PPI), reviewed by Geyer et al. [26]. Data were recorded as potential in voltage and were concatenated for compatibility with Excel software. %PPI was calculated using the following formula:

$$\%\text{PPI} = 100 - \left[\left(\frac{\text{Mean PPI}}{\text{Mean startle only}}\right) \times 100\right]$$

There were 25 sessions in each trial with each trial consisting of 3 tests; there was a 15 min interval between each trial. Each session consisted of 5 types which were randomly placed throughout the trial, but each trial used the same format, depicted below in tabular form (Table 7.1).

Table 7.1 Protocol for delivery of auditory stimuli in the acoustic startle paradigm

Type	No stimulus	Prepulse (dB)	Pulse (dB)
1	75 decibels (broadband)	—	—
2	—	—	120
3	—	95	120
4	—	105	120
5	—	115	120

7.5.6 Data Analysis

Data were analyzed by Graph Pad Software Inc. and PRISM (La Jolla, CA, USA). Statistical significance between and within gender (sex) and dosage of cocaine was delineated via Two-Way Analysis of Variance (ANOVA). Data were considered significant at $p < 0.05$ and above. The expected baseline for acoustic startle response is defined by SR-LAB™ as 90th percentile.

7.5.7 Acoustic Startle Response: Sex Differences: Results/Outcome

Sex differences in responses during the acoustic startle paradigm after cocaine **Group A** (Figure 7.6) showed statistically significant differences in the dose response interaction of cocaine in females and males ($F\ (2,12) = 11.15$, $p < 0.002$). In Group A, in females and males, the 20 mg/kg dose of cocaine showed a maximum effect in pre-pulse attenuation of acoustic startle. **Group B** (Figure 7.7) female *versus* male analysis showed dosage × gender interaction significance ($F\ (2,12) = 6.81$ and $p < 0.01$). Gender alone significance was analyzed, and significance was found ($F\ (1,12) = 11.90$, $p < 0.005$). Cocaine dosage alone was significant as well ($F\ (2,12) = 8.00$, $p < 0.006$). In Group B, in females, the 20 mg/kg i.p. dose of cocaine resulted in a maximum attenuation of pre-pulse inhibition in the injection test. The within gender data gave the following results; Group A males *versus* Group B males showed between group significance, male × male ($F\ (1,12) = 14.91$, $p < 0.002$). Group A females with Group B females, females × females, showed cocaine dosage significance; ($F\ (2,12) = 15.76$, $p < 0.0004$).

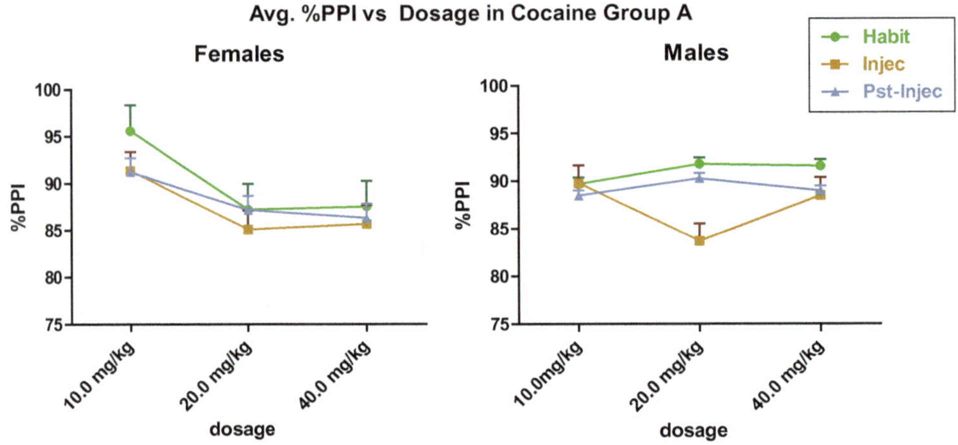

Figure 7.6 (Environment: fall season, Group A) (n = 6) shows the effect of cocaine on females (left) and males (right). Statistically significant differences occurred in the interaction between and within dosage.

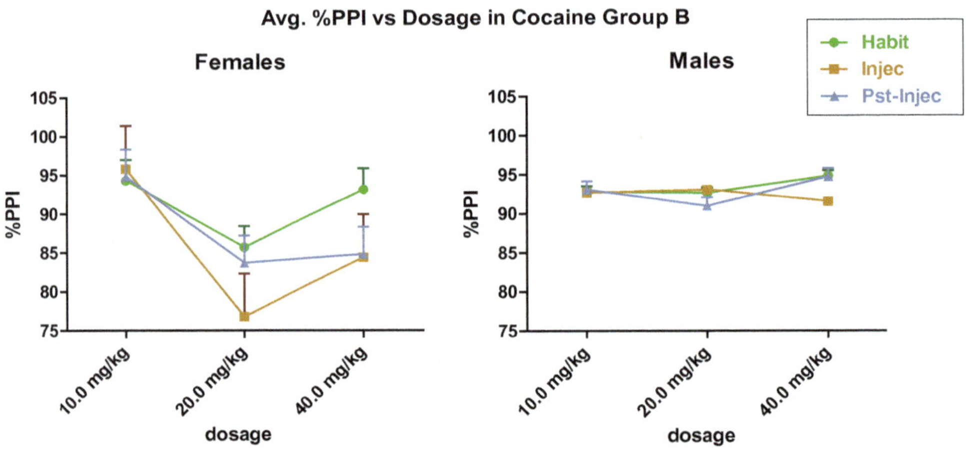

Figure 7.7 (Environment: spring season, Group B) (n = 6) shows the effect of cocaine on females (left) and males (right). Statistically significant differences occurred in the interaction between and within dosage and gender (sex).

A significant dosage effect of cocaine in its attenuation of pre-pulse inhibition in **Group C** (further isolated females; housed in a different room) was seen at 20 mg/kg in the injection and post-injection tests in summation of 4 trials [95% Confidence Limits ($p < 0.05$)] (Figure 7.8).

Given the caveat that different numbers of subjects comprised Group C and Group A females, still no female × female significance occurred. Nonetheless, Group C *versus* Group B, female × female comparison, exhibited a cocaine interaction significance in dosage (F (2,12) = 6.80, $p < 0.01$).

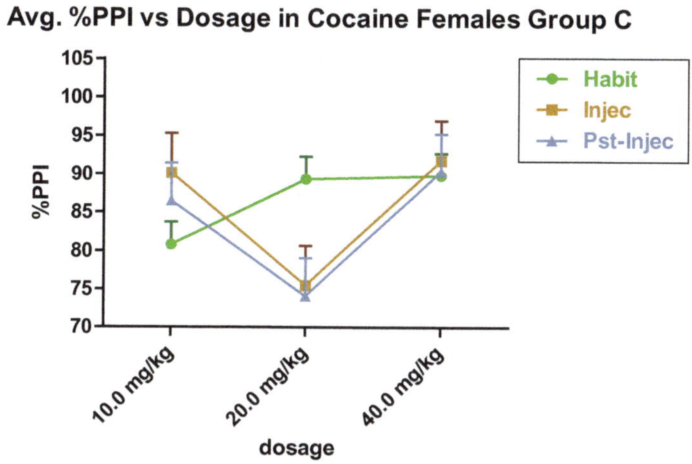

Figure 7.8 Group C (n = 6) shows the effect of cocaine in the acoustic startle paradigm in females. Females were isolated from males during the experimental period.

Saline Control

Figure 7.9 shows the effect of saline administration to animals during auditory acoustic startle.

7.5.8 Acoustic Startle Response: A Paradigm for Atypical Antipsychotic Therapy

Our research into acoustic startle, its relevance and its meaning to the pharmaceuticals of cocaine substance abuse comorbid with psychiatric disorders continues. Atypical antipsychotic agents were used at low doses to compare with higher doses as different dose responses are seen when studying typical versus atypical antipsychotics. Furthermore, subtle changes and variations of responses between sexes at the lower doses may belie the underpinnings of the mechanism of such an atypical medication, risperidone on fear potentiation [31].

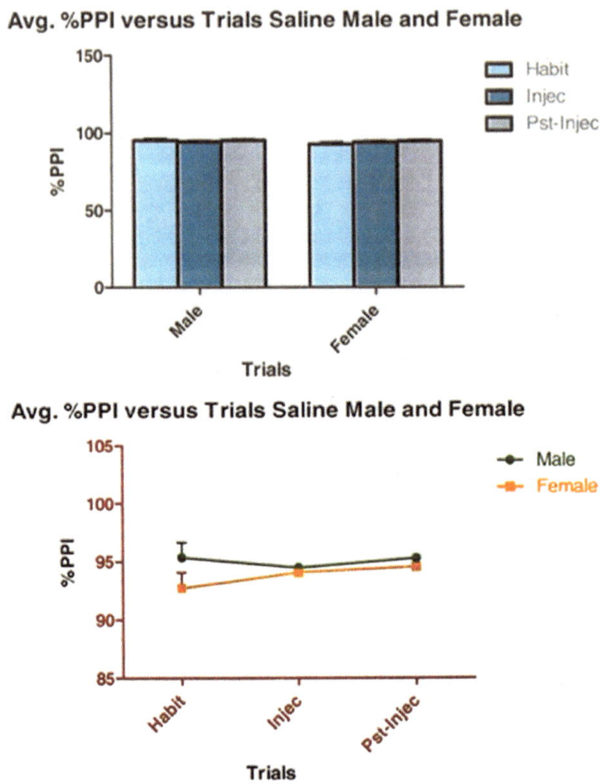

Figure 7.9 The bar graph (Top)/line graph (Bottom) is representative of an n value of 4 female and 4 male animals treated with saline (n = 8). Analysis of the saline data shows an interaction between gender and trials that is **not** significant, (F (2,12) = 0.94, p < 0.42). This represents 10.63% of variance. Gender was **not** significant (F (1,12) = 3.42, p < 0.1). Gender was 19.27% of variance. Trials were **not** significant after administration of saline, (F (2,12) = 0.23, p < 0.80). Trials represent 2.55 % of total variance.

7.5.9 Acoustic Startle Response: Antipsychotics: The Pons Connection

According to TM Lee et al. [41], for example, affective disorders can come from auditory disruption. The inability to have control over shocking or startling sounds causes disruption in affect. The nuances in adapting to sudden startling happenings in life have been intensely studied heuristically, empirically and theoretically. Affective disorders include such schizophrenic subtypes as schizoaffective; the neurotransmitter, serotonin, is critical in the process of affect.

7.5.9.1 Acoustic startle response and seasonal mood: The pons connection

Finally, as we take a glance at data from TM Lee et al. [41], and place it side by side with sex and seasonal effects in sensorimotor gating, further evidence for multifaceted neural connections are illuminated among mood, seasons and sex differences in schizophrenia from cocaine addiction. A seasons variable is interesting since this is rarely thought of in auditory acoustic sensorimotor gating studies. It is well known that seasons are critical factors in mood, schizoaffective disorders and bipolar disorder.

7.5.9.2 Acoustic startle response and SARS/Covid-19 (*SARS-CoV-2)*: The pons connection

***From*: Dr. Broderick**

"COVID-19, also called SARS/Covid-19 and SARS-CoV-2, may well be the major startle response to the human and animal "pons" in the historical universe. Brain reward was blocked in the collective world."

7.6 Summary

- Norepinephrine and Dopamine electroactive signals are separated by the BRODERICK PROBE®
- The pons is a neuroanatomic substrate in the Brain Stem.
- The locus coeruleus is a neuroanatomic substrate in the Brain Stem,
- The *pons* and the *locus coeruleus* are noradrenergically innervated.
- The pons is the nucleus of fear potentiation.

- Norepinephrine is adrenaline.
- Dopamine was found in the *pons* and the *locus coeruleus.*
- A *brain reward* pathway signals through the *Brain Stem* to the *Ventral Tegmentum*.
- Shock and startle block the brain reward pathway in the auditory system of the brain.
- Shock and startle block the dopamine brain reward response in the auditory system of the brain.
- The result is fear and potentiation of fright.
- Schizophrenia causes multiple channels in the brain to open simultaneously.
- One may say, "everything is bigger."
- Acoustic startle is a way to study shock, startle *versus* brain reward.
- Acoustic startle is experienced in the auditory system.
- Sex differences are associated with the administration of psychostimulants.
- Cocaine induces psychosis.
- Antipsychotic medications can block cocaine-induced-psychosis (CIP).
- Atypical antipsychotic medications act differently than typical antipsychotics in fear potentiation.
- Sex differences occur after administration of self-medicating with cocaine.
- Substance abuse is comorbid with psychosis.
- Females are more sensitive than males to substance abuse.
- Females are more sensitive than males to the effects of cocaine.
- Females are more sensitive than males to brain reward.
- *SARS-CoV-2 brought global shock and startle, enabling fear to predominate.*
- *SARS-CoV-2* is the auditory and cortical startle on a global scale.
- *SARS-CoV-2* may well respond to somatostatin medicine.
- It is suggested herein that somatostatin agents are therapy for *SARS-CoV-2.*

7.6.1 Conclusion

The seminal data that the norepinephrine and dopamine electroactive signals can be imaged separately in the brain stem and ventral tegmental area adventures us to a new world. It is a game changer to separate fear from brain reward and open up therapies for anxiety, depression and panic disorders. Finding dopamine in the pons and the locus coeruleus assisted with great aplomb. Startle stimuli and responses were monitored in the mammal's auditory circuit, localizing, a priori, in the brain stem and then is tracked to the midbrain, Ventral Tegmental Area, the brain reward/addiction pathway. Sex differences play a dramatic role in the etiology of affective and substance abuse responses. Brain reward and reinforcement are wonderful feelings and from there, can come the dreaded addiction, "like a thief in the night."

Sex differences in the ***mesocorticolimbic*** *circuit are next.*

References

1. Batki SL, Harris DS (2004) Quantitative drug levels in stimulant psychosis: Relationship to symptom severity, catecholamines and hyperkinesia. *Am J Addict* 13: 461–470.
2. Bleuler M (1954) The concept of schizophrenia. *Am J Psychiatry* 5: 382–383.
3. Boutros NN, Gooding D, Sundaresan K, Burroughs S, Johanson CE (2006) Cocaine-dependence and cocaine-induced paranoia and mid-latency auditory evoked responses and sensory gating. *Psychiatry Res* 145: 147–154.
4. Brady KT, Lydiard RB, Malcolm R, Ballenger JC (1991) Cocaine-induced psychosis. *J Clin Psychiat* 52: 509–512.
5. Broderick PA (1988) Distinguishing in vitro electrochemical signatures for norepinephrine and dopamine. *Neurosci Lett* 95(1–3): 275–280.
6. Broderick PA (1989a) Cathodic Electrochemical Current Arrangement with Telemetric Application. U.S. Patent # 4,883,057, Issued.
7. Broderick PA (1989b) Characterizing stearate probes in vitro for the electrochemical detection of dopamine and serotonin. *Brain Res* 495(1): 115–121.

8. Broderick PA (1995) Microelectrodes and their use in cathodic electrochemical current arrangement with telemetric application. U.S. Patent # 5,433,710, Issued.

9. Broderick PA (1999) Microelectrodes and their use in an electrochemical arrangement with telemetric application. U.S. Patent # 5,938,903, Issued.

10. Broderick PA (2016) Noninvasive Electroactive Photonic Protein Sensor with Polymer Photovoltaic Optics for Memory Transduction Using Organic and Inorganic Elements as Platforms. *US Non–Provisional International Application No. PCT/US 16/68879.2020. US Application No.* 16/074,254, *Notice of Allowance, August 16th, 2021.*

11. Broderick PA (2021) Identification, Diagnosis, and Treatment of Neuropathologies, Neurotoxicities, Tumors, and Brain and Spinal Cord Injuries using Electrodes with Microvoltammetry. *US Patent Serial No. 10,980,460, Issued.*

12. Broderick PA (2017) Noninvasive Electroactive Photonic Sensor with Polymer Memory Transduction Using Organic and Inorganic Elements as Platforms. US Provisional Application No. 62/273,693, December 31, 2015 published, February, 2017.

13. Broderick PA, Pacia SV (2006) Identificaton, diagnosis, and treatment of neuropathologies, neurotoxicities, tumors, and brain and spinal cord injuries using microelectrodes with microvoltammetry. USPTO_U.S. Patent # 7,112,319, Issued.

14. Broderick PA, Rosenbaum T (2013) Sex-Specific Brain Deficits in Auditory Processing in an Animal Model of Cocaine-Related Schizophrenic Disorders. *Brain Sci* 3: 504–520.

15. Buckley PF (1998) Substance abuse in schizophrenia: A review. *J Clin Psychiatry* 59: S26–S30.

16. Caton CL, Samet S, Hasin DS (2000) When acute-stage psychosis and substance use co-occur: Differentiating substance-induced and primary psychotic disorders. *J Psychiatr Pract* 6: 256–266.

17. Chapman T, Freeman T, McGhie A (1959) Clinical research in schizophrenia; the psychotherapeutic approach. *Br J Med Psychol* 32: 75–85.

18. Cocaine and related disorders. Available online: *http://www.minddisorders.com/Br-Del/Cocaine-and-related-disorders.html* (accessed on 28 December 2012).

19. Crochet S, Sakai K (2003) Dopaminergic modulation of behavioral states in mesopontine tegmentum: a reverse microdialysis study in freely moving cats. *SLEEP* 26(7).

20. Davis M (1984) The mammalian startle response. In *Neural Mechanisms of Startle Behavior* Plenum Press, New York, pp. 287–342.

21. Delavenne H, Garcia FD, Lacoste J, Cortese S, Charles-Nicolas A, Ballon N (2012) Psychosis in a cocaine-dependent patient with ADHD during treatment with methylphenidate. *Gen Hosp Psychiatry. http://dx.doi.org/10.1016/j.genhosppsych.2012.05.010.*

22. Fendt M, Koch M, Schnitzler H-U (1996) Somatostatin in the Pontine Reticular Formation Modulates Fear Potentiation of the Acoustic Startle Response: An Anatomical, Electrophysiological, and Behavioral Study. *The Journal of Neuroscience* 16(9): 3097–3103.

23. Fendt M, Li L, Yeomans JS (2001) Brain stem circuits mediating prepulse inhibition of the startle reflex. *Psychopharmacology* (Berl) 156: 216–224.

24. Fiorentini A, Volonteri LS, Dragogna F, Rovera C, Maffini M, Mauri MC, Altamura CA (2011) Substance-Induced Psychoses: A Critical Review of the Literature. *Curr Drug Abuse Rev* 4: 228–240.

25. Freeman T, McGhie A (1957) The relevance of genetic psychology for the psychopathology of schizophrenia. *Br J Med Psychol* 30: 176–187.

26. Geyer MA, Kerbs-Thomson K, Braff DL, Swerdlow NR (2001) Pharmacological studies of prepulse inhibition models of sensorimotor gating deficits in schizophrenia: A decade in review. *Psychopharmacology* 156: 117–154.

27. Harris D, Batki SL (2000) Stimulant psychosis: symptom profile and acute clinical course. *Am J Addict* 9: 28–37.

28. Hetrick WP, Smith DA, Spuriell E, Sandman CA, Bunney WE (1997) The putative phenomenological correlates of sensory gating: Factor analyses and a self-report rating scale. *Schizophr Res* 24: 17.

29. Hoffman HS, Fleshler M (1963) Startle reaction: modification by background acoustic stimulation. *Science* 141: 928–930.

30. Hoffman HS, Ison JR (1980) Reflex modification in the domain of startle: I. Some empirical findings and their implications for how the nervous system processes sensory input. *Psychol Rev* 87: 175–189.

31. Hordof JR, Broderick PA (2020) Acoustic Startle Response and Prepulse Inhibition: Sex Differences with Cocaine. *EC Psychology and Psychiatry* 9.5: 08–16.

32. Ilango A, Shumake J, Wetzel W, Scheich H, Ohl FW (2012) The role of dopamine in the context of aversive stiumuli with particular reference to acoustically signaled avoidance learning. *Front Neurosci* 6: 132.

33. Joseph RG (2000) The Pons of the Brainstem: Reticular Formation, Dopamine, Norepinephrine, Serotonin, Dreaming, REM Sleep. Cranial Nerves. Academic Press.

34. Keating GL, Rye DB (2006) Where you least expect it: Dopamine in the pons and modulation of sleep and REM sleep. *Article in Sleep*, December, Editorial Comment.

35. Koch M (1999) The neurobiology of startle. *Prog Neurobiol* 59: 107–128.

36. Koch M, Fendt M, Kretschmer BD (2000) Role of the substantia nigra pars reticulata in sensorimotor gating, measured by prepulse inhibition of startle in rats. *Behav Brain Res* 117: 153–162.

37. Koch M, Friauf E (1995) Glycine receptors in the caudal pontine reticular formation: are they important for the inhibition of the acoustic startle response? *Brain Res* 671: 63–72.

38. Koch M, Kungel M, Herbert H (1993) Cholinergic neurons in the pedunculopontine tegmental nucleus are involved in the mediation of prepulse inhibition of the acoustic startle response in the rat. *Exp Brain Res* 97: 71–82.

39. Koch M, Schnitzler HU (1997) The acoustic startle response in rats–circuits mediating evocation, inhibition and potentiation. *Behav Brain Res* 89: 35–49.

40. Lang PJ, Davis M (2006) Emotion, motivation, and the brain: Reflex foundations in animal and human research. *Prog Brain Res* 156: 3–29.

41. Lee TM, Sun DL, Wong NML, Shao R, Men W, Ge J, So KF, Gao JH, Chan CCH (2015) A Pontine Region is a Neural Correlate of the Human Affective Processing Network. *EBioMedicine* 2(11): 1799–1805.

42. Leitner DS, Cohen ME (1985) Role of the inferior colliculus in the inhibition of acoustic startle in the rat. *Physiol Behav* 34: 65–70.

43. Leumann L, Sterchi D, Vollenweider F, Ludewig K, Früh H (2001) A neural network approach to the acoustic startle reflex and prepulse inhibition. *Brain Res Bull* 56: 101–110.

44. Lieberman JA, Kinon BI, Loebel AD (1990) Dopaminergic mechanisms in idiopathic and drug-induced psychoses. *Schizophr Bull* 16: 97–110.

45. Luppi PH, Aston-Jones G, Akaoka H, Chovet G, Jouvet M (1995) Afferent projections to the rat locus coeruleus demonstrated by retrograde and anterograde tracing with cholera-toxin B subunit and Phaseolus vulgaris leucoagglutinin. *Neuroscience* 65(1): 119–160.

46. Lysaker P, Bell M, Beam-Goulet J, Milstein R (1994) Relationship of positive and negative symptoms to cocaine abuse in schizophrenia. *J Nerv Ment Dis* 182: 109–112.

47. Maeda T, Kitahama K, Geffard M (1994) Dopaminergic Innervation of Rat Locus Coeruleus: A Light and Electron Microscopic Immunohistochemical Study. *Microscopy Research and Technique* 29: 211–218.

48. McGhie A, Chapman, J (1961) Disorders of attention and perception in early schizophrenia. *British Journal of Medical Psychology* 34: 103–116.

49. McGhie A, Chapman J, Lawson JS (1964) Disturbances in Selective Attention in Schizophrenia. *Proc R Soc Med* 57: 419–422.

50. Mendoza R, Miller BL, Mena I (1992) Emergency room evaluation of cocaine-associated neuropsychiatric disorders. *Recent Dev Alcohol* 10: 73–87.

51. Merlo LJ, Carnes PJ, Gold MS (2008) Response to "Comorbid psychiatric diagnoses and their association with cocaine-induced psychosis in cocaine-dependent subjects". *Am J Addict* 17: 247–248.

52. Miller BL, Mena I, Giombetti R, Villanueve-Meyer J, Djenderdjian AH (1992) Neuropsychiatric effects of cocaine: SPECT Measurements. *J Addict Dis* 11: 47–58.

53. Mitchell J, Vierkant AD (1991) Delusions and hallucinations of cocaine abusers and paranoid schizophrenics: a comparative study. *J Psychol* 125: 301–310.

54. Muesser KT, Nishith P, Tracy H, de Girolamo J, Molinaro M (1995) Expectations and motives for substance use in schizophrenia. *Schizophr Bull* 21: 367–378.

55. Nambudin DE, Young RC (1991) A case of late-onset crack dependence and subsequent psychosis in the elderly. J *Subst Abuse Treat* 8: 253–255.

56. Nunes JV, Broderick PA (2007) Novel research translates to clinical cases of schizophrenic and cocaine psychosis. *Neuropsychiatr Dis Treat* 3: 475–485.

57. Ornstein K, Milon H, McRae-Degueurce A, Alvarez C, Berger B, Würzner HP (1987) Biochemical and radioautographic evidence for dopaminergic afferents of the locus coeruleus originating in the ventral tegmental area. *J Neural Transm* 70(3–4): 183–191.

58. Parker KJ, Hyde SA, Buckmaster CL, Tanaka SM, Brewster KK, Schatzberg AF, Lyons DM, Woodward SH (2011) Somatic and

neuroendocrine responses to standard and biologically salient acoustic startle stimuli in monkeys. *Psychoneuroendocrinology* 36: 547–556.

59. Peak H (1939) Time order error in successive judgments and in reflexes. I. Inhibition of the judgment and the reflex. *J Exp Psychol* 25: 535–565.
60. Roncero C, Daigre C, Gonzalvo B, Valero S, Castells X, Grau-López L, Eiroa-Orosa FJ, Casas M (2013) Risk factors for cocaine-induced psychosis in cocaine-dependent patients. *Eur Psychiatry* 28: 141–146.
61. Rose PK, Scott SH (2003) Sensory-motor control: a long-awaited behavioral correlate of presynaptic inhibition. *Nat Neurosci* 6: 1243–1245.
62. Rosenthal RN, Miner CR (1997) Differential diagnosis of substance-induced psychosis and schizophrenia in patients with substance use disorders. *Schizophr Bull* 23: 187–193.
63. Rosse RB, Collins JP Jr, Fay-McCarthy M, Alim TM, Wyatt RJ, Deutsch SI (1994) Phenomenologic comparison of the idiopathic psychosis of schizophrenia and drug-induced cocaine and phencyclidine psychoses: a retrospective study. *Clin Neuropharmacol* 17: 359–369.
64. Rosse R, Deutsch S, Chilton M (2005) Cocaine addicts prone to cocaine-induced psychosis have lower body mass index than cocaine addicts resistant to cocaine-induced psychosis—Implications for the cocaine model of psychosis proneness. *Isr J Psychiatry Relat Sci* 42: 45–50.
65. Rudomin P (2002) Central control of information transmission through the intraspinal arborizations of sensory fibers examined 100 years after Ramón y Cajal. *Prog Brain Res* 136: 409–421.
66. Satel SL, Edell WS, (1991) Cocaine-induced paranoia and psychosis proneness. *Am J Psychiat* 148: 1708–1711.
67. Satel SL, Seibyl JP, Charney DS (1991) Prolonged cocaine psychosis implies underlying major psychopathology. *J Clin Psychiatry* 52: 349–350.
68. Seki K, Perlmutter SI, Fetz EE (2003) Sensory input to primate spinal cord is presynaptically inhibited during voluntary movement. *Nat Neurosci* 6: 1309–1316.
69. Serper MR, Chou JC, Allen M II, Czobor P, Cancro R (1999) Symptomatic overlap of cocaine intoxication and acute schizophrenia at emergency presentation. *Schizophr Bull* 25: 387–394.

70. Sherer MA, Kumor KM, Cone EJ, Jaffe JH (1998) Suspiciousness induced by four-hour intravenous infusions of cocaine. Preliminary findings. *Arch Gen Psychiat* 45: 673–677.

71. Suzuki M, Hurd YL, Sokoloff P, Schwartz J-C, Sedvall G (1998) D3 dopamine receptor mRNA is widely expressed in the human brain. *Brain Res* 779(1–2): 58–74.

72. Swerdlow NR, Hartman PL, Auerbach PP (1997) Changes in sensorimotor inhibition across the menstrual cycle: implications for neuropsychiatric disorders. *Biol Psychiatry* 41: 452–460.

73. Tang YL, Kranzler HR, Gelernter J, Farrer LA, Cubells JF (2007) Comorbid psychiatric diagnoses and their association with cocaine-induced psychosis in cocaine-dependent subjects. *Am J Addict* 16: 343–351.

74. Tang YL, Kranzler HR, Gelernter J, Farrer LA, Pearson D, Cubells JF (2009) Transient Cocaine-Associated Behavioral Symptoms Rated with a New Instrument, the Scale for Assessment of Positive Symptoms for Cocaine-Induced Psychosis (SAPS-CIP). *Am J Addict* 18: 339–345.

75. Taylor WA, Staby AE (1992) Acute treatment of alcohol and cocaine emergencies. *Recent Dev Alcohol* 10: 179–191.

76. Tueth MJ (1993) High incidence of psychosis in cocaine intoxication and preventing violence. *Am J Emerg Med* 11: 676.

77. Valasamis B, Schmid S (2011) Habituation and Prepulse Inhibition of Acoustic Startle in Rodents. *J Vis Exp* 55: e3446, DOI: 10.3791/3446.

78. Vorspan F, Bloch V, Brousse G, Bellaise L, Gascon J, Lépine JP (2011) Prospective assessment of transient cocaine-induced psychotic symptoms in a clinical setting. *Am J Addict* 20: 535–537.

79. Vorspan F, Brousse G, Bloch V, Bellais L, Romo L, Guillem E, Coeuru P, Lépine JP (2013) Cocaine-induced psychotic symptoms in French cocaine addicts. *Psychiatry Res* 200: 1074–1076.

80. Waters F (2010) Auditory Hallucinations in Psychiatric Illness. *Psychiatric Times* 27(3): 1–7.

Chapter 8

Sex-Specific Synaptic Signals While the Subject Walks: The Mesolimbic Brain Reward Circuit

Sexual dimorphism means sex specificity. Sexual dimorphism is said to be "differential effects on the growth of bone, muscle, and fat at puberty that increase considerably the difference in body composition between the sexes. Males have a greater increase not only in stature but especially in breadth of shoulders; females have a greater relative increase in width of hips. These differences are produced chiefly by the changes that occur during puberty, but other sex differentiations arise before that time. Some, like the external genital difference itself, develop during fetal life. Others develop continuously throughout the whole growth period by a sustained differential growth rate. An example of this is the greater relative length and breadth of the forearm in the male when compared with whole arm length or in whole body length. Part of the sex difference in pelvic shape antedates puberty. Females at birth already have a wider pelvic outlet. Thus, the adaptation for childbearing is present from an early age. The changes at puberty are concerned more with widening the pelvic inlet and broadening the much more noticeable hips" [56].

Sexual dimorphism predicts that females would be more sensitive or inherently susceptible to the untoward effects of drugs of addiction than would males, due to a natural selection on the part of the female to use a kind of prophylactic method against substance abuse in order to protect

Neuroimaging: Sensing Biochemistry in the Brain
Patricia A. Broderick

ISBN 978-981-4613-49-1 (Hardcover), 978-0-429-02974-5 (eBook)
www.jennystanford.com

their offspring. Perhaps even a sensory cue, or an "inner sixth sense" of the neurotoxicity that presumably associates with so many neurobiological aspects of addiction furnish such a caveat to the female sex. The ability to streamline sensory channeling to avoid disaster may be why females inherently know of their sensitivity to the use of substances that exploit. Although the numerical gap between male and female substance abusers seems to be narrowing on the social scene, the hypothetical prediction here is that the sexes will show a physiological dimorphism to drugs of abuse within brain reward neuro-centers.

When the BRODERICK PROBE® *images electroactive signals, profiling the neurochemistry of the male versus the female brain reward neuro-centers in real time and LIVE, sex difference is the empirical case. The meso-corticolimbic (mesolimbic) circuit includes the limbic brain, called the "feeling brain." Feelings differ between the sexes; feelings between the sexes matter. With the* BRODERICK PROBE®, *the neurotransmitter, dopamine, is imaged non-invasively from the feeling brain while males and females walk! While the nanobiosensor is a selective imager for multiple neurotransmitters, the chapter focuses on this classically known brain reward neurotransmitter, the catecholamine, dopamine.*

8.1 Brain Reward Signaling through the Pons to the Nucleus Accumbens

Brain reward communicates through the pons in the brain stem to the nucleus accumbens through the somatodendritic ventral tegmental area, named A10, that are cell bodies that synthesize the biogenic amine neurotransmitter, dopamine. Most importantly, the A10 neurons are in close proximity to the pons. Moreover, in close proximity to the pons, are neuronal terminals such as the nucleus accumbens that instigate, along with the hypothalamus, the septum, the amygdala, the reticulum, and the Reticular Activating System (RAS), a myriad of brain reward mechanisms such as the neurological effects of reinforcement and pleasure that neuromodulate behavior. There is some debate as to whether these pleasure centers are stimulated by motivation of pleasure and not the pleasure itself. The BRODERICK PROBE® sensors come close to settling such cause and effect dilemmas. Nonetheless, abundant evidence suggests that human and murine pleasure/reward/reinforcement happens across a distributed brain circuit in which the sub-cortical as well as cortical regions are involved.

> "A graphic illustration conveys a stronger message than words, as in "*The book jacket is a big selling point—one picture is worth a thousand words.*" This saying was invented by an advertising executive, Fred R. Barnard. To promote his agency's advertisements, he took out an ad in *Printer's Ink* in 1921 with the headline "One Look Is Worth a Thousand Words" and attributed it to an ancient Japanese philosopher. Six years later he changed it to "Chinese Proverb: One Picture Is Worth Ten Thousand Words," illustrated with some Chinese characters. The attribution in both was invented; Barnard simply believed an Asian origin would give it more credibility."
>
> *https://en.wikipedia.org/wiki/A_picture_is_worth_a_thousand_words.*

8.2 The Nanotechnology

Imaging (NMI) is all about pictures, images with novel nanobiosensors enables the selective detection of neurotransmitters *in vivo* within seconds, on line and in real time while the subject walks and behaves. It's been said, "It's incredible *seeing* the BRODERICK PROBE® and *seeing* what it does is mind boggling. Now, it is said, I *see* further its novelty, its value, more than ever!" When one realizes that the Trademark, BRODERICK PROBE® signifies a series of carbon allotrope and protein photodiode nanoprobes, the excitement escalates. When one sees that specific sensors comprised of carbon allotropes exhibit patented semiconductor properties, the body of work is more deeply appreciated. Indeed, further mind-boggling is the established empirical fact that the biosensors remain in place for continuing online imaging over a period of months. This biotechnological advance is based on conventional electrochemistry; the nanobiosensors detect neurotransmitters by, in one example of our nanobiotechnology, electron transfer. Simply stated, nanobiosensors *adsorb* electrons from each neurotransmitter at a specific oxidation potential, also called half-wave potential and the current derived from electron transfer is directly proportional to the concentration of the neurotransmitter or neurochemical. Selective electron transfer properties of these biosensors permit the imaging of neurochemicals, enzymes, nucleic acids, peptides, amino acids and neurotransmitters, molecules galore. The BRODERICK PROBE® images within its protein photodiode and photons are *absorbed by* quantum mechanics kinetics at the same time as the BRODERICK PROBE® images electron kinetics. Information around and about the network of dendritic cell body cables impinge on axons at synapses to generate the LIVE image in real time! Personalized medicine benefits. The healthy physiological neuronal being benefits. The novel, inventive BRODERICK PROBE® nanobiosensors differ in formulation and detection capabilities from electrodes used in conventional electrochemistry/voltammetry. In the present studies, NMI, specifically, the BRODERICK PROBE® laurate biosensor images neurotransmitter signals within mesolimbic neuronal terminals, within the ventrolateral nucleus accumbens (NAcc).

Within the exact same paradigm, experimental design and same subject control, infrared photocells monitor the sensor-brain directed motor responses that enable one to walk and to perform other multiple tasks such as to masticate and groom. These are motor responses highly involved in the study of spinal cord injuries and for example, such orphan diseases as Lesch Nyhan's disease of uncontrollable chewing and biting. One example of the nanobiosensor is schematically shown in Figure 8.1 along with a Neuromolecular LIVE image depicting signally for neurotransmitters and metabolites in mammalian mesolimbic brain while the subject walks.

Let the pictures begin...

Figure 8.1 shows the BRODERICK PROBE® and a typical Neuromolecular LIVE image—follow the legends!

Figure 8.2 presents dopamine (DA) and serotonin (5-HT) in mesocorticolimbic nucleus accumbens (ventrolateral (vl)) in Sprague Dawley laboratory murine animals using immunocytography.

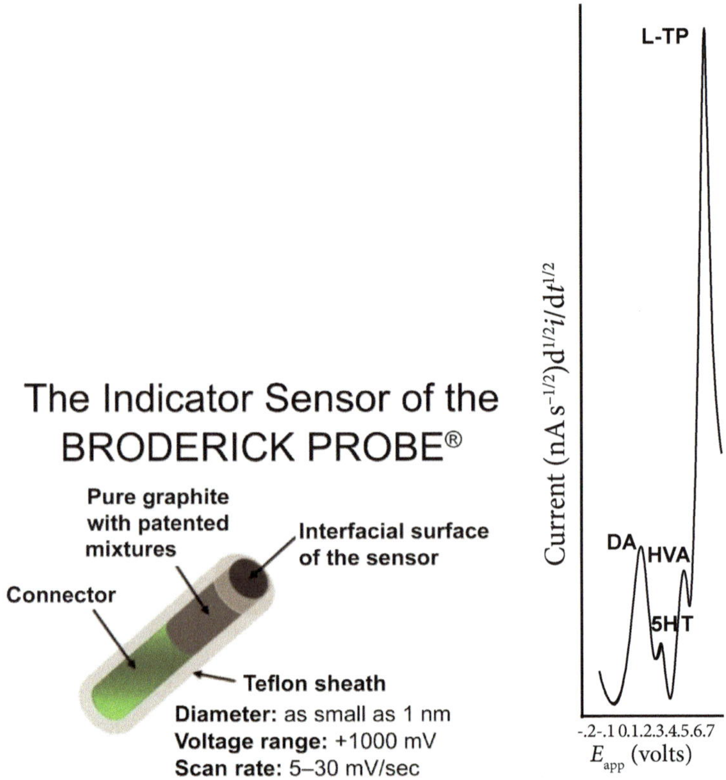

Figure 8.1 (Left) Schematic diagram of the BRODERICK PROBE®. (Right) A typical neuromolecular live image of neurotransmitters and metabolites, derived from the mesocorticolimbic circuitry, in particular, the brain reward circuit, the vlNAcc. *Abbreviations*: DA, dopamine; 5-HT, serotonin; HVA, homovanillic acid; L-TP, L-tryptophan. Each neurochemical is detected at a specific oxidation potential, experimentally delineated.

8.3 Some Background on the Stimulants Studied

Live Imaging can and does dramatically demonstrate the differential sex effect of stimulants on brain reward circuitry. Seeing inside the brain reward circuit, actually sensing the brain in real time is innovation personified. NMI and the BRODERICK PROBE®

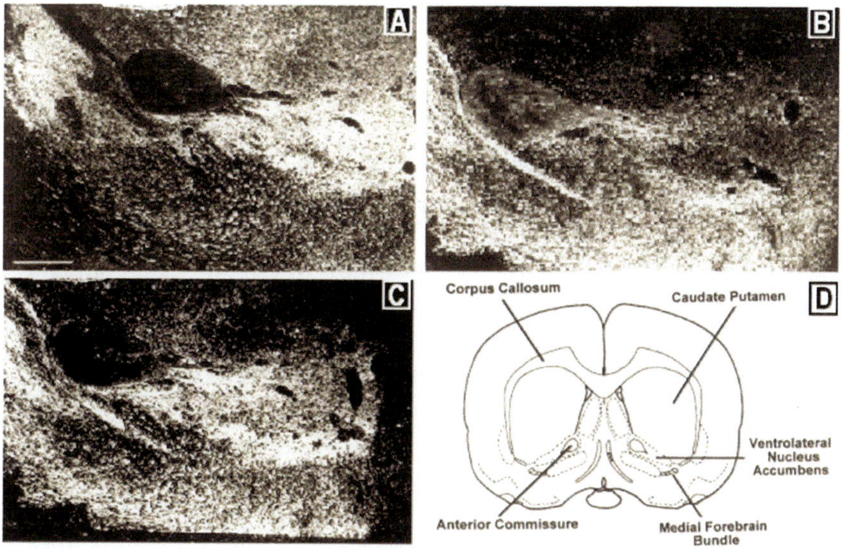

Figure 8.2 Immunocytographs of dopamine (DA) and serotonin (5-HT) in ventrolateral nucleus accumbens (NAcc) in murine mammalians. Dark field photomicrographs show the distribution of (A) DA neurons, stained with tyrosine hydroxylase; two high density patterns of DA are apparent in the medial and lateral core, (B) 5-HT axons in the caudal one-third of NAcc; 5-HT was stained with a sensitive silver intensification procedure. Thus, axons and terminals are black, (C) 5-HT axons in DA neurons in NAcc at the site of the BRODERICK PROBE® laurate biosensor. In B, two low density patterns of 5-HT are apparent in the ventral and ventrolateral NAcc. High density 5-HT is seen in the perimeter around the core. In A, the scale is shown. Scale = 500 um as shown by the horizontal line in (A) [4, 20, 47, 48]. (D) Coronal section of NAcc depicting vlNAcc. The almond-shaped Anterior Commissure, shown in dark black in (A–C) immunocytographically and shown in (D) schematically, provides a marker for neurotransmitter loci. Adapted with permission from Elsevier Limited. Kidlington, Oxford, OX5 1GB.UK.

series of nanobiosensors image stimulant sexual dimorphism as neurotransmitter gender differences are imaged within seconds and sub-seconds online.

Cocaine is known as an alkaloid stimulant, that, through its role as an indirect dopamine agonist, mimics a schizophrenic syndrome as shown, for example, by Nunes and Broderick, 2007 [42] using risperidone in patients and animals in the presence of cocaine. Cocaine causes euphoric feelings by this DA response by way of blocking DA transporters. Importantly, there is another way to enhance DA release at the presynapse and that is by reversal of synaptic DA release, that is, by depolarizing DA neuronal firing in somatodendritic mesocorticolimbic circuitry with gamma-butyrolactone, synaptic DA release is inhibited and cocaine's effects are disenabled as shown empirically by Broderick, 1991 [3]. With Live Imaging, *via* Neuromolecular Imaging, we see inside the brain of the living mammalian animal, the physiologic release of DA from NAcc and the consequent disenabling of cocaine's dysfunctional sensory-motor influences and affect. Thus, NMI sees synaptic release of DA as a presynaptic mechanism of action for the alkaloid, cocaine.

Moreover, cocaine exhibits a true pharmacological dose response curve; its pharmacokinetic linear dose-response curve is clear while DA release in NAcc and instant motoric effects are imaged and videotracked. The basic tenet is: "the higher the dose the greater the response."

Caffeine is also known as an alkaloid stimulant that influences wakefulness as well as induction of increased motoric effects *via* non-selectively antagonizing the adenosine A1 receptors (A1R) and A2A receptors (A2AR). A1 receptors are found throughout the brain but are abundantly found primarily in the hippocampus, cerebral cortex, NAcc and caudate-putamen (Fredholm et al., 1999) [16]. A2AR are localized in dopaminergic-rich areas. The NAcc, where both receptor types are located, is the DA dense brain area involved in reward that also corresponds to motor response. This, then, is a critical neuroanatomic site of interest for both basal ganglia DA release and motor responses that include among others, walking, locomotion, also called, ambulation. The NAcc is a brain reward nucleus and it is essential to the study of the alkaloids such as the psychostimulant, cocaine and the stimulant, caffeine.

It is more than interesting that caffeine exhibits an inverted "U-shape" dose response curve. This dose response curve is not the typical pharmacokinetic dose response that is classically known for drug effects in terms of pharmacological pharmacokinetics. It is not the classical linear dose response that occurs in the case of the alkaloid stimulant, cocaine. The basic pharmacokinetic tenet is, "what is true of cocaine, is not true of caffeine, that is, for caffeine, "the higher the dose- does not equate to -the greater the effect." Caffeine exhibits an inverted U shape response. *At the higher dose, the caffeine response diminishes.*

8.3.1 Sexual Dimorphism: Experimental Design

- **Goal:** Decode the neurochemical differences between the nitrogenous alkaloids, cocaine, and caffeine while the subject walks.
- **Surgery:** Place the BRODERICK PROBE® biosensor (one chosen from the family of patented sensors) in the brain reward nerve terminal, ventrolateral nucleus accumbens (vlNAcc), for example, in the brain of an animal model preferably free of viruses. Use the map for stereotaxic coordinates published by Pellegrino et al., 1979 [46]. The use of Na pentobarbital anesthesia is one suggestion for reliable and safe anesthesia.
- ***In Vivo*-Non-Invasive Studies:** Perform studies after following procedures for recovery from anesthesia. Sending the signal via a commutator, selectively image synaptic DA separately from ascorbic acid release *in vivo*, with NMI using miniature BRODERICK PROBE® biosensors. Monitor walking and various aspects of behavioral functionalities with computerized infrared photobeams at the same time as the subject walks and behaves using culled pre-programmed parameters in the brain-machine interface.
- **Dosage Regimen:** Administer cocaine at doses of 2.5, 5.0 and 10.0 mg/kg intraperitoneally (i.p.) and in separate studies, administer caffeine at 12.5, 25.0 and 50.0 mg/kg *via* the same route of administration. Administer the cocaine/caffeine combination dose regimen using the same paradigm.
- **Interpret the Data:** Discern the novel phenomenon of **neurobehavior** of sexual dimorphism online.

8.4 Switching Stimulant Gears in the Classical Brain Reward Circuit

Neurochemical Sexual dimorphism within males and females in a mammalian animal model is shown in Figures 8.3 to 8.11.

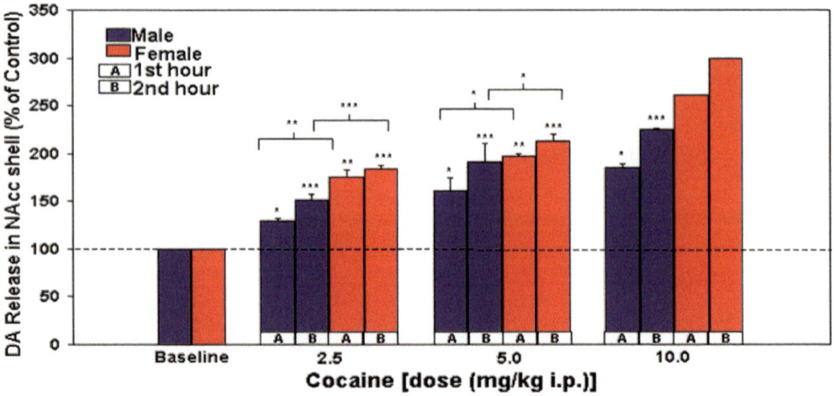

Figure 8.3 Shows in histogram form, synaptic DA release enhancement produced by the psychostimulant, cocaine, *in vivo* and on line. Electroactive signals, selectively imaged for the biogenic amine neurotransmitter, dopamine (DA), are seen in real time as electron transfer redox reactions are sensed in the brain, in the reward centers, videotracked within ventrolateral nucleus accumbens (vlNAcc) as the mammalian males versus females are freely moving about the chamber, unrestrained after intraperitoneal (ip) administration of the alkaloid, cocaine, at specified doses of 2.5 mg/kg, 5.0 mg/kg, and 10.0 mg/kg. The BRODERICK PROBE® nanobiosensor images neurotransmitters selectively within subseconds with excellent spatial and temporal resolution and reliable operational stability. In this example of the nanobiotechnology called, Neuromolecular Imaging (NMI), the neuroprobe is a conductive polymeric carbon/lipid/phosphotidyl matrix smaller than one human hair. Note that the dopaminergic response observed in females [red font] after cocaine was significantly and statistically greater than in males [blue font] (p = 0.001). Linear dose response pharmacokinetics is notable. Interestingly, LIVE IMAGING of dopamine synaptic release from a specific brain reward nucleus, the vlNAcc, is reported for continuous 2 h videotrack and continued enhancement of synaptic dopamine release is readily imaged online and in real time. Data are presented as baseline = 100%. Nonetheless, sexual dimorphism is clear in the natural physiologic state. The term, baseline, signifies that healthy animals are moving about freely, walking, behaving, without drug or medication. Such studies are enabled in the art by the BRODERICK PROBE® nanotechnology and its circuits and designs. The data show that the animals are continued to be videotracked under circumstances of cocaine, that is, during the cocaine-induced drug state. This is the only neurochemical *in vivo*, in real time and online technology that enables the brain of the freely moving healthy subject to be imaged and mapped as well as is the brain of the unhealthy subjects. This is a feat that has not been, to the author's knowledge, previously accomplished.

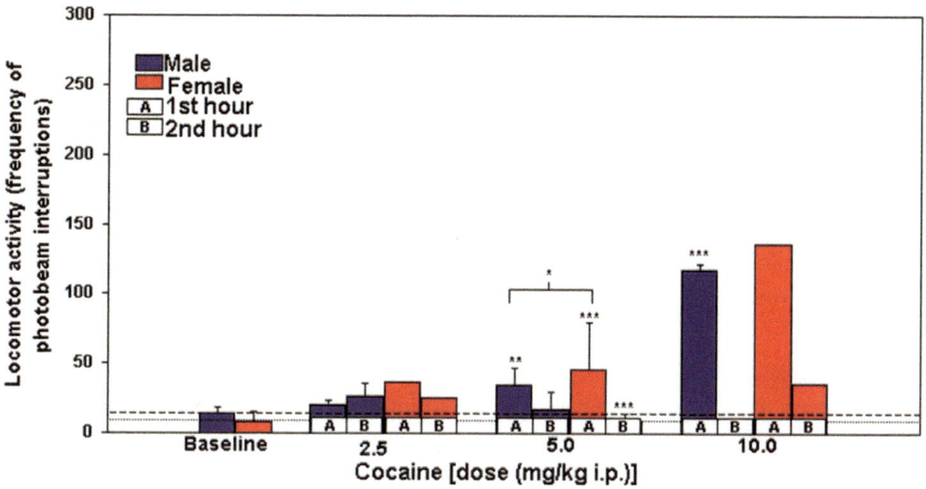

Figure 8.4 Motor effects of cocaine during neuromolecular imaging (shown using the three-dose regimen as described in the experimental design).

Figure 8.4 shows forward walking locomotion as these motor responses take place and are recorded online at exactly the same time during which neuromolecular imaging is taking place. Thus, neurotransmitters are sensed and videotracked while locomotion is monitored with infrared photobeams before and after cocaine and importantly after natural, physiologic NMI and motor responses are imaged. (Photobeams are placed around the chamber in which the animal is freely moving). Therefore, NMI is the unique and breakthrough nanobiotechnology that the BRODERICK PROBE® enables, a feat not previously accomplished. It is important to understand that there is only one set of dose response injections for cocaine pharmacokinetics needed thereby eliminating the need for "hundreds of animals lives lost in vain." Our laboratory is the only laboratory capable of brain LIVE IMAGING during behaviors in addition to brain defect studies on line, in this case, behavior is open-field locomotor activity. We sense the brain, see inside the brain as sensory and motor responses are monitored within subseconds with no lag in either parameter. The data show that females were observed to have a greater behavioral locomotor response to cocaine as compared to males as happened in brain neurochemistry on line. In this animal model of cocaine-induced psychostimulation, we continuously videotrack electroactive signals into the second hour as well and of course, videotracking electrochemical signals for neurotransmitters are imaged online and at the same time as behavior is tagged with photobeams, a most distinct feature of our LIVE IMAGING nanobiotechnology. NMI electroactive signals and motor responses to saline, studied in a separate group of studies, using the same animal subject control, remain insignificant statistically (data not shown here but published frequently). Females show an even more dramatic enhancement of DA release and concomitant motor results. Second-hour data show that DA release increases in the second hour and locomotion decreased in the second hour during online videotracking.

Figure 8.5 shows synaptic DA release in the vlNAcc in males following administration of the alkaloid, caffeine using doses well studied in conventional caffeine doses in the literature. The dose paradigm for caffeine is described in the experiment design and is comprised of 12.5 mg/kg, 25.0 mg/kg, and 50 mg/kg i.p.

As mentioned, it is more than of interest that caffeine exhibits an *"inverted U-shaped"* dose response curve with high dose exhibiting less DA release, concomitant with less behavioral motoric walking responses at the high dose as well. Therefore, dopaminergic synaptic release from brain reward terminals are in synchrony with basal ganglia-driven walking responses. The subjects are walking as the neurotransmitter, dopamine, the brain reward neurotransmitter, is imaged online, in real time and *in vivo*, converging brain with behavior in such a way that that has never been shown previously. This is heuristically seen as a true advance over the prior art.

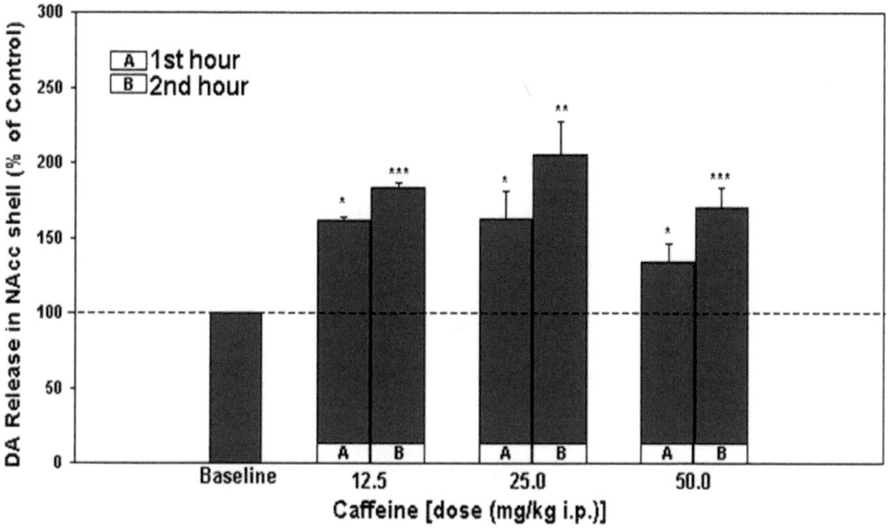

Figure 8.5 Using the same parameters, experimental design and bar codes, the effect of the alkaloid, caffeine is shown in histogram format. Thus, NMI shows caffeine-induced increases in synaptic DA release in NAcc in basal ganglia while stepping up walking at the same time in the same subject. The subset, Figure 8.6 shows that walking is in concert with the neurotransmitter response and the response is fascinating in its U-shaped interesting response curve just as is caffeine's induction of dopamine. Caffeine-induced increase in motor walking and DA release is a cautious "dose-dependent" pharmacokinetic. The author supports the work from Solinas et al. (2002) [55] and Karcz-Kubicha et al. (2003) [23] regarding a general increase in DA release and motor activity albeit the two parameters shown by the previous authors did not show "cause and effect" as the subjects were tested in different paradigms. Moreover, the author confirms the work from the latter teams in that adenosine is certainly involved as caffeine acts *via* adenosine. Adenosine is studied comprehensively by the author's team and by the Solinas and the Karcz-Kubicha teams.

Hypothesis: *"The middle dose effects of caffeine may be the rationale for DA-non-selective caffeine/ adenosine bonds that* ***may*** *clue cocaine seeking."*

Figure 8.7 shows dopamine electron transfer kinetics perpetrated by dopamine and derived from the three- dose regimen of the two-dose stimulant paradigm.

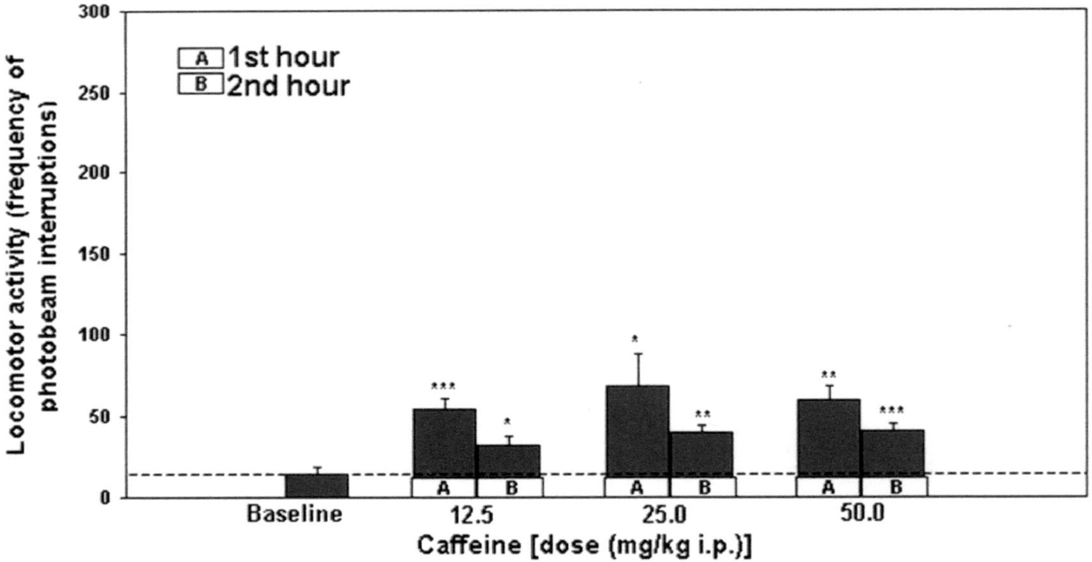

Figure 8.6 Shows motor response in males, online with NMI signaling from NAcc as shown in the previous figure. The classical inverted U-shaped curve for locomotion is seen once again as response diminishes somewhat at the higher dose, given the caveat that the second hour effect does not follow suit. The dose response curve is unusual for caffeine as compared with cocaine as cocaine exhibits a classically linear dose response curve, so appreciated in the field of pharmacology. The caffeine U-shaped data expectedly appear as a significant decrease in DA release at the high dose of caffeine, again so significantly different from the dose response effects of cocaine. The caffeine data shows the highest release of DA and motor activity at the middle dose. Note that we have published low-dose, high-dose, and middle-dose effects for caffeine and cocaine throughout the body of evidence from our laboratory.

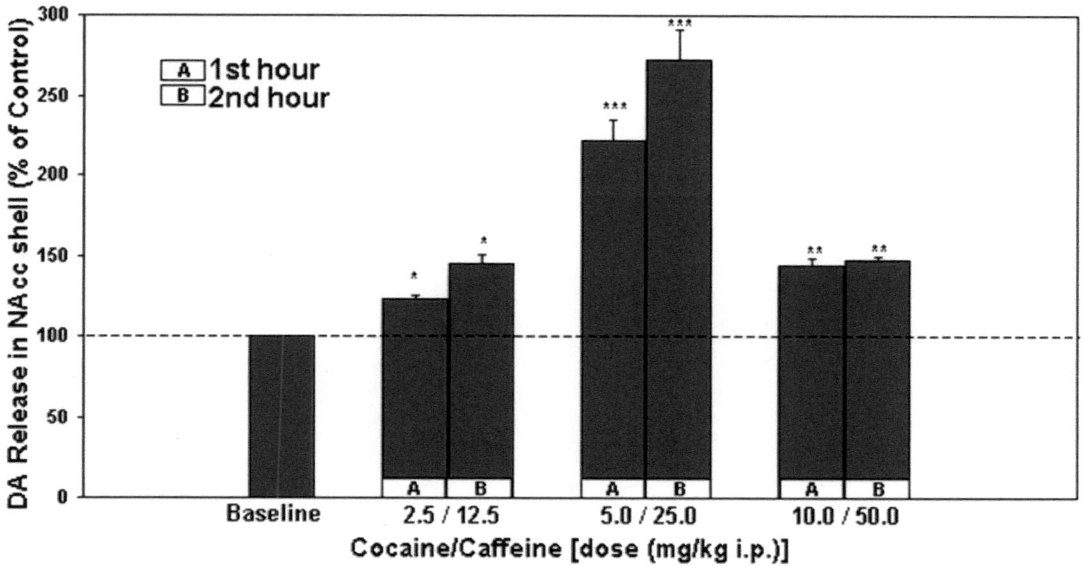

Figure 8.7 When caffeine is administered with cocaine, cocaine's effect on dopaminergic neurotransmission is attenuated. This is in essence, a neuroprotection effect that is provided by caffeine. Cocaine effects on synaptic DA release in NAcc at the low-dose and high-dose regimen for combination caffeine and cocaine administration are neurocehemically protected. Note the U-shaped curve and middle-dose effect. *The subjects are walking*!

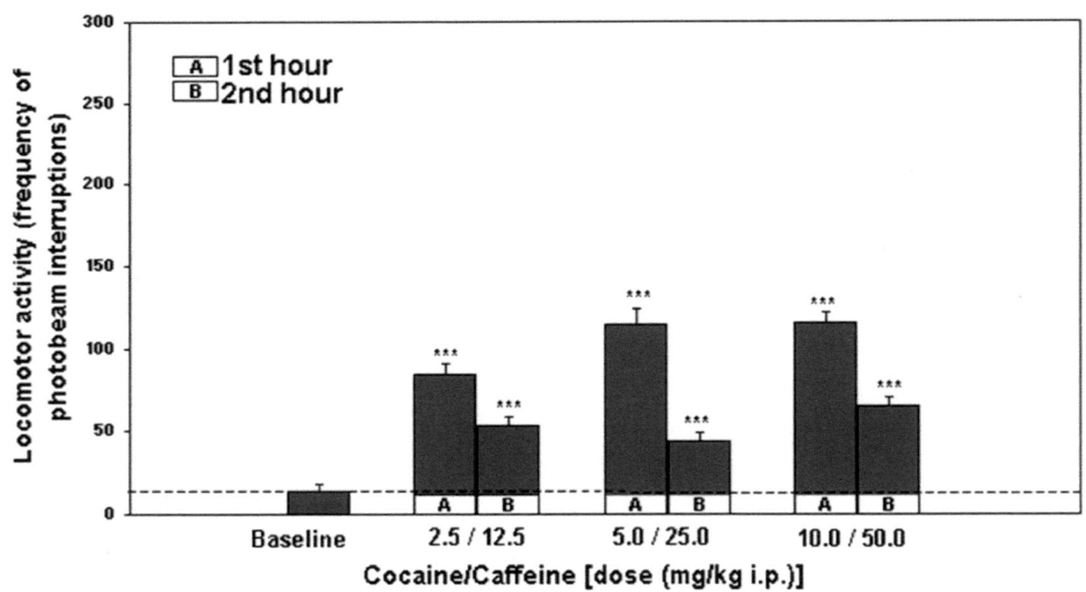

Figure 8.8 Results of the combination, cocaine and caffeine administration at low-, middle-, and high-dose regimens of each stimulant are shown here in histogram form. Here, though, the high dose plateaus with the middle-dose effects. The high dose behavioral effect of cocaine on walking seems compromised or unprotected in the neuronal sense.

Figure 8.9 shows data from males and females after low-dose caffeine and cocaine.

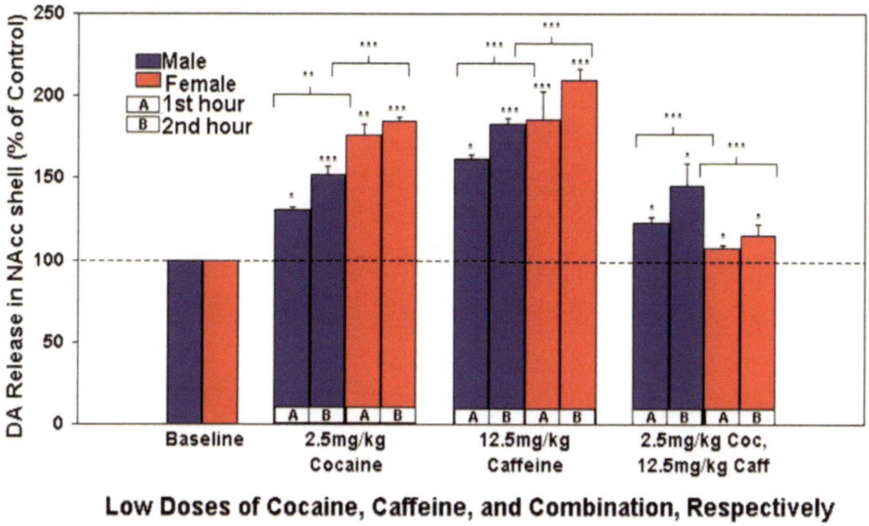

Figure 8.9 Neuroprotection of cocaine effects by caffeine at the low-dose regimen. Females are more sensitive to both alkaloid stimulants than are males. Females exhibit a higher percentage of neuroprotection from cocaine effects than do males. Acute doses of caffeine exhibit greater effects on both genders than does cocaine. Sex-specific dimorphism is seen throughout the studies.

Since neuroprotection from cocaine by caffeine is observed experimentally at low- and high-dose regimens of both alkaloids, and not at the middle dose regimen, we sought to study the middle dose of cocaine further in terms of adenosine receptor activity. Therefore, Figure 8.10 decodes the mechanism of action of caffeine by using cyclopentyltheophylline (CPT), a specific A1 adenosine antagonist. Given the caveat that caffeine is actually a non-specific adenosine antagonist, the data provide interesting clues to "how and when reinforcing behavior becomes an addiction." The data further decode the mechanism of action of cocaine and further decode the mechanism of action on caffeine's neuroprotective effect on the psychostimulant, cocaine.

Figure 8.10 The Neurochemistry of the Middle Dose Regimen. Enter 8-Cyclopentyltheophylline

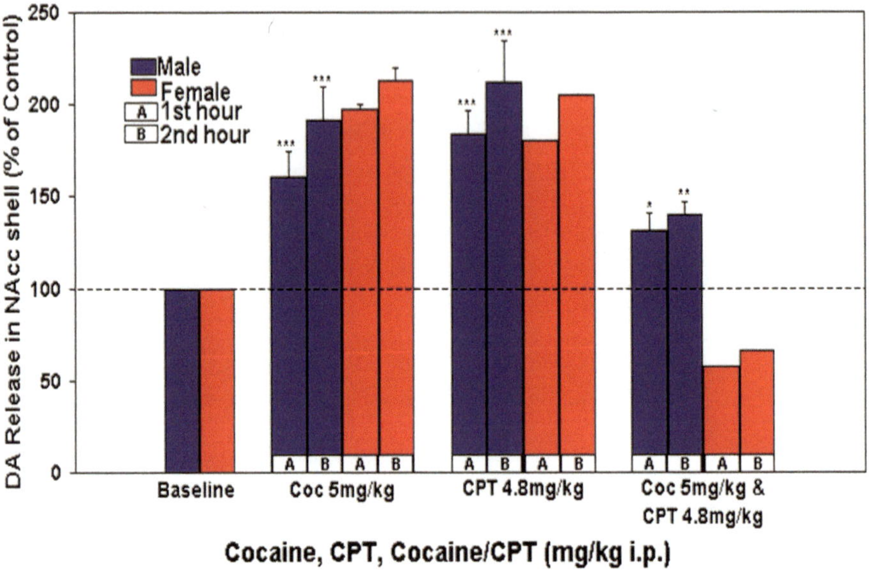

Figure 8.10 The synaptic release of dopamine (DA) by cocaine at the middle dose is compared with cocaine at the middle dose as well as a further comparison with the novel use of 8-cyclopentyltheophylline (8-CPT, a specific A1 adenosine antagonist). The combination of cocaine and 8-CPT compounds were imaged, monitored and the data was interpreted. Deciphering differences between stimulants and psychostimulants is critical to the study of sexual dimorphism. Interestingly, there is the significant implication of the male sensitivity to reinforcing substances seen in this set of histograms. The *specific* A1 antagonist, (8-CPT) shows that males can sometimes be more sensitive to addicting molecules than are females. Subtle nuances in sexual dimorphism are easily detected by the BRODERICK PROBE® and these refinements are essential to the understanding of the complexities of sexual dimorphism.

Cocaine as well as 8-CPT each increased DA synaptic release in NAcc in *simili modo* but males were slightly more so but still significantly affected by 8-CPT than were females. Yet, 8-CPT effects on synaptic DA release in NAcc were less dramatic than those induced by cocaine. These data are amazingly accurate and unexpected because caffeine is seen to frequently cause a greater effect on DA release in the reward accumbens neuronal terminals than does cocaine empirically. This subtlety has always plagued the author and since pursuing the more specific 8-CPT adenosine antagonist in our studies, the clarity of such a phenomenon is delightedly crystal clear. The selective adenosine antagonist, 8-CPT has elucidated the *modus operandi* wonderfully insofar as this author is concerned. Moreover, 8-CPT blocked DA neurotransmission induced by cocaine further confirming the ability of a stimulant to overcome the untoward effects of a psychostimulant and the outcome is dramatic. Caffeine in the form of 8-CPT, so to speak, dramatically blocked DA release induced by cocaine in NAcc neurons. Thus, neuroprotection, under selective adenosine antagonism, may have DA-adenosine receptor mediated neuroprotective properties. Indeed, the observed empirically-derived neuroprotection from a selective adenosine antagonist is drastically and significantly more than twofold greater in females as compared with that seen in males. The existence of female sensitivity to substances of abuse is further confirmed by such LIVE data as that shown herein.

Sexual dimorphism is clear throughout and within each LIVE imaging NMI study. A selective adenosine antagonist has the ability to neuroprotect at the ***middle dose***! When the classical caffeine molecule has been and is seen to save neurons from injury, the high and low doses in the dose regimen shone through to protect neurons from degeneration. Therefore, these enlightening innovative data for the mechanism of "cocaine seeking" *versus* "neuroprotection from cocaine" by caffeine is enabled by NMI LIVE Imaging and the BRODERICK PROBE® at the intersection of adenosine added to the equation. Therefore, the NMI – BRODERICK PROBE® data provide solid evidence that the presence of adenosine is the key factor in solving the puzzle of "the turning point of addiction" since all doses across the board, at least in our considered chosen regimen, are able to keep neurons safe from damage when adenosine-mediation seems waylaid. This is a new look at reinforcement and addiction and a new look at the molecule, adenosine.

Note: 8-CPT also exhibits an inverted "U-shape" dose response curve as does caffeine with peak activity at a lower dose of 4.8 mg/kg equivalent to the caffeine dose regimen.

8.5 Integrating the Data

Studies in our laboratory directly examined synaptic DA release in the NAcc following cocaine, caffeine, and combination of cocaine and caffeine while the subject is walking and behaving. Neuromolecular imaging data were collected non-invasively using the BRODERICK PROBE® nanobiosensor. Cocaine increased DA release linearly with dose, whereas, caffeine exhibited the inverted "U-Shape" curve. Combinations of cocaine and caffeine revealed a neuroprotection of the surge of DA release induced by cocaine. Importantly, this effect was dose-specific and sexually dimorphic specific. Caffeine appeared not enabled to produce such a cocaine-induced DA release at the ***middle doses*** (25 mg/kg and 5 mg/kg, respectively) but was seen at high (50 mg/kg and 10 mg/kg, respectively) and low doses (12.5 mg/kg and 2.5 mg/kg, respectively). Enter 8-cyclopentyltheophylline (8-CPT; 4.8 mg/kg), a selective adenosine A1 antagonist and with such receptor binding, neurons were safe from cocaine (5 mg/kg)-induced DA release! The data support a direct role for adenosine receptor antagonists at the adenosine-1 receptor and in concert with the dopamine, D2 receptor, adenosine is seen to play a neuromodulatory role within the regulating actions of cocaine, especially since cocaine is highly neuromodulated by adenosine as well.

These data point out neuroprotection when a selective A1 antagonist is playing its role against the cocaine response. These data, in fact, in the last figure, are perhaps the most important to understanding the etiologic difference between cocaine seeking with or without caffeine and the phenomenon underlying the caffeine neuroprotection of cocaine. Thinking of adenosine as the well-known break point between waking and sleeping as it is the energy molecule, this begins to teach us the way to discern the ambiguous word, "addiction." Adenosine is the molecule that binds actin with myosin; it is the molecule of life itself with its base in adenine that bridges with thymine in deoxyribose nucleic acid and bridges with uracil in ribose nucleic acid, the DNA and RNA polymers, respectively. Adenosine is a key to unlock the elusive secret of the addicted brain.

Indeed, when H_2O is added to adenosine triphosphate, the phosphate group induces an energetic surge and hydrolysis must bring energy back to the physiologic state. Dopamine DAD1 receptors stimulate cyclic adenosine monophosphate whereas DAD2 receptors inhibit this monophosphate energy molecule. At first glance, the thought came to me that an A1 and DAD2 signaling presynaptic signaling supplied neuroprotection at only the low and high dose regimen. Then, NMI LIVE imaging data with CPT, selective A1 antagonist at adenosine receptors, lends an explanatory note that A1 may not only play a part but may well be the key to blocking cocaine seeking at each dose. Neuroprotection is the molecular contrast to "cocaine seeking" in addition to some of cocaine's many faces such as relapse and recidivism. Perhaps the conceptual and heuristic meaning of "dose," although the bible for the pharmacologist, is not as intriguing as is the trio of sexual dimorphism comprised of dopamine-adenosine-hormonal nuances.

8.5.1 More Important Thoughts

Thus, neuroprotection may be mediated by A1/DAD2 receptor expression, presynaptically in mesocorticolimbic circuitry. Cocaine seeking may be mediated by A1/A2A competition with DAD1 expression. Female sensitivity to drug use and abuse streams into the fetus.

A view of DA-adenosine cocaine exposure in prenatal systems produces direct effects on both the DA and adenosine systems. It attenuates DAD1 receptor function and enhances the DAD2-receptor function thereby tilting the physiological D1-D2 balance in favor of the D2-receptor. Dopamine D2 and adenosine A2A receptors produce antagonistic effects on each other in the embryonic brain. Therefore, the data show that cocaine exposure modulates dopamine and adenosine signaling even in the fetal brain [8, 49)].

8.6 Why Not Look at Caffeine and Its Mediator, Adenosine, in a Different Way?

For this, the following reprint will help...

Citation: Patricia A Broderick. "Caffeine Is A Stimulant but It May Not Be A Psychostimulant." *EC Psychology and Psychiatry 1.1 (2016): 4–13. Reprinted with permission from EC Psychology and Psychiatry.*

According to the *Diagnostic and Statistical Manual of Mental Disorders: Fifth Edition* (DSM-V) [12], there does not appear to be any evidence for caffeine being intimated in substance abuse. In fact, caffeine is not listed in the DSM-V as a substance abuse disorder. Indeed, caffeine is now removed from the Substance Misuse Category of DSM-V, and this provides further evidence for the lack of psychostimulatory action on the part of caffeine since psychosis is comorbid with substance abuse. A partial quote from DSM-V follows:

> **Substance Use Disorder:** In DSM-V, the DSM-IV categories combine with DSM V wherein substance dependence is merged into a single disorder measured on a continuum from mild to severe. Each specific substance (*other than caffeine, which cannot be diagnosed as a substance use disorder*) is addressed as a separate use disorder (e.g., alcohol use disorder, stimulant use disorder, etc.), but nearly all sub-stances are diagnosed based on the same overarching criteria. In this overarching disorder, the criteria have not only been combined, but strengthened. Whereas a diagnosis of substance abuse previously required only one symptom, mild substance use disorder in DSM-5 requires two to three symptoms from a list of eleven. Drug craving will be added to the list and problems with law enforcement will be eliminated because of cultural considerations that make the criteria difficult to apply internationally. In DSM-IV, the distinction between abuse and dependence was based on the concept of abuse as a mild or early phase and dependence as the more severe manifestation. In practice, the abuse criteria were sometimes quite severe. The revised substance use disorder, a single diagnosis, will better match the symptoms that patients experience. Additionally, the diagnosis of dependence caused much confusion. Most people link dependence with "addiction" when in fact dependence can be a normal body response to a substance.

Thus, caffeine, a purine alkaloid, is not one of the stimulants that enables a psychotic condition to actually occur in a subject when the subject is using or being administered "non-excessive" amounts in either the human or murine mammalian species.

Even though this often-debated molecule, called caffeine, has not been empirically or clinically proven to cause psychosis, a few rare cases will be pointed out, *a priori*. These are rare cases in which a caffeine psychosis may or may not have occurred. For example, one case of caffeine psychosis was published in the *New England Journal of Medicine* in October 1936 [33]. In this article, the authors report a case of caffeine psychosis produced by high doses of caffeine, and the effects are described as mental confusion and gastrointestinal diarrhea. The authors go further to describe caffeine intoxication when high doses of caffeine were intravenously administered to the subject and, interestingly, the response to the intravenous dose of caffeine caused psychosis described as "toxic," and further described as "final weakness," "irregularity of the cardiovascular system," a "slight increase in blood pressure," and a "slight blood vessel dilation." Furthermore, the authors describe the human subject who has just been intravenously injected with caffeine as having "clearer ideas," a "more flowing thought process" while fatigue disappears. On the other hand, at lower doses of caffeine that are not described as toxic, the subject is described as having "absolute strength," and the description of the subject includes "peripheral muscle irritability."

Another paper reports that caffeine, when taken at a high dose, is contraindicated in previously diagnosed psychotic patients and this further delineates the importance of interaction with substance abuse and psychosis and further still critically points out the toxicity of very high doses of caffeine, previously emphasized [21]. This was, as reported, a case of "apparent chronic caffeine psychosis" characterized by delusion and paranoia after high caffeine intake. The subject's symptoms resolved in 7 weeks after caffeine intake was lowered, while antipsychotic medication was not medically judged as needed. An article from one other laboratory reports that long-term caffeine use can lead to mild physical dependence, but withdrawal symptoms are unremarkable and these symptoms do not last more than 4 days; neither caffeine use nor a pathological or compulsive form of its use has ever been documented [30]. Thus, the general consensus is that caffeine is not psychoactive and therefore, caffeine does not appear to be a psychostimulant or a substance of abuse as stated in DSM V. The conclusion is clear: The rare exception of toxic symptoms, derived from the caffeine molecule, lies in its excessive and high intake usage.

8.6.1 What Is Adenosine?

The adenosine molecule comprises a purine, as is the molecule caffeine, but the purine in the adenosine molecule is a *nucleotide* attached to a *ribose sugar*, very different indeed. In so doing, the adenosine molecule may be said to spin away from the caffeine molecule, enabling a differential molecular structure and hence, function. Adenosine is ubiquitous in every human cell, and it gives mammalians, for example, a tremendous amount of energy through its monophosphate and triphosphate homeostatic derivatives, (AMP and ATP respectively). When a water molecule is added to ATP, loss of a phosphate group releases high energy; in reverse, the gain of phosphate group from AMP to ATP, energy is released dramatically as well, Adenosine, *per se*, is known as an inhibitory neurotransmitter, it may play a critical role, at least in part, such as is described in *the hypo-function hypothesis of psychosis* [2].

8.6.2 The Adenosine Hypofunction Hypothesis of Psychosis

In fact, there is an adenosine hypothesis of psychosis [9]. In this hypothesis, adenosine is not a psychotic. Rather, in this hypothesis the discussion centers around hypofunction of adenosine, as well as its interaction with other molecules such as glutamate and N-methyl-D-aspartate as these molecules are known to be involved in psychotic processes. Also included in this more recent hypothesis are the biogenic amine neurotransmitters such as dopamine and serotonin. The adenosinergic hypothesis states that the biogenic amine hypothesis of schizophrenia by Schildkraut and Kety, as reviewed by Meltzer and Stahl [35], may not fully answer strategies for the pharmacotherapy of the negative versus the positive symptoms of psychotic behavior from schizophrenia. This recent hypothesis reports that adenosine is a homeostatic, bioenergetics network, a sort of modulator that is able to affect complex circuits synergistically at different levels of receptor-dependent pathways including epigenetics. In this hypothesis of adenosinergic hypofunction, dopamine and glutamate neurotransmitters interact with the adenosine molecule to promise much-needed treatment for negative and positive psychotic polarity.

On the other hand, Brown and Short [9] discuss adenosine as a neuromodulator that has been shown to play a role in reward-related behavior both as an independent mediator and by way of interactions of adenosine receptors with other receptors, since adenosine levels are elevated upon exposure to drugs of abuse. These authors continue on to specify A (2A) receptors as adenosine receptors that are "ideally suited to influence the signaling of neurotransmitters relevant to neuronal responses and neuroplasticity that underlies the development of drug taking and drug-seeking behavior." The rationale for suggesting that A (2A) receptors are excellent targets for treatment of adenosine-induced psychosis from drug abuse is based on findings that show that A(2A) receptors reside in receptor "clusters" with dopamine and glutamate receptors in neuroanatomical regions of brain that are known to be directly related to brain reward and reinforcement behavior. There is an important caveat in this discussion, that is, reinforcement, addiction and psychosis are not the same phenomena.

8.6.3 The Psychosis-Revisited

Let us consider what the condition of psychosis is, so as to realize the consequent detriment to the field of caffeine science when categorizing this caffeine nutrient molecule as a psychostimulant in such a laissez-faire manner. In one view of psychosis, cognitive processes in the cerebral areas of attention and perception are involved [10], whereas the previous definition of psychosis was explained as a breakdown of the ego [1]. Psychosis is currently recognized as cognitive disorganization [31, 32], and one might simply say that psychosis is sensory affect that is undergoing overload. For example, a patient, suffering from psychosis may remark, "too many things are coming into my mind," "the thoughts feel bigger," "the person I am talking to looks bigger." Therefore, one's ability to keep sensory information organized in channels is imperative for control of central nervous system functioning.

8.7 Brain Reward, Pleasure Centers, Reinforcement

The brain reward hypothesis—that is, stimulating specific central nervous system pathways for neurotransmitters and neuromodulators to connect and interact in order to make one "feel good" started with the work of Dr. James Olds at McGill University. Olds et al. [44] electrically stimulated specific neuroanatomic substrates in the brain *in vivo*. Mammalian responses to this neuronal trigger of stimulation demonstrated that pleasure/reward/reinforcement was experienced; the pleasure was reflected in remarkable positive behavioral reactions to the stimuli [43, 44]. This pleasure/brain reward/ reinforcement hypothesis has reached consensus worldwide. The hypothesis of brain reward by Olds has been empirically, continually, and strongly supported. Demonstrations of such have been empirically shown repeatedly by notable pioneers in the brain reward field [25] and those advancing the field by using molecular heteromers technologies [15] and neuromolecular nanobiotechnologies [8].

Now, let us look at the molecular structures to see the differences between drugs that cause psychosis and those do not (Figure 8.11).

These studies may have placed caffeine and adenosine in the realm of brain reward because mesocorticolimbic neuronal dopaminergic circuitry in somatodendrites, ventral tegmentum and nerve terminals, nucleus accumbens along with associated areas such as amygdala and hippocampal somatodendrites, locus coeruleus are likely influenced by caffeine and adenosine and/or adenosine receptors. Channeling sensory auditory signaling through the pons nuclei via the cochlear nucleus is affected by caffeine and adenosine [28]. Indeed, caffeine and adenosine can be reinforcers, but the road from reinforcement to addiction is winding with elusive complexities since in addition to the neuronal sensory circuitry of reward, there are the higher pleasures, directly related to the brain's default neuronal circuits. These circuits existentially enable faculties that the non-prescient mammalian brain cannot likely fathom. These are, for example, cognition and a life of meaning and happiness. The non-hedonic brain with or without positive or negative bias toward the future, may exist as the perennially unhappy brain, the brain that cannot feel pleasure; it is the brain that fast forwards chronic neuropathologies, affective disorders such as depression and addiction, and unyielding pain, peripherally and centrally.

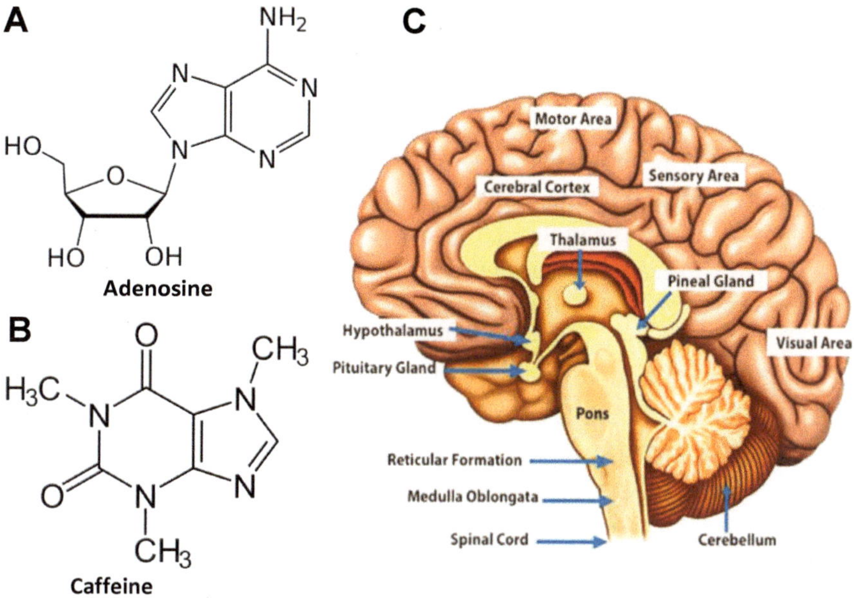

Figure 8.11 (A) The adenosine molecule. Note that this purine *nucleotide* comprises an adenine base which is bonded to the ribose sugar (see the molecular structure on the left side of the molecule); the connection is between the nitrogen on the adenine base and the double bonded oxygen on the first carbon of the ribose sugar. (B) The caffeine molecule, a purine *alkaloid*, is depicted. Caffeine is known as a nutrient substance, a food source. (C) Schematic depiction of the mesocorticolimbic neuronal circuitry and related neuroanatomic substrates involved in brain reward/reinforcement mechanisms as well as in the neuropathology of addiction.

8.8 A Most Notable Difference in the Molecular Structure of Adenosine versus Caffeine

Adenosine is a purine nucleotide; it comprises an adenine base bonded with a ribose sugar, a nucleotide, making the molecular structure of adenosine, a nucleoside. The adenine base in adenosine is bonded to the ribose sugar; the connection is between the nitrogen on the adenine base and the double bonded oxygen on the first carbon of the ribose sugar by an intermolecular dehydration. It is the oxygen bonded to the second carbon of the ribose sugar that distinguishes between the ribose in ribose nucleic acid also known as ribonucleic acid (RNA) and the deoxyribose in deoxyribose nucleic acid, also known as deoxyribosenucleic acid (DNA). It is beyond the scope of this chapter to discuss whether or not a tropane, tropine or tropane-like molecular structure is or is not related to the molecular structure of adenosine.

The author has studied the differences in the distinctive properties of these nitrogenous compounds, the alkaloids, and the author, through the literature and experimental derived data, is led to believe that the study of these differences may shed some light on the reinforcement path to addiction. It is simply suggested that the difference in the molecular structures between caffeine and cocaine, in particular, the cocaine **tropane** alkaloid, further elucidates the tropane ring as a possible key to open a door to reveal reinforcement mechanisms versus the mechanisms for the neuropathology of addiction compared with psychosis. Indeed, there are known dual diagnostic therapies for psychosis and addiction. This interrogative perspective, written to enlighten and appreciate some degrees of reward experienced in our daily lives, addresses only the cocaine tropane alkaloid as having such dual properties of psychosis and addiction whether or not these present in the patient separately or combined.

Figure 8.12 shows the structure of tropane.

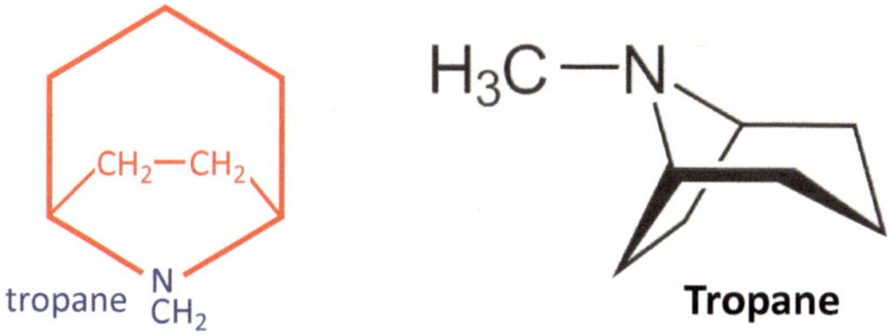

Figure 8.12 The tropane ring, in two structural views, a fascinating, naturally occur-ring bicyclic chemical ring system that lends itself to many pharmaceutical and therapeutic applications. The role of the tropane skeleton in drug research has been pioneered by Laszlo Gyermek, MD, PhD. Professor of Anesthesiology (*in memoriam, 2017*), University of California at Los Angeles, Harbor UCLA Campus, and Los Angeles Biomedical Research Institute, Torrance, CA. The bicyclic ring system of tropane can be construed as a condensation product of a piperidine and pyrrolidine ring with a shared N atom. Thus, the exact chemical name of tropane is 8-azabicyclo[3.2.1]octan-3-one. This name also characterizes the manner in which this bicyclic ring system is branched; the branch has distinct pharmacological significance [19].

8.8.1 Cocaine Is a Psychostimulant and Is Known to Cause Psychosis

In fact, reports of cocaine induced psychotic patients are happening at an alarming rate. The first cases were reported in abundance in the 1990s, and presently, there is still no scarcity of clinical cases of cocaine-induced psychosis [22, 27]. Cocaine-induced psychosis is so prevalent that is has its own abbreviation, CIP [50]. In the Medscape report [22], the authors use the DSM-IV-TR to describe 10 more cases of CIP, and these clinicians state that cocaine psychoses cause hallucinations.

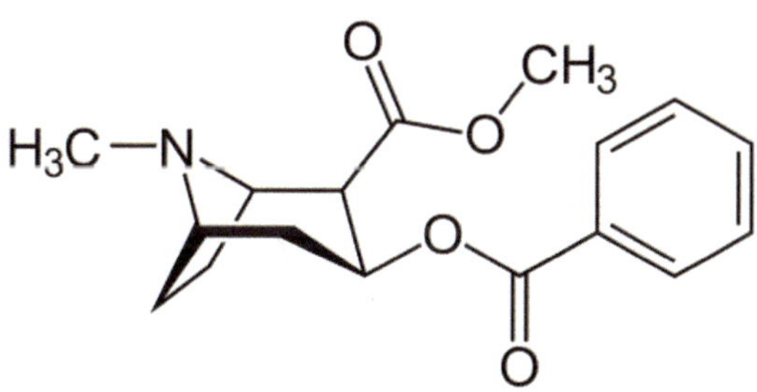

Figure 8.13 The cocaine molecule. Note the tropane ring in bold. Cocaine is a Schedule II drug listed in the Federal and New York State Drug Enforcement Agency Codes.

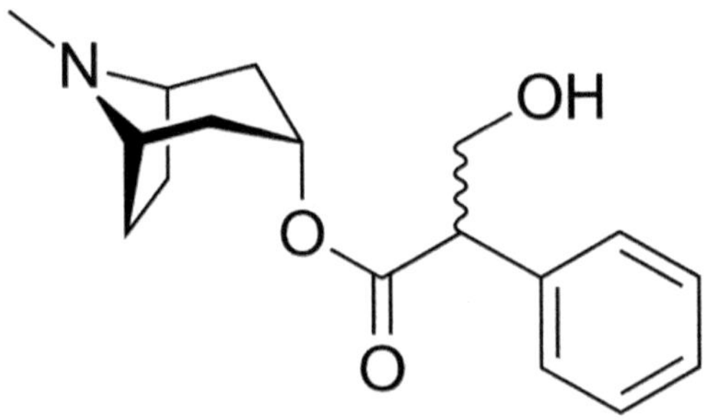

Figure 8.14 The atropine molecule. Note the tropane ring in bold. Atropine, although not a schedule II drug listed in the Federal and New York State Drug Enforcement Agency Codes, is reported to be a "possibly addicting drug." Atropine addiction occurs especially if it is used with another drug that has inherent addicting properties.

A thorough history pertaining to the type of symptoms experienced by the patient and the timing of these symptoms in association with cocaine abuse is necessary to make each diagnosis. Other cocaine-induced medical problems may be present and/or coexistent with any of these ten cocaine-induced psychiatric conditions. For example, a patient may present with specific clinical criteria that lead to the diagnosis of a cocaine-induced psychotic disorder with hallucinations. That same patient also may report chest pain, a symptom that could be associated with cocaine-induced acute coronary syndrome, pneumothorax, or pulmonary edema. The report continues on to cite CIP dysfunctions as intoxication, withdrawal, intoxication delirium, CIP with delirium, CIP with hallucinations, CIP mood disorder, CIP anxiety disorder, CIP sexual dysfunction, and CIP sleep disorder.

The broad category of cocaine related disorders can be subdivided into cocaine use and cocaine disorders, for example delusions (hanging on to beliefs that are clearly nonexistent) and hallucinations (hearing and seeing things) (14). A clinical case of cocaine psychosis presenting with hallucinations and delusions is reported (42). This article reports a patient in the emergency department who had binged on crack cocaine. The patient heard "voices in his head who were telling him to kill himself and the patient continued to believe there was a plot against him." These auditory hallucinations have been given the definition "strange perception of sounds" (57).

8.8.1.1 Cocaine causes stimulant psychosis

Stimulant psychosis is a psychotic disorder that occurs in some people who use stimulant drugs. Stimulant psychosis commonly occurs in people who abuse stimulants, but it also occurs in some patients taking therapeutic doses of stimulant drugs under medical supervision [11]. The most common causative agents are substituted amphetamines and cocaine [13].

8.8.1.2 Cocaine is a tropane alkaloid

Cocaine is a tropane alkaloid; it has a tropane ring in its molecular structure. The structure activity relationship within cocaine and caffeine molecules is dramatically different. It is true that cocaine and caffeine are alkaloids, but cocaine is a tropane alkaloid and caffeine is a purine alkaloid. Both alkaloids are molecular compounds derived from nitrogenous metabolism in plants, and it is true that both molecules are reinforcing; it is known that both molecules bring about a feeling of *joie de vivre*. However, cocaine is a psychostimulant. Caffeine, in contrast, may not be a psychostimulant albeit it is a stimulant known throughout the ages. To repeat, psychosis and addiction are not the same; in the next section, only addiction and the cocaine alkaloid are discussed.

8.8.1.3 Addiction and the cocaine alkaloid

In the science of pharmacology, cocaine is listed in formula comprised of a tropane ring and is, moreover, the only one that has been shown to cause addiction. Thus, the correlation between tropane and addiction may be, as pointed out before, a simple key to unveiling the complexities of addiction. Indeed, empirical studies have reported that molecules with the tropane ring are primarily the ones that show the ability to cause what we might call addiction again, given the caveat that the neuropathology of addiction remains to be solved. The suggestion in the literature, though, is that the tropane ring may well be the addicting part of any molecule, and this fact, then, can lead us to insightfulness in each intricate step that brings one from hedonia/reinforcement/pleasure/reward to an addiction/hedonia/neuropathological state. There is a class of drugs called "local anesthetics." Most interestingly, cocaine is the only drug in this class of local anesthetics that has a chemical tropane ring.

8.8.1.4 Cocaine and atropine

Cocaine is an ester of benzoic acid and methyl ecgonine; ecgonine is a base, an amino acid alcohol closely related to tropine, the amino alcohol of atropine. Atropine is sometimes called "possibly addicting." This is a surprising concept in the conventional sense of pharmacology. One would not expect atropine, a belladonna alkaloid, to possess addicting properties because it is an important physiologic anticholinergic in the autonomic nervous system, and it is not generally reported as psychogenic. Atropine is possibly addicting [18, 41] because the atropine molecule has a molecular structure that is comprised of the tropane ring. The similarity in the chemical formulas in atropine and cocaine molecules again emphasizes the critical role of the tropane ring in the neuropathology of addiction. Molecular structures for cocaine and atropine are shown in Figures 8.13 and 8.14, parts of which have been previously described with data from this laboratory [5].

Then, to emphasize the role of cocaine as a local anesthetic that is addicting, cocaine does have the same fundamental structure as the other classical synthetic local anesthetics, but the dramatic difference in the molecular structure of cocaine versus all the other local anesthetics lies in the cocaine tropane ring. Indeed, it may be true that the tropane ring may be the source of cocaine's ability to lead the path from controlled reinforcement, enjoyment, to the unfortunate maladies associated with addiction pathology. It is important to note that each molecular structure of each of the other local anesthetics are not comprised of the tropane ring and are not known to be addicting.

Finally, it is the dopamine transporter studies that have encouraged intense thinking and writing about the role of the have published the critical nature of the tropane ring in addiction [18, 24]. The dopamine transporter is believed to perform the mechanism by which dopamine remains in the synapse. The work of the scientists in the field of molecular pharmacology has experimentally shown the correlations among cocaine, tropane, and addiction [24, 45].

It may be the work in the field of the dopamine transporter and molecular pharmacology that reinforces the scientific belief that the tropane ring highly influences the addiction properties of cocaine and, indeed, in at least some of the processes involved in the mechanisms of addicting drugs.

8.9 Take Home Messages

Caffeine is a stimulant; it produces reinforcement/brain reward, but it may not be a psychostimulant because it is not psychoactive given the caveat that all stimulants may not produce psychoses *per se.* Caffeine is not psychoactive because it has not been shown to produce psychosis. The molecular structure of caffeine is not comprised of a tropane ring; it is a purine alkaloid. Caffeine blocks cocaine-induced dysfunctional neurochemistry and behavior via adenosine receptor antagonists; the responses are based on sex differences [29]. Caffeine has also been shown to restore to normal estrus cycle changes induced by cocaine [6]. Caffeine is reported to be medically beneficial in the treatment of Parkinson's disease [17].

Adenosine and adenosinergic receptor interactions and derivatives may reinforce either alone or through their adenosine derivatives. Adenosine is a purine alkaloid, comprised of an adenine base and ribose sugar; its molecular structure is dramatically different from that of caffeine. Caffeine/adenosine agonist and antagonist interactions and DA-adenosine antagonists provide the etiology of the cocaine seeking and neuroprotective actions of cocaine. The nitrogenous bases of the adenosine molecule are the chemical operative key to the deoxyribose nucleic acid and the ribose nucleic acid DNA and RNA polymers. Adenosine is medically beneficial in many instances, for example in cardiovascular disease as a vasodilator [54].

Note: Derivatives of adenosine comprise tropane rings [34].

Cocaine is a psychostimulant; it is psychoactive because it has been shown to be causative in psychosis. Cocaine is a tropane alkaloid. Cocaine produces reinforcement/brain reward, and it has been shown to do so to the extreme, to be causative in addiction neuropathology. Cocaine disrupts the sensory processing in the auditory neuronal pathway through the cochlear nucleus, and the effects are sex differentiated (5). Cocaine has been a controlled substance under the United States Federal Drug Enforcement Code since 1970. Cocaine is medically beneficial in pharmacology as a local anesthetic. It is prescribed only in limited cases (18).

Atropine is a tropane alkaloid, and it is a belladonna alkaloid; it dilates the pupils and is anecdotally referred to as the "beautiful woman" alkaloid. Atropine is possibly addicting likely due to its tropane molecular structure. Atropine has a wide spectrum of use clinically ranging from influencing gastrointestinal tract function to use in ophthalmic preparation by paralyzing the iris sphincter to cause cycloplegia, a necessary paralysis used successfully in everyday cataract eye surgeries [18].

To conclude, then, it is submitted that caffeine may not be a psychostimulant and cocaine is a psychostimulant and may be causative in psychosis. Adenosine-dopamine interactions during cocaine and caffeine interactions are critical in delineating the breaking point between wakefulness and sleep between cocaine seeking and mesocortical limbic neuronal inhibitory activity, nigrostriatal circuitry and neurochemical and behavioral properties of adenosine as neuromodulator of caffeine. The role of selective and non-selective adenosine antagonists in their expression of functional A1/A2A and DAD1/DAD2 receptor affinities in delineating the path from reinforcement to addiction cannot be underestimated.

Thus, clinical psychogenesis reports can be relied on to decipher stimulant psychosis. Furthermore, an in-depth study of molecular structures of stimulants and other reinforcers will provide strategic discoveries in the drug addiction field. Clues from stimulant molecular structures can assist in the analyses of unique mechanisms and subsequent pharmacotherapeutic.

The adenosine-dopamine-estrogen trio cannot be underestimated in this trio's role in the phenomenon of the sexual dimorphism of caffeine! Chapter 9 depicts the role of adenosine and dopamine in the estrous cycle.

References

1. Bleuler M. (1954) The concept of schizophrenia. *Am J Psychiatry* 5: 382–383.
2. Boison D, Singer P, Shen HY, Feldon J, Yee BK (2012) Adenosine hypothesis of schizophrenia—opportunities for pharmacotherapy. *Neuropharmacology* 62: 1527–1543.
3. Broderick PA (1991) *In Vivo* Voltammetric Studies on the release mechanisms for cocaine with gamma butyrolactone. *Pharmacology Biochemistry and Behavior.* 40: 969–975.
4. Broderick PA, Phelix, C F (1997). I. Serotonin (5-HT) within dopamine reward circuits signals open-field behavior: II. Basis for 5-HT-DA interaction in cocaine dysfunctional behavior. *Neuroscience and Biobehavioral Reviews,* 21(3), 227–260.
5. Broderick PA, Rosenbaum T (2013) Sex specific brain deficits in auditory processing in an animal model of cocaine-related schizophrenic disorders. *Brain Sciences* 3: 504–520.
6. Broderick PA, Malave LB (2014) Cocaine shifts the estrus cycle out of phase and caffeine restores it. *J Caff Res* 4: 109–113.
7. Broderick PA (2016) Caffeine Is A Stimulant But It May Not Be A Psychostimulant. *EC Psychology and Psychiatry* (1.1) 4–13.
8. Broderick PA (2016) Cocaine and Neuromolecular Imaging of Neurotransmitters in the Brain: BRODERICK PROBE® Laurate Nanobiosensors in Mesocorticolimbic Neurons and the Nucleus Accumbens. In: Preedy VR (ed). The Neuropathology of Drug Addiction and Substance Misuse. Amsterdam, The Netherlands. pp. 74–85.
9. Brown RMI, Short JL (2008) Adenosine A(2A) receptors and their role in drug addiction. *J Pharm Pharmacol* 60: 1409–1430.
10. Chapman T, Freeman T, McGhie A (1959) Clinical research in schizophrenia; the psychotherapeutic approach. *Br J Med Psychol* 32: 75–85.
11. Curran C, Byrappa N, McBride A. (2004) Stimulant psychosis: systematic review. *Br J Psychiatry* 185: 196–204.
12. Diagnostic and Statistical Manual for Mental Disorders, fifth ed. Arlington, VA: American Psychiatric Association, 2013.
13. Elliott A, Mahmood T, Smalligan RD (2012) Cocaine bugs: a case report of cocaine-induced delusions of parasitosis. *Am J Addict* 21: 180–181.

14. Encyclopedia of Mental Disorders. Cocaine and related disorders. Available at www.minddisorders.com/Br-Del/Cocaine-and-related-disorders.html. Accessed August 2, 2015.

15. Ferré S, Baler R, Bouvier M, et al. (2009) Building a new conceptual framework for receptor heteromers. *Nat Chem Biol* 5: 131–134.

16. Fredholm BB, Battig K, Holmen J, Nehlig A, Zvartou EE (1999) Actions of caffeine in the brain with special reference to factors that contribute to its widespread use. *Pharmacol Rev* 51(1): 83–133.

17. Golembiowska KI, Wardas J, Noworyta-Soko1owska K, Kamińska K, Górska A (2013) Effects of adenosine receptor antagonists on the *in vivo* LPS-induced inflammation model of Parkinson's disease. *Neurotox Res* 24: 29–40.

18. Goodman LS, Gilman AG, Limbird LE, Milinoff PB, Ruddon RW. Goodman and Gilman's: The Pharmacological Basis of Therapeutics, ninth ed. New York: McGraw-Hill, 1995.

19. Gyermek L The role of the tropane skeleton in drug research. URL: inhn.org/fileadmin/archives_new/Gyermek/Gyermek_TheRoleof Tropane Skeleton.pdf.

20. Handbook of Neurotoxicology, Broderick PA Microsensors Detect Neuroadaptation by Cocaine. chapter 13; volume 2, Massaro EJ, Broderick, PA, Mattsson, JL, Schardein, JL. Schlaepfer, TE (eds.) Springer Science and Business Media, March 20, 2002.

21. Hodges DW, Woon FL, Hoopes SP (2009) Caffeine-induced psychosis. *CNS Spectr* 14: 127–129.

22. Holstege CP, Bienenfeld D (2013) Cocaine-related psychiatric disorders: clinical presentation. *Medscape*, October 17.

23. Karcz-Kubicha M, Antoniouk K, Terasmaa A, Quarta D, Solinas M, Justinova Z, Pezzola A, Reggio R, Muller CE, Fuxe K, Goldberg SR, Popoli P, Ferre' S (2003) Involvement of adenosine A1 and A2A receptors in the motor effects of caffeine after its acute and chronic administration. *Neuropsychopharmacology* 28(7): 1281–1291.

24. Kopaitic TA, Lau Y, Surratt CK, Donavan DM, Newman AH, Katz JL (2010) Dopamine transporter-dependent and -independent striatal binding of the benztropine analog JHW 007, a cocaine antagonist with low abuse liability. *J Pharmacol Exp Ther* 335: 703–714.

25. Kringelbach ML, Berridge KC (2010) The functional Neuroanatomy of pleasure and happiness. *Discov Med* 9: 579–587.

26. Leard-Hansson J, Guttmacher L (2006) Cocaine-Induced Psychosis and Schizophrenia. *Clin Psychiat News* vol. 34, No.2.

27. Lysaker P, Bell M, Beam-Goulet J, Milstein R (1994) Relationship of positive and negative symptoms to cocaine abuse in schizophrenia. *J Nerv Ment Dis* 182: 109–112.

28. Malave LB. Caffeine's attenuation of cocaine-induced deficiency in acoustic startle response by inhibition of adenosine in a sex and dose dependent manner. [M.S. thesis]. The City College of New York, The City University of New York, NY, 2014.

29. Malave L, Broderick PA (2014) Caffeine's attenuation of cocaine-induced dopamine release. *J Caff Res* 4: 1–6.

30. Malenka RC, Nestler EJ, Hyman SE. Reinforcement and addictive disorders. In: Sydor A, Brown RY (eds). Molecular Neuropharmacology: A Foundation for Clinical Neuroscience, second ed. New York: McGraw-Hill Medical, 2009.

31. McGhie A, Chapman J (1961) Disorders of attention and perception in early schizophrenia. *Br J Med* Psychol 34: 103–116.

32. McGhie A, Chapman J, Lawson JS (1964) Disturbances in selective attention in schizophrenia. *Proc R Soc Med* 57: 419–422.

33. McManamy MC, Schube PG (1936) Caffeine intoxication report of a case the symptoms of which amounted to a psychosis. *N Engl J Med* 215: 616–620.

34. Medical tropane alkaloids Tropanes are derived from adenosine triphosphate. URL: https://www.meriam-webster.com/medical/tropane%20alkaloid.

35. Meltzer HY, Stahl SM (1976) The dopamine hypothesis of schizophrenia: a review. *Schizophr Bull* 2: 19–76.

36. Mendoza R, Miller BL, Mena I (1992) Emergency room evaluation of cocaine-associated neuropsychiatric disorders. *Recent Dev Alcohol* 10: 73–87.

37. Miller BL, Mena I, Giombetti R, Villanueve-Meyer J, Djenderdjian AH (1992) Neuropsychiatric effects of cocaine: *SPECT measurements. J Addict Dis* 11: 47–58.

38. Mitchell J, Vierkant AD (1991) Delusions and hallucinations of cocaine abusers and paranoid schizophrenics: a comparative study. *J Psychol* 125: 301–310.

39. Morton WA (1999) Cocaine and psychiatric symptoms. Prim Care Companion *J Clin Psychiatry* 1: 109–113.

40. Nambudin DE, Young RC (1991) A case of late-onset crack dependence and subsequent psychosis in the elderly. *J Subst Abuse Treat* 8: 253–255.

41. New World Encyclopedia. Atropine. Available at www.newworldencyclopedia.org/entry/Atropine. Accessed on December 28, 2012.

42. Nunes JV, Broderick PA (2007) Novel research translates to clinical cases of schizophrenic and cocaine psychosis. *Neuropsychiatry Dis Treat* 3: 475–485.

43. Olds J (1956) Pleasure centers in the brain. *Sci Am* 195: 105–116.

44. Olds J, Milner P (1954) Positive reinforcement produced by electrical stimulation of the septal area and other regions of rat brain. *J Comp Physiol Psychol* 47: 419–427.

45. Parnas ML, Gaffaney JD, Zou MF, Lever JR, Newman AH, Vaughn RA (2008) Labeling of dopamine transporter transmembrane domain 1 with the tropane ligand N-[4- Azido-3-[125I]iodophenyl)butyl]-2b-carbomethoxy-3b-(4-chlorophenyl)tropane implicates proximity of cocaine and substrate active sites. *Mol Pharmacol* 73: 1141–1150.

46. Pellegrino LJ, Pellegrino AS, Cushman AJ (1979) A stereotaxic atlas of the rat brain. Second edition. (Plenum, New York).

47. Phelix CF, Broderick PA (1995) Light microscopic immunocytochemical evidence of converging serotonin and dopamine terminals in ventrolateral nucleus accumbens. *Brain Research Bulletin* 37(1): 37–40.

48. Phelix CF, Rodriguez VR, Rodriguez JS, Broderick PA (2001) Comparative morphometric analysis of serotonin axons in the nucleus accumbens subterritories of albino rat, *Rattus Norvegicus*, and white-footed mouse, *Peromycus Leucopus. Soc. Neurosci. Abstr*, 1927.

49. Regina C, Kubrusly C, Bhidew PG (2010) Cocaine exposure modulates dopamine and adenosine signaling in the fetal brain. *Neuropharmacology* 58(2): 436–443.

50. Roncera C, Dangre C, Gonzalvo B, et al. (2013) Risk factors for cocaine-induced psychosis in cocaine-dependent patients. *Eur Psychiatry* 28: 141–146.

51. Rosenthal RN, Miner CR (1997) Differential diagnosis of substance-induced psychosis and schizophrenia in patients with substance use disorders. *Schizophr Bull* 23: 187–193.

52. Rosse RB, Collins JP Jr, Fay-McCarthy M, Alim TM, Wyatt RJ, Deutsch SI (1994) Phenomenologic comparison of the idiopathic psychosis of schizophrenia and drug induced cocaine and phencyclidine psychoses: a retrospective study. *Clin Neuropharmacol* 17: 359–369.

53. Satel SL, Edell WS (1991) Cocaine-induced paranoia and psychosis proneness. *Am J Psychiatry* 148: 1708–1171.
54. Sato A (2005) Mechanism of vasodilation to adenosine in coronary arterioles from patients with heart disease. *Am J Physiol Heart Circ Physiol* 288:H1633–1640.
55. Solinas M, Ferre' S, You ZB, Karcz-Kubicha M, Popoli P, Goldberg SR (2002) Caffeine induces dopamine and glutamate release in the shell of the nucleus accumbens. *J Neurosci* 22(15): 6321–6324.
56. Tanner JM (2010) Human Development. Encyclopedia Britannica Inc. URL: http://www.britannica.com/science/human-development/Development-of-the-reproductive-organs-and-secondary-sex-characteristics.
57. Waters F (2010) Auditory hallucinations in psychiatric illness. *Psychiatr Times* 27: 3.

Chapter 9

Sex Specific Intricacies in Brain Reward: Adenosine Streams Estrus

9.1 Sexual Dimorphism Steps Forward from the Past: Neuronal Rewiring

It is not surprising that our past experiences shape the present and the future. But what is most relevant to Neuromolecular LIVE Imaging (NMI) within the monoaminergic neural circuitry is the work done by Bayer [1] which is written in the quote below, for the readers' interest. Within this quote, you will read that the authors discuss the rewiring of monoaminergic circuits during our past lives; the studies were done by pruning synaptic connectivity. Indeed, we show herein in this chapter, that this rewiring of neural circuits in echoic/auditory and mesocorticolimbic brain reward, happens. We can see it happening as it is imaged, LIVE, online and in vivo in mammalian animal models, freely moving about in their own element, their own environtome. Too, in the process of rewiring neuronal monoaminergic circuits, we are also either subconsciously or consciously rewiring its effects on endocrinology in the form of hormones within the sexes. Thus, males will always play a part as in pruning synaptic connections. But since females are generally more sensitive than males to drugs including medications, sexual dimorphism steps forward from the past by way of what is called, the female "estrous cycle."

Neuroimaging: Sensing Biochemistry in the Brain
Patricia A. Broderick

ISBN 978-981-4613-49-1 (Hardcover), 978-0-429-02974-5 (eBook)
www.jennystanford.com

"Past experience shapes sexually dimorphic neuronal wiring through monoaminergic signaling. Differences between female and male brains exist across the animal kingdom and extend from molecular to anatomical features. In the nematode *Caenorhabditis elegans*, sexual differentiation entails the sex-specific pruning of synaptic connections between neurons that are shared by both sexes, (then) giving rise to sexually dimorphic circuits in adult animals. Bayer discovered that starvation during juvenile stages is memorized in males to suppress the emergence of sexually dimorphic synaptic connectivity. These circuit changes result in increased chemosensory responsiveness in adult males following juvenile starvation. Bayer found that an octopamine-mediated starvation signal dampens the production of serotonin (5-HT) to convey the memory of starvation. Serotonin production is monitored by a 5-HT1A serotonin receptor homologue that acts cell-autonomously to promote the pruning of sexually dimorphic synaptic connectivity under well-fed conditions" [1].

9.1.1 Neuronal Rewiring: Neurochemical Sexual Dimorphism is Linked with Adenosine

It is postulated from sound evidence that brain and behavior sexually dimorphistic neurotransmission circuit changes are continually shaped online by addictive/neuroprotective brain circuits that are well known in the neuroscience community to involve a "brake" in the cocaine-seeking circuit mediated by A1/A2A competition co-expressed with DAD_2 receptor expression. The empirical LIVE data from the BRODERICK PROBE® sensing device shows that cocaine-seeking comes about at the well-designed low and high doses of cocaine/caffeine combination studies and cutting-edge discoveries with the BRODERICK PROBE® further show, as the subject walks freely, that cocaine seeking can derive from a combination of these two stimulants at any dose or circumstance where the inherent adenosine "brake" is hindered by adenosine/DA co-expression receptor imbalance when the adenosine is no longer in control!

These further illuminating data were seen online when using the selective adenosine antagonist, 8-CPT. Therefore, LIVE neurotransmitter reflection of protection from cocaine-seeking by caffeine highly supports the scientific constructs that expression of adenosine may well be the key to seek cocaine or not to seek cocaine and/or other types of implacable reinforcement. Then, adenosine may be the link for caffeine to neuroprotect or not to neuroprotect. The word "neuromodulation" is critical. Adenosine neuromodulates; the mighty molecule can put its foot on the brake pedal to signal reinforcement forward and/or it can make the brake pedal come to a stop, signaling reinforcement to be held at the proverbial traffic light, even perhaps cross signally clandestinely to the addictive process, unannounced to the purveyor.

9.1.2 Caffeine Acts at Adenosine Receptors

Caffeine is a nonselective adenosine inhibitor and adenosine itself is an inhibitory neurotransmitter. Thus, the author's team report an enhancement of synaptic release of the neurotransmitter dopamine associated with caffeine and with adenosine as well in brain reward nerve terminals such as the nucleus accumbens. These data are empirically shown here with LIVE Imaging from within the mesocorticolimbic brain of the conscious and freely in motion mammalian animal model.

It may seem odd but a blocker/inhibitory molecule is blocking/antagonizing another blocker, enabling signal amplification as the exciting outcome. The author suggests that this may be the *modus operandi*, the mechanism of action, for the stimulant, caffeine. Based on the novel work from the Laboratories of Drs. Arvid Carlsson, Goteborg, Sweden with Drs. Monty Piercey, Michigan, USA and Patricia Broderick, NY, USA, the mechanics may resemble a type of presynaptic autoreceptor terminal neurotransmitter enhancement [43, 44]. Satisfying such a concept are data from the Valenti et al., [57] laboratory that show that an electrophysiologic depression of somatodendritic ventral tegmental neurons that derive from caffeine stimulation.

9.1.3 Caffeine Amplifies Biogenic Amines, Metabolites and Precursors *via* Adenosine

Sensing Caffeine While the Subject Walks:
See Caffeine Effects Online

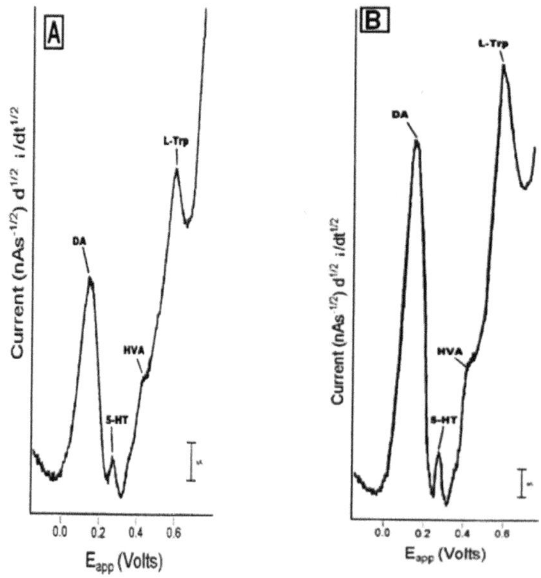

Figure 9.1 NMI LIVE Imaging data from within the neuroanatomic substrate, the Nucleus Accumbens (NAcc) in the freely moving, behaving and walking animal model. The data are from Broderick's Lab using the BRODERICK PROBE® biosensors. LIVE recordings show in **A.** the physiologic videotrack of neurotransmitters and metabolites as the animal is comfortable, unrestrained and walking in its evirontome. In the same animal model, a murine model, **B.** the data shows the effect of caffeine at the low dose of 12.5 mg/kg ip. Injections are expertly done without removing the animal from the comfort of the chamber and perfect timing is essential to this online data collection, which is not invasive, nor is it manipulated. No centrifuging, no cannulas, no heavy resistor head sets are needed in the BRODERICK PROBE®, a carbon lipid polymer. The *x*-axis is *E* applied in mv; the *y*-axis is current as semiderivative or semidifferentiation of the linear circuit. Original raw data are shown. These raw data caffeine scans are published for the first time within this Chapter of the Neuroimaging Book. As previously not enabled until the NMI nanotechnology, the data show neurotransmitters videotracked in the physiologic normal state **A.** and then the data show the dramatic online videotracked response of the brain neurotransmitters to caffeine, **B.** in real time within seconds, *in vivo* and online. Thus, raw data are collected non-invasively as the subject walks, advancing the field of caffeine studies.

Figure 9.1 shows this caffeine phenomenon performing *via* adenosine with LIVE Neuromolecular Imaging (NMI). Monoaminergic neurotransmission in addition to metabolites and precursors are shown in each LIVE electroactive signal.

9.2 Caffeine Neuroprotects the Neurochemical Effects of Cocaine *via* Adenosine

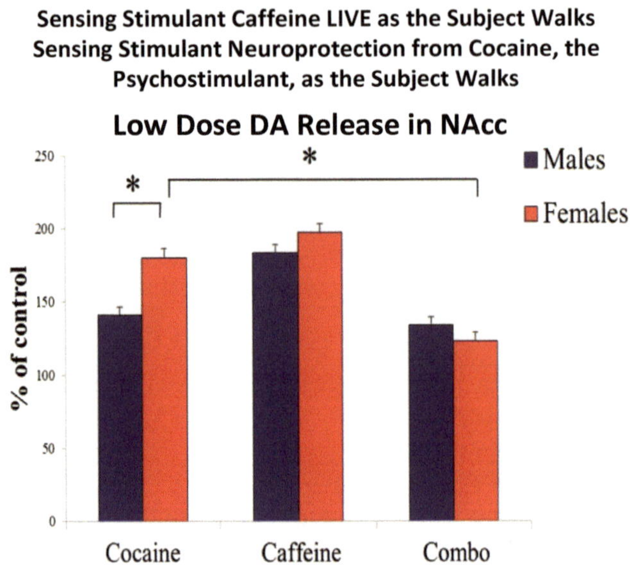

Figure 9.2 Shows, in histogram form, specific sexual dimorphic excess of DA induced by cocaine as well as sexually dimorphic DA excess induced by caffeine using the low dose regimen of both substances (12.5 mg/kg ip caffeine and 2.5 mg/kg ip of cocaine) as signaled online with the BRODERICK PROBE®. Both stimulants enhance the electroactive signals for dopamine; females are more sensitive to both stimulant substances than are males. Yet, together, when these stimulating substances are administered in combination (combo), the expression of adenosine and dopamine receptors allow for their molecular inhibitory roles. Sexual dimorphism is seen online and is dramatically seen, in fact. Moreover, sex differences show that females are protected from cocaine by caffeine and these empirical data show that females are protected from cocaine dysfunction by the molecule, caffeine, to a greater degree than are males. This is interesting as females are especially amenable to the effects of both cocaine and caffeine to a higher degree and further, neuroprotection from cocaine by caffeine is dramatically greater in females than it is in males.

9.3 *Adenosine is a Neuromodulator*

9.3.1 What Is Adenosine?—*a deja vu*

The adenosine molecule comprises a purine, as does the molecule caffeine, but the purine in the adenosine molecule is a *nucleotide* attached to a *ribose sugar*, very different indeed. In so doing, the adenosine molecule may be said to spin away from the caffeine molecule, enabling a differential molecular structure and hence, function. Adenosine is ubiquitous in every human cell, and it gives mammalians, for example, a tremendous amount of energy through its monophosphate and triphosphate homeostatic derivatives, (AMP and ATP respectively). When a ***water*** molecule is added to ATP, **loss of a phosphate** group releases high energy; in reverse, the **gain of phosphate** groups from AMP to ATP, energy is released dramatically as well, regardless of charge. Adenosine, *per se*, is known as an inhibitory neurotransmitter.

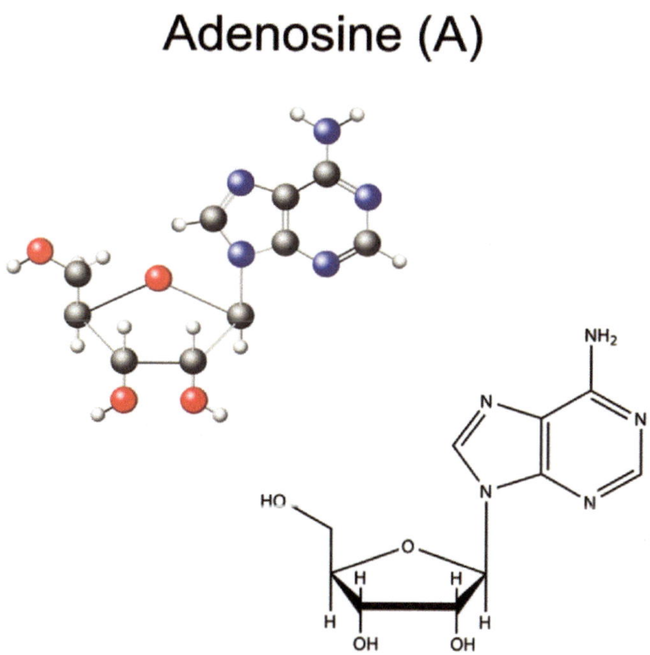

Figure 9.3 Molecular and structural formula of adenosine. From Vector Stock (4007508).

The metaphor *"**It bears repeating,**" fits next. The modus operandi of adenosine control of brain reward coupling/uncoupling to the traumatic psyche of addiction is shown in Figure 9.4. These NMI LIVE data using the innovative tiny* BRODERICK PROBE® nanosensors *fast-forward adenosine A1/dopaminergic influence to the apex of cerebral decision, the point at which cocaine enjoyment and reinforcement progresses to an act of addiction.*

Sensing 8-Cyclopentyltheophylline (8-CPT) While the Subject Walks. Sensing (8-CPT) Neuroprotection from Cocaine-the Psychostimulant While the Subject Walks.

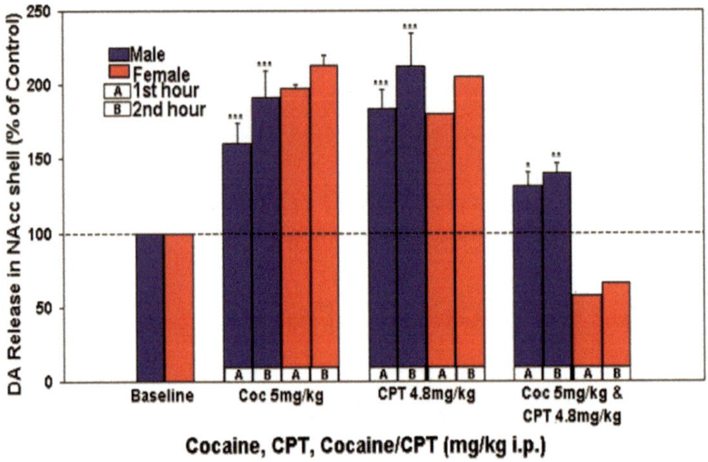

Figure 9.4 The synaptic release of dopamine (DA) by cocaine in nucleus accumbent at the middle dose is shown here with a further comparison with the novel use of 8-cyclopentyltheophylline (8-CPT), a specific A1 adenosine antagonist. The combination of cocaine and 8-CPT compounds were imaged, monitored and the data was interpreted as neuroprotection from cocaine by the specific Adenosine A1 receptor. Deciphering differences between stimulants and psychostimulants is critical to the study of sexual dimorphism. Interestingly, there is the significant implication of the male sensitivity to reinforcing substances seen in this set of histograms. The *specific* A1 antagonist, (8-CPT) shows that males can sometimes be more sensitive to addicting molecules than are females. Subtle nuances in sexual dimorphism are easily detected by the BRODERICK PROBE® and these refinements are essential to the understanding of the complexities of sexual dimorphism. The adenosine-dopamine bond is made clear with *in vivo* NMI data for the first time. The work was published in abstract form at the *Society for Neuroscience*, 2006 [15] and is in further preparation for publication.

9.4 Stimulant 8-Cyclopentyltheophylline Neuroprotects the Neurochemical Effects of Cocaine

9.4.1 Adenosine Streams Estrus

Then, adenosine may be the real molecule that is censorial to streaming estrus. Adenosine is indeed a neuromodulator and most importantly, it is comprised of adenine, a base pair with thymine in the deoxyribosenucleic acid (DNA) molecule, the molecule of life *per se*; it is the energy molecule that forcefully works *via* mitochondria within neuronal entities. Indeed, more and more evidence has accumulated to show that adenosine is involved in regulating dopamine (DA) release within caffeine mechanics.

By the way, caffeine is a true stimulant and its task is to promote wakefulness by non-selectively antagonizing adenosine A1 receptors and A2A receptors. There are observed and critical interactions between the adenosine system and dopaminergic systems. Therefore, the adenosine system regulates cocaine and caffeine reinforcement, substance abuse, addiction processes *per se* and those comorbid with psychiatric conditions with limbic upset that cause the psyche to go awry. A major role for adenosine to neuromodulate hormonal influences on the estrous cycle and the menstrual cycle certainly appears imminent.

Moreover, the CPT NMI LIVE electrochemical data from the BRODERICK PROBE® that reflect subtle and significant male gender influences in adenosine sexual dimorphism [15] (Figure 9.4) solidly lend an explanatory note of support to adenosine data from LIVE electrochemical graphite fiber data [3]. In the latter studies, males were shown to be significantly more sensitive than females to adenosine as studied in the murine medial prefrontal cortex using fast cyclic voltammetry. The studies were performed during estrous cycles and proestrus was found to be the site of the most adenosine signally when estrogen and progesterone are highest as Ramikie and Ressler, 2018 [49] have reported as follows, "female hormone levels change over the stages of estrous with concentrations of estrogen and progesterone peaking during proestrus, then dropping during estrus and the beginning of metestrus and rising again during late metestrus and diestrus."

Therefore, the data from the **Venton** team [3] that markedly departs from the predicted is that of diestrus following proestrus within adenosine signaling within the estrous cycle. Other factors governing the prefrontal cortex that suggest medial prefrontal neurons as mediators of rapid behavioral responses as well as slow adaptive responses [30] provide rationale for the reported swift transition from proestrus to diestrus [3]. Nonetheless, a precipitous and abrupt drop in superficial epithelial cells on exfoliate cytology examination can happen and such a phenomenon can in fact shift the estrus phase to a stage of diestrus. Adenosine may have caused this abrupt shift in prefrontal cortical neurons in the fast-cyclic voltammetry data [3]. Finally, sexual dimorphism to stimulants and stimulant-like substances is clear.

9.4.2 The Adenosine-Dopamine-Estrogen Trio in Psychiatry: *Post-Partum* Depression

> "Mesolimbic dopamine (DA), particularly in the nucleus accumbens, importantly regulates activational aspects of maternal responsiveness. DA antagonism and accumbens DA depletions interfere with early postpartum maternal motivation by selectively affecting most forms of active maternal behaviors, while leaving nursing behavior relatively intact. Considerable evidence indicates that there is a functional interaction between DA D2 and adenosine A(2A) receptors in striatal areas, including the ventral striatum, the nucleus accumbens. This study was conducted to determine if adenosine A(2A) receptor antagonism could reverse the effects of DA receptor antagonism on early postpartum maternal behavior. The adenosine A(2A) receptor antagonist MSX-3 (0.25–2.0 mg/kg, IP) was investigated for its ability to reverse the effects of the DA D2 receptor antagonist haloperidol (0.1 mg/kg, IP) on the maternal behavior of early postpartum female murine animals. Haloperidol severely impaired the expression of active maternal components, including retrieval and grouping the pups at the nest site, pup licking, and nest building. Co-administration of MSX-3 (0.25–2.0 mg/kg, IP) with haloperidol produced a dose-related attenuation of the haloperidol-induced behavioral deficits in early postpartum

females. Doses of MSX-3 that effectively reversed the effects of haloperidol (0.5, 1.0 mg/kg), when administered in the absence of haloperidol, did not affect maternal responding or locomotor activity. Adenosine and DA systems interact to regulate early postpartum maternal responsiveness. This research may potentially contribute to the development of strategies for treatments of psychiatric disorders during the postpartum period, with particular emphasis in maintaining or restoring the mother-infant relationship" [42].

9.4.3 The Adenosine-Dopamine-Estrogen Trio in Psychiatry: Sex and Sensory Gating

In *Sex as a Moderating Factor in the Relationship Between Hippocampal Volume and Sensory Gating in Patients with Schizophrenia* [29], we revert to the auditory system and sex differences in sensory gating *via* the acoustic startle response. The study found a hippocampal involvement in the P50 aspect of schizophrenia disorders; a correlation between hippocampal volume and P50, which implies a defective pattern in pre-attentional filtering of sensory information, was elucidated. The correlations between hippocampal volume and P50 were analyzed in males and females separately, corrected for potential confounders, including age, duration of illness, and antipsychotic medication. Different associative patterns of hippocampal volume with P50 gating ratio and with P50 difference were found by sex, suggesting sex plays a role in the physiology of sensory gating [29].

9.4.4 The Adenosine-Dopamine-Estrogen Trio in Psychiatry: Sex and Anxiety

A broad definition of depression can be studied anthropomorphically using animal models. Adolescence is a period of development and networking for social and emotional growth in the central nervous system, particularly brain. As this period is especially hormonally influenced, the estrous cycle needs to modulate toward a healthy balance. Various classical behavioral tests such as the elevated plus maze behavioral assay were studied. The Wistar Kyoto (WKY) model has been proposed as an animal model of adolescent depression with various face, construct, and predictive validities for clinical depression having been established. Female adolescent WKY were studied during different phases of the estrous cycle with inbred, age-, and phase-matched Wistar animals as controls. Wistars in proestrus-estrus demonstrated reduced anxiety levels. WKY also demonstrated increased open arm time and entries and closed arm time, but less than Wistars, and as closed arm entries remained unaffected, it did not translate into a lowering of the anxiety levels. Risk taking and risk assessment behaviors were not affected by estrous phases in WKY, though exploratory behavior was reduced in proestrus-estrus. In Wistars, increased risk taking and decreased risk assessment behaviors were observed during proestrus-estrus. Increased immobility in the forced swim test (FST), indicative of Hans Selye's learned helplessness was not influenced by phase in the WKY, which was at variance with Wistars that demonstrated phase-specific differences. The relevance of these studies is that estrous phase-specific time windows are needed for pharmacotherapeutic interventions of anxiolytic and anti-depressant psychiatric disturbances [data from reference 20].

9.4.5 Estrous Cycle Disorders Can Be Neurovascular: Estrogen Protection: Sars-Cov-2

Angiotensin Converting Enzymes block the conversion of Angiotensin 1 to Angiotensin 11 and pivotal in the mechanism of action of Angiotensin 11 is the cerebrovascular dysfunction that is induced by this blood pressure molecule as it ***varies across the estrous cycle.*** **Diovan® is an example of such a medicin or drug.** The dysfunction is not observed in proestrus and estrus but is present in diestrus. The protection from the cerebrovascular actions of Ang II in proestrus is abolished by the nonselective estrogen receptor antagonist ICI182,780, suggesting the involvement of estrogen in the mechanisms of the protection. Like the cerebrovascular effects, Angiotensin II-induced Reactive Oxygen Species (ROS) production is also estrous cycle dependent and is maximal in diestrus and attenuated in proestrus and estrus. ICI182,780 enhances the ROS increase in proestrus, also attesting to a role of estrogen receptors in the Angiotensin II-induced ROS generation. These findings, collectively, unveil a previously unrecognized dependence of the cerebrovascular effects of Angiotensin II on the estrous cycle and raise the possibility that the vulnerability of the female brain to cerebrovascular injury varies cyclically across the estrous cycle.[1]

Not to digress but to integrate, the author will briefly introduce this ACE molecule in relationship to the current pandemic called Sars-Cov-2, the virus that attacks organ systems galore *via* ACE-11.[2] Broderick is thinking, protection from ACE 11 (ACE-2) Sars-Cov-2 **viral** attack is probable by way of a steady state blood level of the ACE-11 (ACE-2) inhibitor, Diovan® for one example of this class of drugs or medicines.

Note: The Generic name for Diovan® is Losartan, Candesartan or Valsartan, for example.

Thus, the estrous cycle becomes a moat impressive factor in pharmacotherapeutics in so many instances.

9.4.6 Integrating Sex Hormones with the Estrous Cycle and Biogenic Amines: An Historical View

Carter talks about neuroendocrinology as it relates to sexual behavior in the female gender in Becker JB, Breedlove SM, Crews D (eds). *Behavioral Endocrinology* [17].

Sex hormones, such as estrogen, are regarded as the underlying cause. Estrogen levels fluctuate throughout the four stages of the estrus cycle [50]. Changes in the estrous cycle change

[1]Capone C, Anrather J, Milner TA, Iadecola C (2009) Neurovascular dysfunction induced by angiotensin II in the mouse neocortex. *Hypertension* 54: 302–307. https://doi.org/10.1161/HYPERTENSIONAHA.109.133249.

[2]Broderick, PA (2021) The COVID brain: A cytokine crisis: Major depressive disorder and suicide. *Preferred Health Magazine*, Winter Edition, https://www.blurb.com/b/10558458-preferred-health-magazine-winter-2021.

behavior dependent on ovarian hormones [53]. Laboratory animals with higher estrogen levels also have higher sensitivity to cocaine. Thus, during the estrous cycle stages when estrogen levels peak, females are more susceptible to cocaine [45–47]. For instance, females show higher stimulatory response to lower doses of cocaine compared to men. The level of sex hormones fluctuates throughout the four stages of the estrous cycle, which alters sensitivity to cocaine [50]. Human studies showed higher cocaine stimulatory response during the luteal phase (moderate estrogen levels, high progesterone) and not the follicular phase (low estrogen levels and progesterone) [21, 22]. Throughout the estrous cycle in murine animals, the highest estrogen levels are during the proestrus phase and drops during the estrus and diestrus phase and ovariectomy (OVX) procedures significantly diminished these sex differences [2].

Adenosine is associated with the sleep-wake cycle and is most important in regulating neuronal activity within this cycle [18]. Thus, more and more empirical evidence leads to its involvement in regulating DA release [26]. As mentioned, the stimulant, caffeine, works *via* non-selectively antagonizing the adenosine A1 receptors (A1R) and A2 receptors (A2AR). There are observed interactions between the adenosine system and DA [59]. Therefore, the adenosine system may regulate cocaine reinforcement.

The psychostimulant, cocaine, significantly increases neurotransmitters, such as dopamine (DA) and serotonin (5-HT) in the brain reward pathway of the brain [4–7, 10–12]. More and more empirical evidence is pointing to the increased vulnerability of cocaine abuse in women compared to men and research in laboratory animals, females self-administer cocaine more rapidly, become dependent earlier, and relapse faster compared to males [45–47]. The reason for these sex differences are becoming more clarified.

Furthermore, studies using the BRODERICK PROBE® biosensor show that caffeine combined with cocaine reduced the significant increase of dopamine that is associated with cocaine [10]. The biosensors were surgically implanted in the Nucleus Accumbens, which involves brain reward areas. Dopamine (DA) is a key factor in cocaine-seeking behavior causing the euphoric feeling of reward. This effect was more sensitive for females and occurred at lower doses compared to males, supporting the increased vulnerability associated with females [10]. Caffeine enhances biogenic amine, dopamine and serotonin release and caffeine-co-administration with cocaine decreased the cocaine-induced increase in the biogenic amines [8–10, 14–16].

The selective A1 antagonist, 8-Cyclopentyltheophylline (8-CPT) increases synaptic biogenic amine release online *in vivo* while subjects walk and groom [15] and also counteracts the psychostimulant cocaine response when co-administered. Adenosine signals throughout the estrous cycle [3]. Female sensitivity abounds but male sensitivity appears subtly, with reliability and significantly in some instances [3, 15]. Still, 8-CPT blocks cocaine-induced dopamine release more significantly in females than in males [15]. Since caffeine acts on adenosine receptors in the brain, this highly suggests a role for drug therapies that target adenosine receptors for treating cocaine addiction. These findings continue to show that caffeine works through adenosine and then some of the references mentioned [Broderick Laboratory] go on to report that adenosine may protect the limbic neurons from the psychostimulant properties of cocaine [8–11, 14, 15, 18, 24, 25, 32, 38, 39, 48, 55, 56].

9.5 Tenets of the Estrus Cycle and the Menstrual Cycle: Close Relatives

9.5.1 The Estrous Cycle and the Menstrual Cycle are Critical for Procreation: Brain Reward

The estrous cycle can be divided into four stages: proestrus, estrus, metestrus, and diestrus. During proestrus, progesterone declines) and a pre-ovulatory follicle undergoes its final growth phase (estradiol increases). Ovulation usually occurs during estrus.

There are four phases to the menstrual cycle with each phase associated with hormones, e.g., luteal phase. This phase begins after ovulation. It lasts about fourteen days (unless fertilization occurs) and ends just before a menstrual period. In this phase, the ruptured follicle closes after releasing the egg and forms a structure called a corpus luteum, which produces increasing quantities of progesterone. Estrogen is generally the hormone associated with incentive to use drugs of abuse. Progesterone is generally the hormone associated with diminished incentive to use drugs of abuse.

9.5.2 The Estrous Cycle and the Menstrual Cycle Work in Tandem Across species: Murine Model

> In **proestrus**, follicles develop rapidly until ovulation at the end of proestrus. At this stage, two to four large follicles can be observed in a unilateral ovarian section. When **estrus** starts at ovulation, follicles are transformed into corpora lutea, the luteal cells of which are more basophilic and smaller than in later phases of the **menstrual cycle**. These corpora lutea begin to regress at **metestrus** and continue their regression into **diestrus**. This pattern of development and regression then repeats in the next **cycle of estrous**. Given that regressive corpora lutea may persist for some 12 to 14 days, ovaries of murine animals tend to contain a large number of corpora lutea [52].

9.6 Classical and Innovative Paradigms in the Study of the Estrous Cycle

Thus, in order to delve into sex differences in substance abuse, psychology and psychiatry including reinforcement and addiction, the non-prescient mammalian estrous cycle was examined.

During proestrus, estrogen and estradiol reach peak amounts as progesterone may or may not decline during ovarian follicle growth dependent on the pregnancy status given the caveat that murine models are polyestrous. During proestrus then, on vaginal smear, nucleated epithelial cells, including *parabasal* cells, may be visualized. During estrus, Luteinizing hormone (LH) increases as estrogen decreases, Progesterone remains low. These hormonal changes produce mature anucleated cornified cells on vaginal smear. During metestrus, estrogen remains low, while LH and Follicle Stimulating Hormone (FSH) increases.

On vaginal smear, anucleated cornified cells, leukocytes, and nucleated epithelial cells are predominant. The subsequent diestrus stage may be divided into two stages as there is an increase in the number of nucleated cells towards the end of this stage. During diestrus, progesterone is high and FSH is low, which results in leukocytes on vaginal smear. As diestrus transitions into proestrus, FSH increases as ovarian follicles grow.

9.7 The Study of Estrous Cycle Cells

> "Parabasal cells are the smallest epithelial cells seen on a typical vaginal smear. They are round or nearly round and have a high nuclear to cytoplasmic ratio. Parabasal cells are prevalent on smears taken during diestrus and anestrus, and not uncommon during early **proestrus**. Parabasal cells are conspicuously absent during estrus.
>
> A majority of cells observed in a normal vaginal smear are, not surprisingly, vaginal epithelial cells. In addition, varying numbers of leukocytes, erythrocytes and bacteria are usually evident, and small numbers of other contaminating cells and microorganisms are sometimes observed.
>
> Analyzing a vaginal smear is largely an exercise in classifying the epithelial cells into one of three fundamental types: *parabasal, intermediate or superficial cells.* Keep in mind, however, that the epithelial cells reflect a developmental continuum; some of the cells you observe will not fit perfectly into these rigidly-defined categories.
>
> All of the figures on this page are presented at the same magnification and are from a single dog" [cf. *Normal Cytology: An Illustrated, Practical Guide*, 1st ed. (Eds. M. Lew, J. Pang, L. Pantanowitz), Springer, 2022].

9.7.1 Parabasal Cells

Parabasal cells are the smallest epithelial cells seen on a typical vaginal smear. They are round or nearly round and have a high nuclear to cytoplasmic ratio.

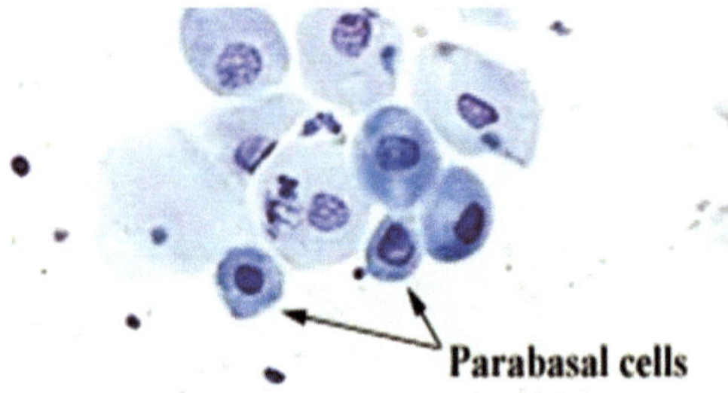

Figure 9.5a Parabasal cells are prevalent on smears taken during diestrus and anestrus, and not uncommon during early proestrus. Parabasal cells are conspicuously absent during estrus.

9.7.2 Intermediate Cells

Intermediate cells vary in size and shape, but typically have a diameter two to three times that of parabasal cells. Many cytologists subclassify these cells into:

- *small intermediates*: nearly round or oval shape with large, prominent nuclei
- *large intermediates*: polygonal shape with a small nuclear/ cytoplasmic ratio

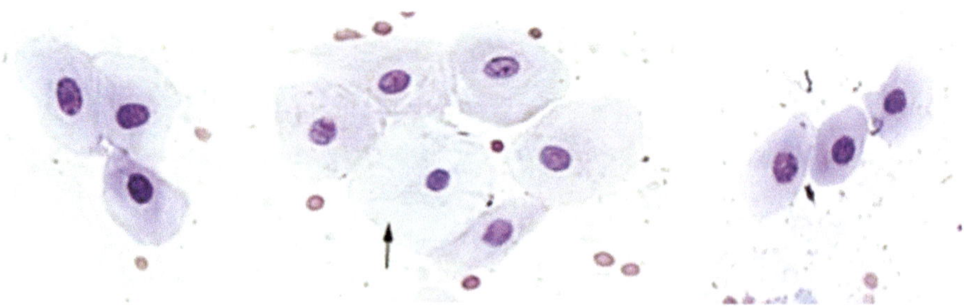

Figure 9.5b Intermediate cells are prevalent during all stages of the cycle except estrus.

In the figures that follow, all of the cells are typical intermediates except for the one cell in the middle panel (arrow), which might be classified as a superficial cell (small, dark nucleus). Note that the intermediate cells vary in size and that some have rounded outlines (small intermediates), while others have a polygonal shape (large intermediates).

9.7.3 Superficial Cells

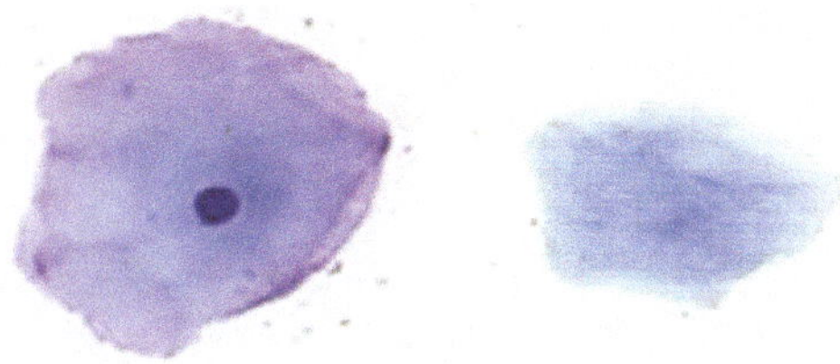

Figure 9.5c Superficial cells are often seen in large sheets or strings, as seen below with fully cornified cells.

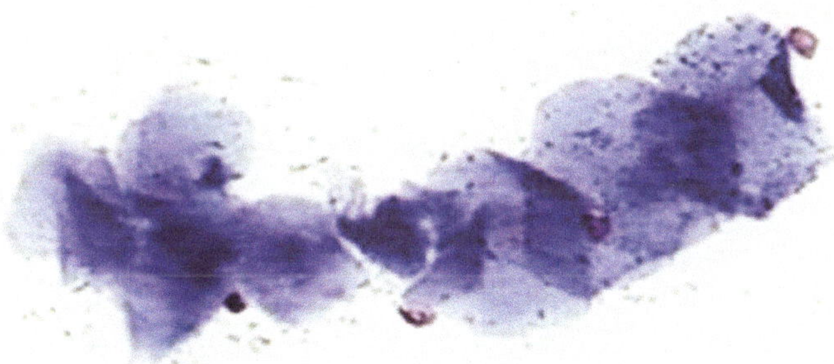

Figure 9.5d Superficial cells are not normally seen during anestrus and increase in prevalence during proestrus. The presence of large numbers of superficial cells or only superficial cells is the defining characteristic of cytologic estrus, and their abrupt and precipitous decline marks the onset of diestrus [cf. *Normal Cytology: An Illustrated, Practical Guide*, 1st ed. (Eds. M. Lew, J. Pang, L. Pantanowitz), Springer, 2022].

Superficial cells are the largest cells seen on a vaginal smear. They are polygonal in shape and distinctly flat, sometimes having the appearance of being rolled up. Their nuclei are either absent or pyknotic (very small and dark). Superficial cells without nuclei are often referred to as being "fully cornified".

9.8 The Experimental Design and the Data Derived

9.8.1 Subjects

Adult female Sprague-Dawley rats (*rattus norvegicus*), 2 months of age were purchased from Charles River Laboratories, North Carolina, USA. Animals were housed in the Animal Care Facility under the auspices of the City College Institutional Animal Care and Use Committee (IACUC) in compliance with National Institute of Health (NIH) guidelines. Food (Purina Rat Chow) and water was available *ad libitum.* Females were housed separately from males in airtight seclusion, three or two per chamber.

9.8.2 Drugs and Doses

Studies were performed in three groups: Cocaine, Caffeine, or combination of cocaine and caffeine with low, middle and high doses. Each group had a sample size of 8. Within the three groups each animal received each dose as a within-subject control. The different doses were given at least 48 hours after the previous one to eliminate residual drug effects. Doses were given as described in table 1 in pseudo-random order. Solutions for cocaine HCl (Sigma-Aldrich, St. Louis, MO) and caffeine (Sigma-Aldrich, St. Louis, MO) were dissolved in distilled water and injected intraperitoneally (i.p). Saline (0.9% NaCl) solution was used as a control. Doses were chosen based on peak effects from the literature and previous NMI experiments done in this laboratory to correlate behavior with biochemistry as the subject walks.

9.8.3 Vaginal Smears Using Exfoliate Cytology

Samples were collected pre- and post- injections within a 30-minute period. Females were swabbed with a saline soaked cotton tip and dotted on a microscope slide. Samples were observed at 40× objective lenses using an Invert Research Microscope (Nikon Eclipse TI) Zeiss Axioplan 2 Imaging and Carl Zeiss Vision GmbH. The vaginal smears were stained with a Diff staining kit (IMEB INC, San Marcos, CA) and analyzed according to four stages: proestrus, estrus, metestrus and diestrus. Samples were assessed based on the proportions of the type of cells to determine the appropriate stages. There were three types of cells: round nucleated epithelial cells, irregular shaped anucleated cornified cells and small round leukocyte cells. Proestrus phase had prominently epithelial cells. Estrus phase had cornified cells. Metestrus phase had an equal amount of all three cells. Finally, the diestrus phase consisted of predominantly leukocytes.

9.8.4 Vaginal Voltage Using Electrical Resistance

Estrus cycle stages were studied by measuring the electrical resistance of the vaginal mucosa (EC40, Fine Science Tools, Foster City, CA). Subjects were held in hand on back and a probe was placed in the vagina opening as a recorder calculated the electrical resistance. This was performed pre and post 2.5 mg/kg cocaine and saline injections and prior to the vaginal smear swab. The cocaine low dose data is shown in Table 9.1 and saline data in Table 9.2. Statistical analysis was performed on the difference between pre- and post-electrical resistance for the two treatments. A one-way analysis of ANOVA and post hoc *Bonferroni* test were performed. Alpha significance was set at $p < 0.005$. Preliminary data has been done on caffeine and combination groups using electrical resistance showing a trend consistent with vaginal smears but more data is needed.

9.8.5 Vaginal Smears and Vaginal Voltage: Tabulation

Table 9.1 Vaginal smears and electrical readings were taken following a low dose of cocaine pre- and post-injection; both physiology and electrical readings differed after an injection of cocaine

Cocaine (2.5 mg/kg ip)			
Exfoliate cytology		**Electrical reading**	
Pre	**Post**	**Pre**	**Post**
Metestrus	Diestrus	1.8	0.8
Diestrus	Proestrus	2.2	1.4
Early Diestrus	Metestrus	2.6	0.8
Metestrus	Diestrus	2.6	1.4
Proestrus	Estrus	2.1	3.7
Metestrus	Estrus	1.4	2.1
Metestrus	Estrus	2.2	4.8
Metestrus	Diestrus	0.7	0.8*
Metestrus	Diestrus	1.7	1.1
Metestrus	Proestrus	2.3	0.6
Proestrus	Estrus	2.4	1.5
Estrus	Diestrus	3.2	1.8
Late Diestrus	Late Proestrus	1.5	2.3

Table 9.2 Vaginal smears and electrical readings were taken following saline; physiology or electrical readings did not change after saline injection

Saline (body weight/kg ip)			
Exfoliate cytology		**Electrical reading**	
Pre	**Post**	**Pre**	**Post**
Diestrus	Diestrus	0.8	0.6
Diestrus	Diestrus	0.7	0.6
Late Estrus	Late Estrus	2.5	2.3
Early Diestrus	Late Diestrus	1.3	1.3
Early Metestrus	Late Metestrus	1.6	1.6
Estrus	Estrus	2.9	3.0
Estrus	Estrus	3.1	2.9
Estrus	Estrus	2.5	1.8*

9.9 Exfoliate Cytology: First Group

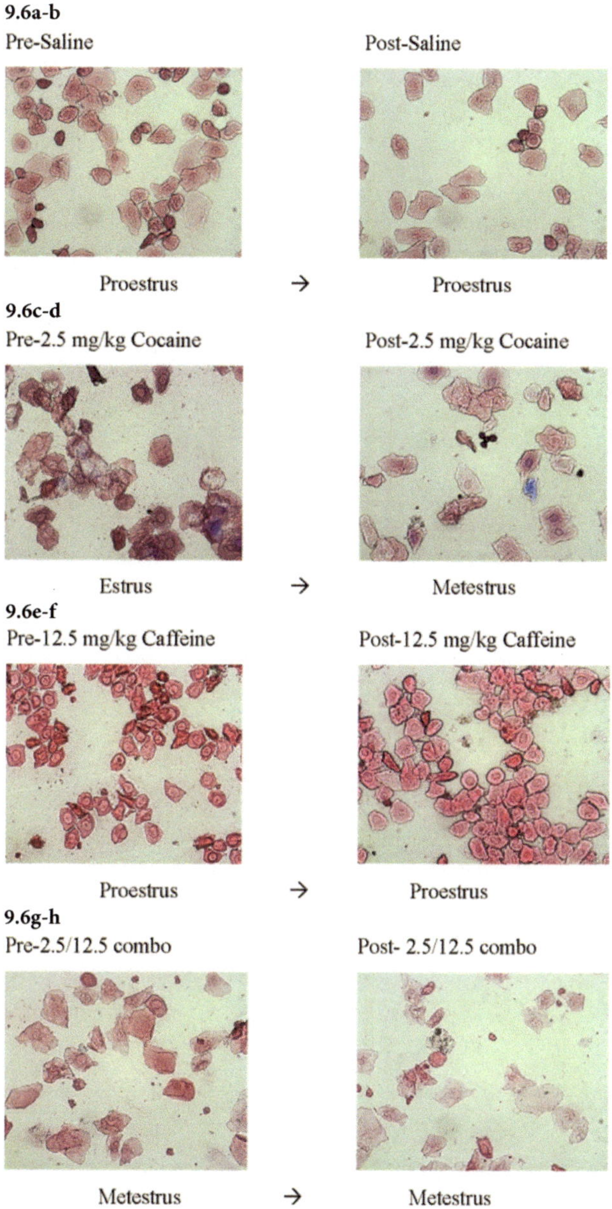

Figure 9.6 (a–h) Studies on the estrous cycle in females show that cocaine induces changes in the estrous cycle that is not seen with caffeine, nor with the combination of cocaine and caffeine or saline. These changes are expressed at the low dose.

9.10 Exfoliate Cytology-Second Group

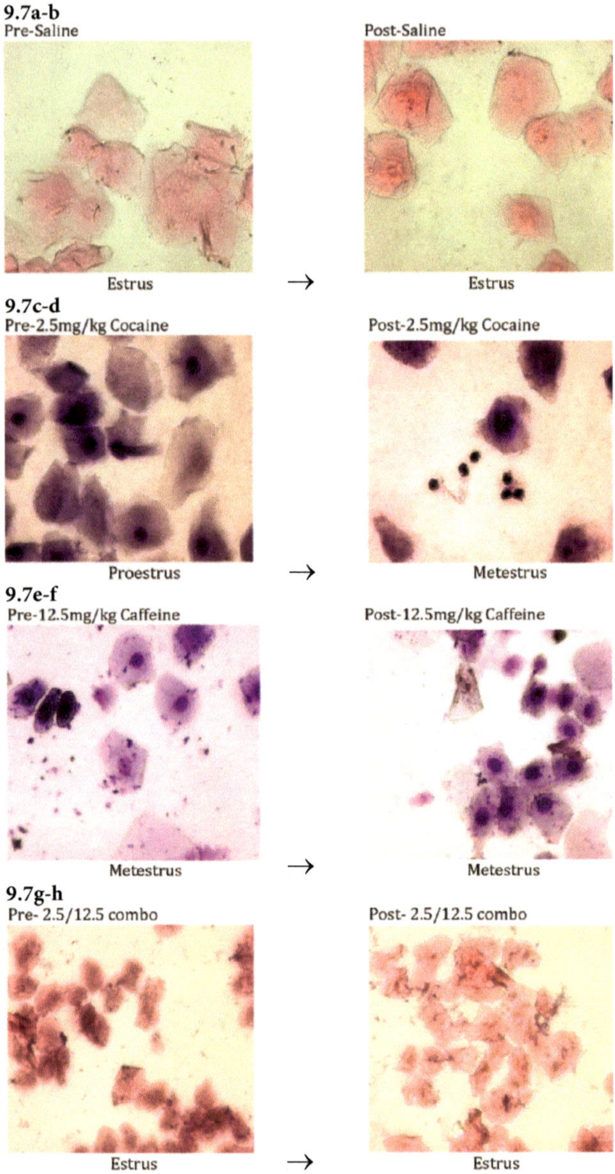

Figure 9.7(a–h) Studies regarding the estrous cycle in females show that cocaine induces changes in the estrous cycle that is not seen with caffeine, nor with the combination of cocaine and caffeine or saline. These changes are expressed at the low dose. Here, exfoliate cytology studies were repeated using the same design.

9.10.1 Vaginal Smears and Vaginal Voltage Statistical Significance

Vaginal smears were assessed using a microscope and scored based on cells described in the methods section (Figures 9.6a–h and 9.7a–h). For all doses of cocaine, female smears showed differences in estrus cycle phase between pre- and post-injection within 30 minutes. Cocaine induced changes in the estrus cycle (Figures 9.6c,d and 9.7c,d). This was not seen following caffeine administration (Figures 9.6e,f and 9.7e,f). Furthermore, combination of cocaine and caffeine did not alter the estrus cycle (Figures 9.6g,h and 9.7g,h). The cocaine-induced changes were blocked by caffeine.

Cocaine injections showed significant differences in the exfoliate cytology data as well as the data from electrical resistance of the vaginal mucosa; the changes were not seen following saline injections **(Tables 9.1 and 9.2)**. A one-way ANOVA showed a treatment significance between the cocaine and saline group ($F\ (1,20) = 16.498$, $p < 0.000619$). A Bonferroni test revealed cocaine significantly changed the electrical resistance pre- and post-injection that was not seen following saline ($p = 0.000619$). These results were consistent with the cocaine-induced estrus cycle stage changes. Caffeine did not change the estrous cycle nor did saline.

9.11 Discussing the Data

Finally, we will discuss further the importance of the abundance of research on the neuroendocrinological influence of sex hormones on sexual behavior in the female sex [17] as well as neuroendocrinology in cocaine reinforcement and/or addiction. Laboratory animals with higher estrogen levels were shown to have higher sensitivity to cocaine [2, 47]. Estradiol studies have shown an increase in cocaine effects in the locomotor and motivational aspect in cocaine addiction [50, 33]. Thus, the most assumed conventional theory is that estrogen is potentially essential in the reinforcement and/or addiction to cocaine.

However, there are still conflicting results in the literature. Some report cocaine responses are increased during estrus phase [28, 45] had the most self-administration during proestrus phase when progesterone was low in their studies [23] in contrast to [3, 49] who found high progesterone levels in proestrus. Finally, to compound the felony that there were no differences between the estrus or proestrus phases [53, 58] during cocaine stimuli.

Broderick Laboratory findings showed that cocaine directly changed the estrus cycle, and thus changed estrogen levels for females affecting their sensitivity. Contrary to the effects of cocaine, caffeine did not cause such changes and blocked the effects of cocaine when taken together. This pattern of neuroprotection *via* caffeine was similar to responses in previous studies of DA release [38], conditioned place preference [39], locomotor activity [11, 48] and acoustic startle response [16] suggesting that dopamine (DA), adenosine, and estrogen strongly interact with each other causing increased sensitivity to cocaine that was reversed when the molecule, caffeine, was administered. The adenosine-dopamine-estrogen trio is clear.

LIVE data in the adenosine system come from the Broderick Laboratory [15] and the Venton Laboratory [3]. The LIVE data agree and both are electrochemical. It is so interesting as one uses a polymeric biomimetic carbon/protein polymer [15] and the other uses a graphite fiber [3]. One uses a semiderivative/semidifferential circuit [15] and the other uses a cyclic voltammetry circuit [3]. Science and medicine agree and the field is advanced.

We approached the estrous cycle studies with a novel method of analyzing the estrous cycle phase changes. This method interestingly reads the electrical resistance of the vaginal mucosa. Results showed a correlation between estrus cycle phase and voltage [39]. Cocaine changed the electrical resistance (kilohm) at low dose, whereas, saline did not significantly change the electrical resistance as expected. Interestingly preliminary data show that caffeine and combination of cocaine with caffeine does not change the electrical resistance as well, but more data is needed in this aspect. The electrical resistance voltage and vaginal smear exfoliate cytology show paralleling results. However, the volt reader gives a conclusive number immediately after, whereas a vaginal smear is subjectively scored. Therefore, for the first time this report on the electrical resistance of the vaginal mucosa is a new tool to better test/evaluate estrus cycle changes.

In addition to hormone effects, there may be other factors involved in these sex differences [47]. Differences in signaling caused by sexually dimorphic regulation of DA or other neuronal regulators may be the origin of these differences. For example, N-Methyl-D-Aspartate (NMDA) and gammaaminobutyricacid$_B$ (GABA$_B$) likely act presynaptically, affecting the overall cell excitability for vesicular release. The ability to regulate adenosine with NMDA and GABA$_B$ receptors will help control the modulatory effects of transient adenosine release [40]. Several neuroanatomic substrates also contribute and amygdala has come to pass in prominence [34].

9.12 Integrating the Data: Where Do We Go from Here

- The author advises not to treat cocaine addiction with coffee. Caffeine and coffee are two distinct entities as the author's student has written in "Coffee Is to a Square as Caffeine Is to a Rectangle": Part 1–3 in the Journal of Caffeine Research (JCR) [35–37].

  Note: The author served as the Editor-in-Chief of *JCR*, New Rochelle, NY, USA.

- We are downright proud that our work made *The Daily Mail* in the United Kingdom [19] but T*he Daily Mail* interpreted the data as "coffee cures cocaine." Actually, it is the captious link between caffeine and adenosine that may help pilot us to ameliorate the unorthodoxy of cocaine's mastery over temporal synchrony.

- Cocaine changed estrogen and caffeine restores it; caffeine does not change estrogen as it simply shifts estrogen back to the stage from which it came, supporting our neuroprotection data for caffeine [14].

9.12.1 Estrogen and Estrus

Nicotine changes the estrous cycle to return to estrus every time, immaterial to which stage it started. One might think that estrogen and estrus are the same since estrogen is in high concentrations in estrus and this is the fertile stage. This would mean that nicotine shifts the estrous cycle continually to the fertile phase from any other cycle [60].

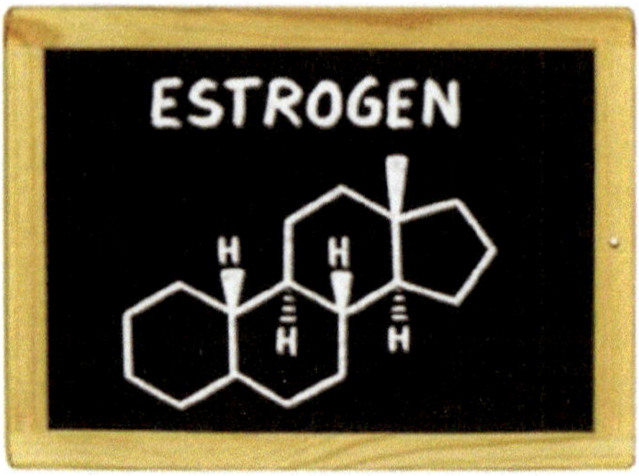

Image credit: https://www.shutterstock.com/image-photo/chemical-formula-estrogen-on-blackboard-125928743.

9.12.2 Sex, Sexuality and the Estrous Cycle

- Estrous cycle may have little to do with prediction of or the essence of sexuality since there is a loss of fecundity for estrus over evolution and female's desires seem extended beyond estrus, that is, that female sexual desires are no longer relegated only to the estrus stage. But the fertile phase is still the fertile phase even though the desires continue to be apparent through the stages of the estrous cycle.
- Analysis of the sexual behavior of nonhuman primates and humans indicates, however, that constant receptivity for the female is not unique to humans and that human sexual behavior is not independent of the phases of the menstrual cycle and in-pair and extra-pair relationships are also factors changing sexuality *ad hoc* so to speak [41]. Pawlowski is a proponent of natural selection.

- Darwin may be called one of the original proponents of sexual selection.
- Natural selection of mates depends on attraction between the sexes and sexual selection depends on fecundity of sexual organs as openly apparent visually.
- Gangestad et al are proponents of natural selection [27].
- "There is a strong correlation between estradiol and testosterone preference specifically during the luteal phase and this further suggests that women's mate preferences may track their fertility between different cycles in addition to being calibrated to the timing of ovulation within individual cycles." Roney and Simmons [51]; these authors are proponents of natural selection.
- Jones, Gangestad and colleagues [31] although proponents of natural selection, disagree with Roney and Simmons [51].
- Thus, the estrous cycle and sex differences seem to be independent of sex and sexuality.

"Behaviourally, oestrus has been thought to entail enhanced proceptivity and receptivity to males when females are fertile. In its stead, women evolved to exhibit 'continuous' sexuality—a sexual nature that changes, if at all, in only minor ways across the cycle, such that women are near-equally sexually proceptive throughout it. Typically, scholars have thought loss of oestrus functions to conceal ovulation.

Recent evidence and theory indicate that this fact is fiction. Women have not lost oestrus. Although women are continuously receptive, their sexuality is not accurately characterized as continuous. Rather, women possess 'dual sexuality,' consisting of a phase of oestrus and a phase of extended (non-fertile) sexuality in their cycles. Loss of oestrus in humans serves no function, as it simply has not occurred, though there is an important sense in which women have evolved to conceal cycle fertility (or, in our terms, their oestrus)" [27].

Gangestad [27] calls the loss of oestrus a myth.

More myths are dispelled in the next chapter. Stay tuned.

References

1. Bayer EA, Hobert O (2018) Past experience shapes sexually dimorphic neuronal wiring through monoaminergic signaling. *Nature* 561(7721): 117–121.
2. Becker JB, Hu M (2008) Sex differences in drug abuse. *Front. Neuroendocrinol* 29: 36–47.
3. Borgus JR, Puthongkham B, Venton JB (2020) Complex sex and estrous cycle differences in spontaneous transient adenosine. *J Neurochem* 153(2): 216–229. https://doi.org/10.1111/jnc.14981.
4. Broderick PA (1992) Cocaine's colocalized effects on synaptic serotonin and dopamine in ventral tegmentum in a reinforcement paradigm. *Pharmacol Biochem Behav* 42: 889–898.
5. Broderick PA (1992) Distinguishing effects of cocaine (IV) and (SC) on mesoaccumbens dopamine and serotonin release with chloral hydrate anesthesia. *Pharmacol Biochem Behav* 43: 929–937.
6. Broderick PA (1993) *In vivo* electrochemical studies of gradient effects of (SC) cocaine on dopamine and serotonin release in dorsal striatum of conscious rats. *Pharmacol Biochem Behav* 46: 973–984.
7. Broderick PA (1993) *On line* 5-HT release patterns in reward circuits signal cocaine-induced DA dysfunction: Implications for therapy. *The Committee on Problems of Drug Dependence, NIDA Res. Mono. Series* 141: 128.
8. Broderick PA (2014) Caffeine counters cocaine's effects on women's estrus cycles *J Caff Res* (press release).
9. Broderick P.A. (2015) Caffeine is Hot-People Want to Know About It. *Alumnus*, The City College of New York, June Issue.
10. Broderick PA, Ho H, Huang J, et al. (2007) Caffeine may neuroprotect against cocaine-induced dopamine release in accumbens of female rats. *Soc Neurosci Abstr* 33 Program Poster 767.7.
11. Broderick P, Ho H, Wat K, Murthy V (2008) Laurate biosensors image brain neurotransmitters *in vivo*: can an antihypertensive medication alter psychostimulant behavior? *Sensors* 8: 4033–4061.
12. Broderick PA, Kornak EP, Eng F, Wechsler RW (1993) Real time detection of acute (IP) cocaine-enhanced dopamine and serotonin release in ventrolateral nucleus accumbens of the behaving Norway rat. *Pharmacol Biochem Behav* 46: 715–722.
13. Broderick PA, Malave LB (2014) Cocaine Shifts the Estrus Cycle Out of Phase and Caffeine Restore It. *J. Caff Res* 4: 109–113.

14. Broderick PA, Malave LB (2015) Caffeine, Adenosine, Sex and the Estrous Cycle. *Soc. Neurosci*, Chicago, Il., Abstract and Presentation (Associated Press Award).

15. Broderick PA, Murthy V et al., (2006) Adenosine A1 antagonist 8-cyclopentyltheophylline) mediates cocaine interactions: a view from the nucleus accumbens. *Soc. Neuroscience Abstract* 32: Atlanta, GA October 11–18.

16. Broderick PA, Rosenbaum T (2013) Sex-specific brain deficits in auditory processing in an animal model of cocaine-related schizophrenic disorders. *Brain Sci* 3: 504–520.

17. Carter CS (1992) Neuroendocrinology of sexual behavior in the female. In: Becker JB, Breedlove SM, Crews D (eds). *Behav Endocrinol*, Cambridge, MA: MIT Press, 71–95.

18. Daly JW (1993) Mechanism of action of caffeine. In: Garattini S (ed.). Caffeine, Coffee, and Health. New York: Raven Press, 97–150.

19. Davies M (2014) Could COFFEE help treat a cocaine addiction? Caffeine 'blocks drug-induced changes in the brain'. *The Daily Mail.* PUBLISHED: 09:11 EST, 21 November 2014 | UPDATED: 10:00 EST, 21 November 2014.

20. D'Souza D, Sadananda M (2017) Estrous cycle phase-dependent changes in anxiety and depression-like profiles in late adolescent Wistar-Kyoto rat. *Ann Neurosci* 24: 136–145. *https://doi.org/10.1159/000477151*.

21. Evans S, Haney M, Foltin R (2002) The effects of smoked cocaine during the follicular and luteal phases of the menstrual cycle in women. *Psychopharmacology* 159: 397–406.

22. Evans S (2007) The role of estradiol and progesterone in modulating the subjective effects of stimulants in humans. *Exp Clin Psychopharmacol* 15: 418–426.

23. Feltenstein MW, See RE (2007) Plasma progesterone levels and cocaine-seeking in freely cycling female rat across the estrous cycle. *Drug Alcohol Depend* 89: 183–189.

24. Ferrè S, von Euler G, Johansson B, Fredholm BB, Fuxe K (1991). Stimulation of high-affinity adenosine A2 receptors decreases the affinity of dopamine D2 receptors in rat striatal membranes. *Proc Natl Acad Sci U S A* 88: 7238–7241.

25. Ferré S, Bonaventura J, Tomasi D, Navarro G, Moreno E, Cortés A, et al. (2016) Allosteric mechanisms within the adenosine A2A–dopamine D2 receptor heterotetramer. *Neuropharmacology* 104: 154–160.

26. Fredholm BB, Battig K, Holme′n J, Nehlig A, Zvartau E (1999) Actions of caffeine in the brain with special reference to factors that contribute to its widespread use. *Pharmacol Rev* 51: 83–133.

27. Gangestad SW, Thornhill R (2008) Human Oestrus. *Proc Biol Sci.* 275(1638): 991–1000. doi: 10.1098/rspb.2007.1425.

28. Hecht GS, Spear NE, Spear LP (1999) Changes in progressive ratio responding for intravenous cocaine throughout the reproductive process in female rats. *Dev Psychobiol* 35: 136–145.

29. Huang W-L, Liu C-Y, Liu H-M, Liu C -M, Hsieh MH (2019) Sex as a moderating factor in the relationship between hippocampal volume and sensory gating in patients with schizophrenia. *Clin EEG Neurosci* 50(4): 227–230. doi: 10.1177/1550059419830562. Epub 2019 Mar 12.

30. Jackson ME, Moghaddam B (2006) Distinct patterns of plasticity in prefrontal cortex neurons that encode slow and fast responses to stress. *Eur J Neurosci* 24: 1702–1710.

31. Jones, BC, Hahn AC, Fisher CI (2018) No Compelling Evidence that Preferences for Facial Masculinity Track Changes in Women's Hormonal Status. 29(6): 996–1005. https://doi.org/10.1177/0956797618760197.

32. Karcz-Kubicha M, Antoniou K, Terasmaa A, et al (2003) Involvement of adenosine A1 and A2A receptors in the motor effects of caffeine after its acute and chronic administration. *Neuropsychopharmacology* 28: 1281–1291.

33. Kippin T, Fuchs R, Mehta R, et al (2005) Potentiation of cocaine primed reinstatement of drug seeking in female rats during estrus. *Psychopharmacology* 182: 245–252.

34. Koob GF, Nestler E (1997) The neurobiology of drug addiction. *J Neuropsychiatry Clin Neurosci*: 482–497.

35. Lauture J, Broderick PA (2014) Coffee is to a Square as Caffeine is to a Rectangle: Part 1. *J Caff Res* 4 (2): 33–34.

36. Lauture J, Broderick PA (2014) Coffee is to a Square as Caffeine is to a Rectangle: Part 2. *J Caff Res* 4 (3), 67–68.

37. Lauture J, Broderick PA (2014) Coffee is to a Square as Caffeine is to a Rectangle: Part 3 *J Caff Res* 4(4): 105–106.

38. Malave L, Broderick PA (2014) Caffeine's attenuation of cocaine-induced dopamine release. *J Caff Res* 4: 1–6.

39. Nesbitt J, Raju C, Vidal L, Alimova A, Steiner J, Broderick PA, Friedman E (2008) Sexual Dimorphism Studies of Cocaine and

Caffeine in Conditioned Place Preference: Estrous Cycle Stage and Open-Field Behaviors. Washington, DC: Society for Neuroscience.

40. Nguyen, MD, Wang Y, Ganesana M, Venton BJ (2017) Transient adenosine release Is modulated by NMDA and $GABA_B$ receptors. *Chem. Neurosci* 8(2): 376–385.
41. Pawlowski B (1999) Loss of oestrus and concealed ovulation in human evolution: The case against the Sexual-Selection Hypothesis, *Curr Anthropol* 40(3): 257–276.
42. Pereira M, Farrar AM, Hockemeyer J, Muller CE, Salamone, JD Morrell JI (2011) Effect of the adenosine A2A receptor antagonist MSX-3 on motivational disruptions of maternal behavior induced by dopamine antagonism in the early postpartum rat. *Psychopharmacology* 213 (1): 69–79. doi: 10.1007/s00213-010-2015-4. Epub 2010 Sep 17.
43. Piercey MF, Broderick PA, Hoffmann WE, Vogelsang, GD (1987) U-66444B and U-68553B, potent autoreceptor agonists at dopaminergic cell bodies and terminals. *Soc Neurosci Abstr* 13.
44. Piercey MF, Svensson KA, Lin CH, Haadsman-Svensson SR, McCall RB, Smith MW, Broderick PA, Carlsson A (1993) U-93385, a high intrinsic activity 5-HT_{1A} agonist with good oral activity. *Soc Neurosci Abstr* 19: 1866 (Broderick-Nobel Laureate Candidate-2022; Carlsson-Nobel Laureate- 2000).
45. Quinones-Jenab V, Ho A, Schlussman S, Franck J, Kreek M (1999) Estrous cycle differences in cocaine-induced stereotypic and locomotor behaviors in Fischer rats. *Behav Brain Res* 101: 15–20.
46. Quinones-Jenab V, Perrotti L, Fabian S, Chin J, Russo S, Jenab S (2001) Endocrinological basis of sex differences in cocaine-induced behavioral responses. *Ann N Y Acad Sci* 937: 140–171.
47. Quinones-Jenab V, Jenab S (2012) Influence of sex differences and gonadal hormones on cocaine addiction. *ILAR J* 53: 14–22.
48. Raju C, Nesbitt J, Yusuf A, Voutsinas N, Alimova A, Steiner JC, Broderick PA (2009) Caffeine protects against estrous cycle changes produced by cocaine: Open-Field behavioral studies. Washington, DC: *Soc for Neurosci.*
49. Ramikie TS, Ressler KJ (2018). Mechanisms of sex differences in fear and posttraumatic stress disorder. *Biol Psychiatry*, 83, 876–885. DOI: 10.1016/j.biopsych.2017.11.016.
50. Roberts DC, Bennet SA, Vickers GJ (1989) The estrous cycle affects cocaine self-administration on a progressive ratio schedule in rats. *Psychopharmacology* 3: 408–411.

51. Roney R, Simmons ZL (2008) Women's estradiol predicts preference for facial cues of men's testosterone. *Hormones and Behavior* 53(1): 14–19. https://doi.org/10.1016/j.yhbeh.2007.09.008.

52. Sato J, Nasu M, Tsuchitani M (2016) Comparative histopathology of the estrous or menstrual cycle in laboratory animals. *J Toxicol Pathol.* 2(3): 155–162. doi: 10.1293/tox.2016-0021.

53. Sell SL, Scalzitti J, Thomas M, Cunningham K (2000) Influence of ovarian hormones and estrous cycle on the behavioral response to cocaine in female rats. *J Pharmacol Exp Ther* 293: 879–886.

54. Sircar R, Kim D (1999) Female gonadal hormones differentially modulate cocaine-induced behavioral sensitization in Fischer, Lewis, and Sprague-Dawley rats. *J Pharmacol Exp Ther* 289: 54–65.

55. Solinas M, Ferré S, You Z, Karcz-Kubicha M, Popoli P, Goldberg S (2002) Caffeine induces dopamine and glutamate release in the shell of the nucleus accumbens. *J Neurosci* 22: 6321–6324.

56. Solinas M, Ferré S, Antoniou K, et al (2005) Involvement of adenosine A1 receptors in the discriminative-stimulus effects of caffeine in rats. *Psychopharmacology* 179: 576–586.

57. Valenti O, Zambon A, Boehm S (2021) Orchestration of dopamine neuron population activity in the ventral tegmental area by caffeine: comparison with amphetamine. *Int J Neuropsychopharmacol* 24(10): 832–841.

58. Walker QD, Francis R, Cabassa J, Kuhn C (2001) Effect of ovarian hormones and estrous cycle on stimulation of the hypothalamo-pituitary-adrenal axis by cocaine. *J Pharmacol Exp Ther* 297: 291–298.

59. Wang JH, Short J, Ledent C, Lawrence AJ, van den Buuse M (2003) Reduced startle habituation and prepulse inhibition in mice lacking the adenosine A2A receptor. *Behav Brain Res* 143: 201–207.

60. Wenning L, Broderick PA, Philip R (2017) Nicotine incites a transformation to estrus from any Estrous Cycle Stage. *EC Psychol Psychiatry* 3 (6): 200–201.

Chapter 10

Myths Dispelled! A Synchronously Rhythmic Brain Defines Wellness Distinct from Disease: Sensing Mobility—The Decode

Translating from Murine to Human

Synchronous brain rhythms make sense and ultimately dispel the myths that the brain uses neurotransmitters statically. Neuromolecular Imaging and BRODERICK PROBE® *nanobiosensors reveal a time sensitive synchrony in brain rhythms during neural transmission online with movement designs during natural physiology. But, when the brain is imaged online in the same subject during pathology, time sensitive synchrony is disrupted. Thus, a rhythmic brain, disrupted, defines disease: The caveat is the distinction between rhythm versus synchrony. One exists within the other. These distinct neurochemical signals from the* BRODERICK PROBE® distinguish subtleties *for cerebral health versus disease and are translatable from animal to human.*

A rhythmic, synchronous brain, disrupted, may well define brain disease. Translated across species, these phenomena from murine animal to human in a dual species contrast, the data demonstrate empirically and heuristically, the translational and existential nature of the brain's synchronous or asynchronous rhythmicity in its critical role of defining the etiology of brain disease. The modus operandi is shown here via sensing. Indeed, for the first time, original sensing nanotechnology enables

Neuroimaging: Sensing Biochemistry in the Brain
Patricia A. Broderick

ISBN 978-981-4613-49-1 (Hardcover), 978-0-429-02974-5 (eBook)
www.jennystanford.com

direct imaging inside the brain, a look inside, a continuous neurotrack a videotrace, never accomplished before. Herein, sensing is wedded across species to brain reward/drug addiction and epileptic seizures.

10.1 The Myths Dispelled

The myths dispelled? Patricia is thinking that current standards of using static data to describe the movement of neurotransmitters is best dispelled. The evidence is clear that data assume a higher degree of soundness when empirically placed online within the context of mobility patterns and/or designs. The art enlightens the science and then, the medicine. Conceptually described here, "sensing brain neurotransmitters in motion" is opposed to the apparently misleading standard of the *"all too simple"* notion that "neurotransmitters increase or decrease". Previously, this notion defined disease and perhaps somewhat mistakenly or dismissively then, defined the therapy? Dynamic data intuitively provides a more complete portrait of the temporal harmony of physiology; it differentiates intensity of a disease or drug-induced temporal cacophony *via* signal processing in online waveform patterns within continuous states of motions. *Temporal Synchrony* is the nomenclature [11].

Temporal Synchrony was discovered in our laboratory using Live Imaging, Neuromolecular Imaging (NMI), and the BRODERICK PROBE® series of nanobiosensors. This novel nanotechnology in the brain began with imaging neurotransmitter release online with specific behavioral components, walking, for example, imaged in the same mammalian murine animal in an open-field paradigm. Notably though, it is the neuroimaging, the nanosensing of each neurotransmitter online at the same time in the same subject at the same temporal and spatial resolution that brings us a nanotechnology that sheds a focus on brain disease that's a game changer. NMI and the BRODERICK PROBE® has provided the way to study the healthy brain paralleled with motor coordination. Never before accomplished, *Temporal Synchrony* reveals a distinctive rhythmic regularity between neurotransmitter synaptic release and movement frequency in the natural, physiologic state and compares the healthy state with the unhealthy state in the same subject.

Using this innovative nanotechnology of NMI and BRODERICK PROBE® nanobiosensors, a point-by-point temporal synchrony was imaged between the synaptic concentration of serotonin (5-HT) released from the specific neuroanatomic structures, dorsal striatum A_9 and ventral striatum A_{10} neuronal terminals and A_{10}, somatodendrites, that is, cell bodies. The dorsal striatum A_9 and ventral striatum A_{10} neuronal terminals are the striatum and ventrolateral nucleus accumbens (vlNAcc). Serotonin release is imaged in seconds in concert with natural open-field behaviors such as walking and finer motor skills, grooming, chewing and the like; these are fine movements, motor responses of the hand and the mouth. These fine movements assume a critically important place in diseases of movement such as *Lesch-Nyhan Syndrome* for one example.

With NMI and the BRODERICK PROBE®, serotonin (5-HT) synaptic release is traced in parallel and simultaneously with the rise and fall of the frequency of movement behaviors in the freely moving murine paradigm. Thus, temporal **synchrony** occurs when the symphony of physiologic neurochemistry and behavior is uninterrupted by drugs, disease, trauma or injury.

In contrast, neurotransmitter release within movement frequency may not act in concert when monitoring psychostimulant-induced behaviors that are produced by cocaine. In fact, after cocaine administration, NMI and BRODERICK PROBE® nanobiosensors reveal temporal ***asynchrony*** between endogenous 5-HT release within A_{10} nerve terminals, nucleus accumbens as well as their connecting somatodendritic A_{10} cell bodies, online and in real time during motor coordination sensing studies in freely moving, murine models.

Intriguingly, the psychostimulant effect of cocaine on serotonin (5-HT) release within dopaminergic accumbens neuronal terminals at A_{10} as well as A_{10} cell bodies, the ventral tegmental area (VTA), the mobility videotraces of walking and the neural finer motor skill patterns lose their synchronous properties; video-trace shows that temporal synchrony is lost after just a single dose! Interestingly, some rhythmic motoric aspects are kept. Indeed, even when utilizing a higher dose of cocaine, cell bodies in A_{10}, the ventral tegmental area (VTA) are disrupted synchronously; yet, the finer motor skills of the hand and of the mouth remain lesser affected insofar as rhythm is the concern.

These precise LIVE data can, to date, only be performed by the BRODERICK PROBE®.

Thus, the neuroanatomic substrates in the dopamine brain reward pathway are affected to different degrees and accordingly to the amount of the drug used or administered. It is apparent that the somatodendrites, the cell bodies, are directing the disruption along the dopamine neural path to its nerve terminals! The data show that cell bodies can be affected differently from nerve terminals! With the BRODERICK PROBE®, sensing mobility provides the algorithmic decode for artificial intelligence to decipher the big data derived from the tiny sensor. The data importantly point out that walking can be disrupted by cocaine while hand and mouth mobility sensing may have a better prognosis after trauma from drugs of abuse such as cocaine. The data read into the mind of the cocaine user, the brain reward seeker. The algorithms can be applied to movement disorders in multiple fields of neurology and psychiatry that include application to the quadriplegic person whether the person is with or without an epidural spinal cord implant.

So, rhythm in neuronal dopaminergic terminals can remain but synchrony is lost! The subtle differences that NMI brings to the table are similar in perspective to the strategy performed by this nanoprobe that is able to sense mobility so discretely. Each drug and/or medicine shows such subtleties that are or can be unearthed by the tiny nanoprobe, enabling personalized medicine when using this sensor, the miniature BRODERICK PROBE®. Thus, psychostimulant effects that appear to produce temporal asynchrony occur with dose dependent patterns of intensity at cell body and nerve terminal nuclei.

As the dendrites form a cable network of sophisticated communication around the brain, signaling axons to make more decisions, our data show that cocaine traumatizes the A_{10} dendrites while still, their distal nerve terminals survive after the acute dose at least in the aspects of rhythmicity. After administration of the higher dose of cocaine, sensing walking mobility is affected in the ventral tegmental area, that is the somatodendritic cell bodies while finer motor skills of hand and mouth are affected but to a lesser degree. The data see into strategies for specified protocols and highly advanced pharmacotherapies and physiotherapies for patients with compromised mobility.

Therefore, the need for NMI and the BRODERICK PROBE® to advance to the market becomes more and more imminent. ***The algorithms for walking and finer motor skills are decoded.*** *Artificial intelligence for walking and finer movement skills of the hand and mouth derive from these data.*

Temporal ***asynchrony*** was also seen during intraoperative *in vivo* surgical studies of neurotransmitter release in the anterior temporal lobe of the epilepsy patient. It is likely that time sensitive asynchrony occurred due to the chemical movement of neurotransmitters away from catecholamine and indoleamine neurotransmitters to excitatory neural transmission *via* the neuropeptide, dynorphin 1-17, which is readily imaged by NMI. Dynorphin 1-17, in fact, is a neuropeptide comprised of the widely known excitatory neurotransmitter, glutamate, capable of exciting neurons to the seizure state. Thus, NMI scans with laminar biocompatible, polymeric carbon-based BRODERICK PROBE® nanobiosensors are taken at a cortical depth of less than 2 mm for 20 min in the epilepsy patient during resection surgery intraoperatively. ***The algorithms for epileptic seizures are decoded.*** *Artificial intelligence for epileptic seizing derives from these data.*

BRODERICK PROBE® sensors are γ-irradiated within a range of 11.6–12.7 kGy. Another valuable property of this miniature sensor is its property of remaining free of bacterial infection. Studies performed in the laboratory of Dr. Philip Tierno in New York University Medical Center show that the BRODERICK PROBE® does not promote bacterial growth when sterilized with γ-irradiation. Thus, intraoperative surgery using the tiny nanoprobe in the temporal lobe of the epilepsy patient is deemed safe from bacterial infection. Further studies show that these tiny nanoprobes do not enable bacterial infection when not sterilized, enabling safe studies in the preclinical murine animal market as well.

Importantly, L-tryptophan (L-TP) was imaged online during the operation showing the lack of the ability of the asynchronous brain to produce 5-HT, the hallmark of the synchronous natural state of physiology. Therefore, L-tryptophan and neuropeptides such as excitatory dynorphins may serve as biomarkers for pharmaco- and/or gene therapy to discern neurodegenerative processes.

Within and among the multiple studies that our laboratory has performed, *in vivo, in situ* and *in vitro,* within manifold neuroanatomic layers of specific neural substrates, algorithms galore are available for discernment and enlightenment. In this manner, temporal patterns may be used to create a dynamic data profile in the clinical setting. Although static neurotransmitter levels are currently the primary standard, these static parameters will become more valuable when empirically studied within the context of movement rather than solely focusing on whether a neurotransmitter *level* has increased or decreased. Not only does this BRODERICK PROBE® provide these algorithmic data dynamically, the nanoprobe brings to the world, a more complete portrait of the temporal harmony of natural physiology. The sensing nanoprobe may also be used to distinguish the intensity of disease or drug-induced temporal cacophony on specific parts of microneuroanatomy.

10.1.1 Translating from Animal to Human

Figure 10.1 shows striking similarities in the human and murine animal brain.

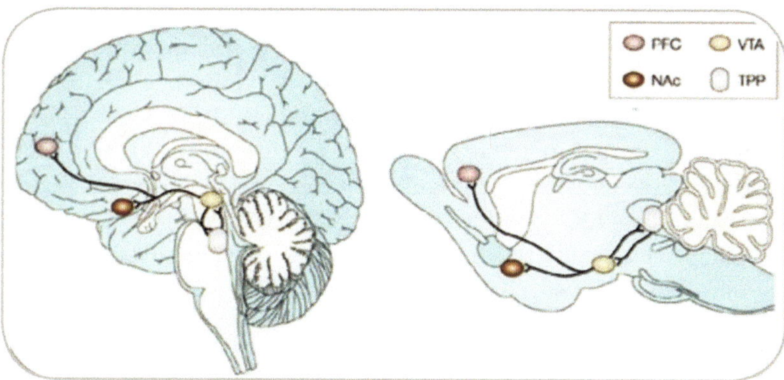

Figure 10.1 Human (left) and murine animal brain (right) depicting mesolimbic DA pathways which originate in VTA and send ascending projections to NAcc and Prefrontal Cortex (PFC). Feelings of reward as well as aversion are derived in DA and non-DA mediated reward signaling. Adapted with permission from Macmillan Pub. Ltd., *Nature Reviews/Neurosci.* 2004, 5, 55–65. http://www.nature.com/nrn/journal/v5/n1/fig_tab/nrn1298_F2.html.

10.2 Brain Reward/Drug Addiction and Epileptic Seizures

The 2016 National Survey on Drug Use and Health reported that cocaine usage has remained at a relatively constant rate since 2009 [25], and the 2017 United Nations on Drugs and Crime Report states that cocaine trafficking in both North America and Europe is on the rise [37].[1] Indeed, cocaine use is prevalent. As with other drugs of abuse, cocaine's reinforcing effects are associated with an increase of dopamine (DA) and serotonin in the reward terminal, the Nucleus Accumbens [19]. The dopaminergic DA-associated enhancement regarding reinforcement has become the conventional signature for cocaine in multiple neuroanatomic substrates [1, 18, 19, 27, 29, 33, 35, 36]. Moreover, serotonergic systems have played a tremendously important role in the conventional signature for cocaine and the role that serotonin plays may relate to both neuronal terminals and cell bodies (somatodendrites) [11, 19, 26, 28, 34, 35].

Similarly, in seizures, there is also a disruption in chemical neurotransmitter concentration, leading to abrupt, disproportionately hypersynchronous depolarizations in cortical neurons. In epilepsy, these hypersynchronous depolarizations are recurrent and unprovoked by acute central nervous system or systemic insult. Typically, a lower concentration of the inhibitory neurotransmitter gamma-aminobutyric-acid (GABA) leads to seizure activity. Moreover, in the presence of $GABA_B$ receptor agonists (i.e. Baclofen®) that are associated with second messengers, GABA concentrations are decreased presynaptically, leading to hyperpolarization *via* opening of K^+ channels [24]. Thus, a decrease in presynaptic GABA has become the conventional signature for seizure activity.

Indeed, a static DA increase is associated with cocaine usage, while a static decrease in GABA is associated with epilepsy. Although these signatures for disease states are useful, they do not convey the full pattern of neurotransmitter activity in the disease state. In this chapter, we examine a point-by-point temporal

[1]UN News, (June 26, 2024), reports a spike in cocaine use with violent behavior. NEW YORK (WABC), (August 13, 2024) reports the emergence of "pink cocaine," a mixture of a small amount of cocaine likely combined with varying amounts of amphetamine, ketamine and/or fentanyl.

synchrony between synaptic neurotransmitter concentration and movement in the untreated and innocent state, so to speak, the natural physiologic state. In the physiologic state, synaptic neurotransmitter release reflects synchronous movement frequency that is rhythmic with neural transmission.

In particular, this point-by-point temporal synchrony as imaged while sensing dexterous hand and mouth skills and agile walking skills as neurotransmission is imaged, is indeed, the forte of the BRODERICK PROBE® sensor. Together with psychostimulant asynchrony found by the BRODERICK PROBE®, the BRODERICK PROBE® algorithms on its interfacial surface provides artificial intelligence for cutting-edge diagnosis and treatment. The nanoprobe is patented and trademarked worldwide to enable the epidural spinal cord person to walk with celerity and to help the autistic epileptic person to speak. The autistic person will rather enjoy speaking without neuronal excessive synchrony in the interruptive art of epilepsy.

Jacobs and Azmitia posit that serotonin's (5-HT's) primary function in CNS neuronal circuitry is to facilitate motor output [31]. As discussed by Broderick and Phelix [23], 5-HT somatodendritic cell bodies exhibit automaticity as they depolarize with regularity by using central pattern generators (CPGs) to produce plateau potentials. The repetitive discharge characteristics of serotonin (5-HT) neurons occurs in the somatodendritic raphe nuclei before movement onset such as increased muscle tone; and this 5-HT neuronal cell firing is maintained during sustained movements [30]. Moreover, serotoninergic raphe nuclei cell firing may be phase-locked to repetitive behavioral stereotypic finer motor skill responses of the hand and of the mouth, activated preferentially and in association with chewing, locomotion, and CPG-stimulated stereotypic behaviors [32].

Thus, in the disease state, we examine what happens to synchrony and/or rhythm after usage of an acute (single) dose of cocaine and we examine temporal asynchrony between synaptic neurotransmitter concentration and seizure activity in the epileptic state using NMI and BRODERICK PROBE® nanobiosensors. In the case of the anesthetized human epilepsy patient, physical movement was not possible at this time, but the use of the newest laser, fiber optic BRODERICK PROBE® voltaic photonic protein probe [16] enables this goal.

Nonetheless, temporal asynchrony and/or synchrony in terms of movement designs such as chemical movement from non-toxic neurotransmission to toxic neurotransmission is established with NMI and BRODERICK PROBE® biosensors. ***The* BRODERICK PROBE® *is a brain transmitter not limited to the brain***. Hence, for the first time, the neurochemistry underlying normal, compared with pathologic neuronal function in the intact brain of the epilepsy patient as studied intra-subject, during surgery, is reported. The work is demonstrated *in vivo* at NYU Tisch Hospital under Institutional Review Board (IRB) approval with the BRODERICK PROBE® nanobiosensors. The work presented herein then, further advances the field of nanotechnology and epilepsy because empirical imaging is enabled in the epilepsy patient intraoperatively for the first time [17, 20–22].

10.2.1 Sensing Mobility: Murine Studies

In adult murine mammalians, NMI with the BRODERICK PROBE® was used to study the continuous empirical relationship between endogenous 5-HT release in basal ganglia and natural movement behavior, which is directed by basal ganglia *per se*. Specific studies imaging 5-HT release in dopaminergic (DA-ergic) motor/reward pathways in the nerve terminals, A_9, dorsal striatum (DStr), nerve terminals, A_{10}, ventral striatum (nucleus accumbens) (NAcc) and somatodendrites (cell bodies), Ventral Tegmental Area (VTA), were performed, *in vivo*. NMI enables empirical studies in each murine subject as its own control. Movement was monitored with computerized infrared photobeam cells online with imaging neurochemistry, indigenous to each neuroanatomic substrate.

A highly specialized and customized plexiglass chamber enclosed with copper was manufactured by the Albert Einstein College of Medicine Machine Shop (Bronx, NY), further modified by San Diego Instruments (San Diego, CA) and still further modified by the Broderick Laboratory (Manhattan, NY).

For the sensing mobility studies, our laboratory installed a novel commutator circuit to connect with the Broderick instant inventive detector to sense neurotransmission LIVE while the computerized infrared photobeams surrounding the outside of the chamber assessed each mobility signal in the subject LIVE.

Each movement type is monitored at the same time as LIVE neuromolecular signals are videotracked directly from each brain substrate online within seconds and sub-seconds. Thus, the copper did not touch or hamper the animals' free movement, exploratory or habituated.

Copper is used around the specialized chamber to avoid possible white light; brain signaling was uninterrupted by noise. Electrical systems were engineered grounded for animals' safety and untampered data entry and collection. The author is the designer of her laboratory as well as the three "side by side" laboratories on the first floor (previously called, the Plaza Level) of the Marshak Building at the site of the City College of New York and the CUNY School of Medicine. The experimental setup is shown in Figure 10.2.

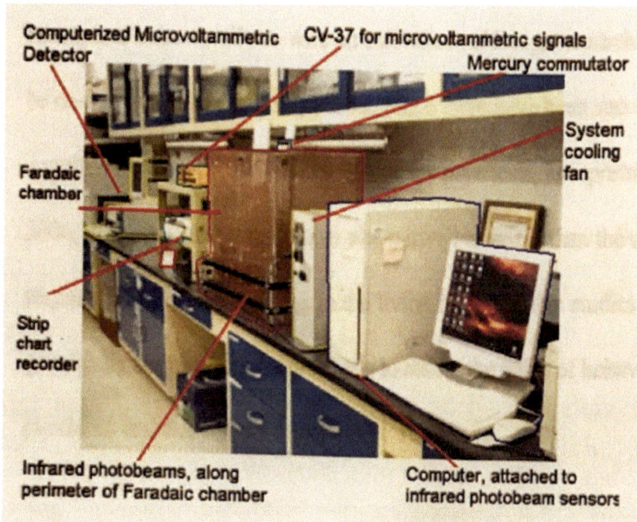

Figure 10.2 Digital photographs of the NMI analytical devices and behavioral (open-field) equipment used for the simultaneous study of neurochemistry and behavior.

Therefore, these studies have two control systems. The first control system is the inventive NMI paradigm wherein the murine model of animals is their own control; this property is enabled for the first time by the tiny Broderick nanoprobe. In the second control system, the murine freely moving and behaving animal model does not receive cocaine and is imaged for one hour of normal/natural behavior during which the non-prescient mammal explore. Subsequently, the subject receives an injection of saline and this takes place in the second hour without disturbing the animal; this is another property enabled by the tiny Broderick nanoprobe. Decoding and sensing mobility provides algorithms and LIVE tethered telemetry takes place without pause. Next, the subject receives a drug, In this case, it is cocaine (10 and 20 mg/kg) and decoding sensing immobility, impaired by psychostimulation, is accomplished, thereby computerizing algorithms for temporal asynchrony for machine learning, aka Artificial Intelligence (AI).

The BRODERICK PROBE® may well be the *only* technology to date that enables imaging the natural state for comparison with the disease state in the same subject and this is why intrasubject imaging of temporal synchrony and temporal asynchrony advance the science of medicine dramatically. Importantly, work in our laboratory progresses to use such LIVE online algorithmic data to encode artificial intelligence (AI) instructions on the interfacial surface of the BRODERICK PROBE®. Thus, the tiny probe provides another instant inventive medical advance to help, for one example, the quadriplegic person to walk.

In studies with murine models, using carbon-based BRODERICK PROBE® biosensors with our patented detector [3, 7, 8] configured by the author and described in chapter 4 herein, the sensitivity parameter can be, for one example, 5 nanoamperes/volt at 10 millivolts/second at a 5 second time constant or a 1 second time constant as well will provide reliable operational stability for the sensor. The BRODERICK PROBE® is patented in the United States and trademarked worldwide for both murine and human use [3, 7, 8, 12–16, 21].

It is important to note that in the recent patent issued in December, 2021 [16], quantum mechanics with the BRODERICK PROBE® purple protein biosensor, rhodopsin, inventive lasers and fiber optic voltaics are leading the way in imaging technologies. Intriguing is the utter novelty that the brain is not opened at all! Further intriguing is the patent by Broderick [14] issued in April, 2021, wherein nanoscale science of advanced materials with fullerene nanotechnology uses semiconductor inventive art to specialize deeper into the exciting exploration of the elusive brain.

10.2.2 Sensing Mobility: Human Studies

Using semiderivative and linear voltammetric circuits, NMI, and BRODERICK PROBE® biosensors, the selective detection of specific neurotransmitters and neurochemicals in discrete parts of the intact brain are imaged within a temporal resolution of seconds in a 42-year-old female patient with temporal lobe epilepsy and generalized tonic-clonic seizures. This patient underwent partial cortical resection with intraoperative *in vivo* studies using NMI and BRODERICK PROBE® biosensors. These studies began with electroencephalographic (EEG) monitoring wherein the site of cortical resection was defined with subdural grid epilepsy electrodes, placed on the surface of the brain with epilepsy strip electrodes placed deeper in the temporal lobe (i.e. hippocampus).

Then, biosensors, with a diameter 5 times less than epilepsy depth electrodes for invasive EEG were placed by direct visualization in the exposed cortical region with and without epileptic spike activity in regions destined for resection. About 6 to 10 recordings with γ-irradiated (11.6–12.7 kilograys), laminar biocompatible carbon-based BRODERICK PROBE® laurate biosensors were taken at a cortical depth of microns to less than 2 mm for a 20–30 min time period. Glial Fibrillary Acidic Protein (GFAP) stains of tissue were post-surgically conducted in order to assess gliosis or other neuronal damage after electrode placement. ***Based on the pathology report from the NYU surgical team, "the*** BRODERICK PROBE® ***biosensors did not cause gliosis, sclerosis, brain/neuronal damage, or histiocyte infiltration and are, thus, a safe biomedical tool."***

10.3 BRODERICK PROBE® Biosensors

Figures 10.3 and 10.4 illustrate the BRODERICK PROBE® and its use in nanotechnology to study neural transmission in both the murine and human brain.

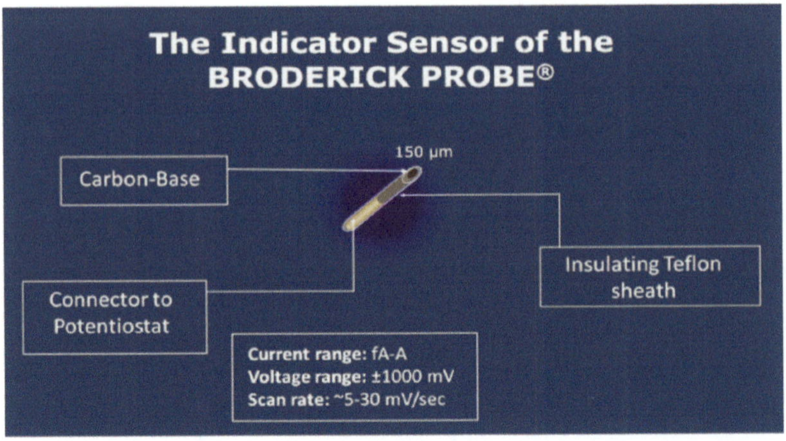

Figure 10.3 Schematic drawing of just one of several designs for the BRODERICK PROBE® series of nanobiosensors.

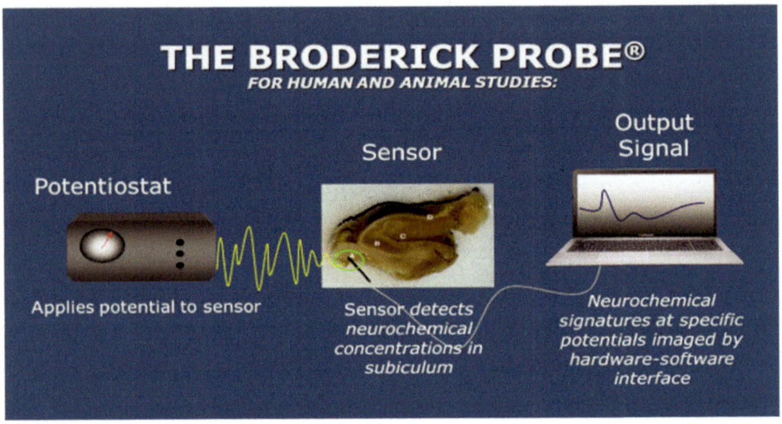

Figure 10.4 The diagram emphasizes the tiny Broderick nanoprobe sensing *a* longitudinal section of the human hippocampus resected *en bloc* from the epilepsy patient *post* NMI intraoperatively. A: subiculum; B: pyramidal cell layer; C: granular cell layer of the dentate gyrus (DG); D: alveus.

The BRODERICK PROBE® encompasses several formulations of carbon and protein-based biosensors. The patents are held by *Eazysense Nanotechnologies Incorporated* in New York State, USA as assignee under an Option Agreement with the CUNY Technology Commercialization Office (TCO) of the City University of New York. Sensors are tested in research paradigms, protocols and seminal publications available in our laboratory and herein [2, 4, 5, 9, 10, 18]. A description of specific components for each biosensor formulation, as well as the use, design and applications for these nanobiosensors, are published; detector/potentiostat, electrical circuits and associated NMI biotechnology are also patented. As mentioned, the patents describe novel inventive constructions and formulations of the Broderick biosensor [3, 7, 8, 12–16, 21].

The sensor selects an image for a specific neurochemical at an electroactive oxidation/half wave potential using electron transfer kinetics. Subunits of volts and amperes are used to measure the electroactive signature for each neurochemical and is reliant on the electronic circuitry paired with the detector [3, 7, 8]. In many of these studies, a semi-derivative reduction electrical circuit and a linear circuit were chosen. Several neurotransmitters and neurochemicals can be separately imaged within less than 1 min with each neurochemical imaged within sub-seconds, and recordings can be repeated continuously for hours or longer periods of time up to months without gliosis and bacteria formation.

Specific sensing properties including the hydrophobicity and hydrophilicity of biosensor formulations within the context of specific anion/cation exchange interactions with neurotransmitters is used to derive current from electron transfer kinetics. In the following research studies, neurotransmitter profiles are imaged using the BRODERICK PROBE® laurate biosensor. Further, γ-butyrolactone investigations are used to confirm imaging of release mechanisms at the pre-synapse of the neuron [5]. Indeed, γ-butyrolactone is used to block neuronal depolarization in specific neuroanatomic sites to prove this release neuronal *modus operandi* empirically and the data are added to the already burgeoning data showing that reuptake inhibition at the synapse is included.

Without wishing to be bound by theory, the author believes that between the surface of the sensor and the biomolecule *per se,* weak covalent bonds are formed by first adsorption of the biomolecule onto the surface of the instant inventive sensor; the sensor is comprised of adsorptive carbon or carbon allotropes, semiconductor and protein-based configurations of highly polar covalent and hydrogen/covalent bonds, intra- and inter-molecularly.

Also, the author wishes to indicate that, although highly polar covalent bonding that includes hybrid hydrogen/covalent bonds provide rationale for the memory or reflective image of the biomolecule produced by the BRODERICK PROBE® sensor, it is plausible that a number of other interactions may occur between the anion-cation biomolecule and the instant inventions. For one example, chemisorption is an adsorptive interaction between a molecule and a surface in which electron density is shared by the adsorbed molecule or molecules and the sensor surface. Physisorption is another plausible explanation for the reflective adsorptive image on the instant inventions. In this mechanism, a biomolecule does not actually undergo a specific chemical bonding. Physisorption involves an electrostatic attraction or dipole-dipole interaction but no real chemical bonds are made. Instead, there is a sharing of electrons between the biomolecule and the instant inventive sensor material, that is to say, as the adsorbate, *i.e.,* the biomolecule migrates closer to the surface of the sensor, the dipole moment of the adsorbate induces an ***image dipole***. *Since dipoles attract each other and generate an electromagnetic field on the surface, the electromagnetic field may well factor into reflection or imaging of the biomolecule or biomolecules on the part of the instant sensor invention.*

10.4 Algorithms for Sensing Mobility in Healthy Subjects Are Decoded: Wellness-Temporal Synchrony

NMI studies were conducted on freely moving and behaving non-prescient mammals. Concurrently, infrared photocell studies of open-field behavior in a novel environment were performed. Thus, each neurochemical data point had a corresponding, correlative behavioral data point. In the control, natural and physiological condition, basal 5-HT release was detected at synaptic environments at

- A_{10} somatodendrites, Ventral Tegmental Area (VTA)
- A_{10} terminal field, Ventrolateral Nucleus Accumbent (vlNAcc)
- A_9 terminal field, Dorsal Striatum (DStr)

10.4.1 Decoding the Algorithm for Walking from the Brain Reward Cell Bodies: VTA

Figure 10.5 demonstrates the release of 5-HT in **VTA** (A_{10} DA cell body neurons) as the subject walks.

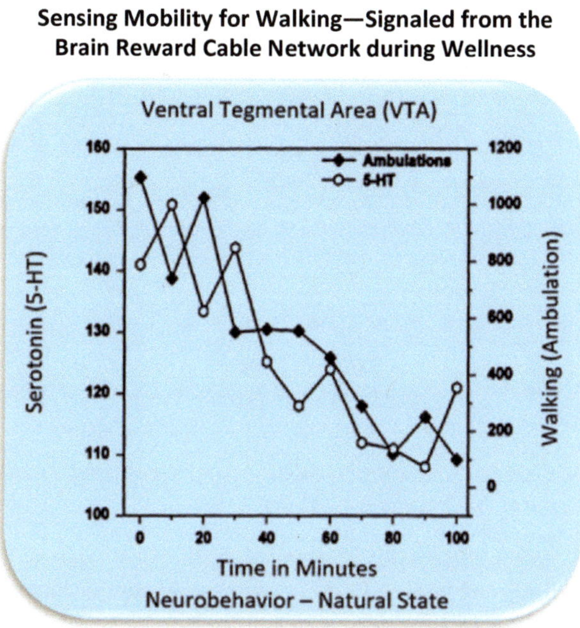

Figure 10.5 *Sensing Walking During Wellness. Natural Neurobehavior*: The *y* coordinate axis videotraces the instant synaptic release of the neurotransmitter, serotonin, (5-HT) (open circles) as imaged with the BRODERICK PROBE® at ventral tegmental brain reward A_{10} cell bodies, **VTA,** in real time while the subject is walking (closed circles). The *x* axis denotes tempo at points in time (first hour-exploratory) and subsequent *post-saline* injection (second hour-habituation). *Note the time sensitive juxtaposition of algorithmic data that show brain sensing neurotransmission online with behavior.*

10.4.2 Decoding the Algorithm for Hand and Mouth Skills: The Brain Reward Cell Bodies—VTA

Figure 10.6 demonstrates the release of 5-HT in **VTA** (A_{10} DA cell body neurons) as the subject performs finer motor responses of the hand and mouth.

Sensing Mobility for Motor Skills of the Hand and Mouth—Signaled from the Brain Reward Cable Network During Wellness

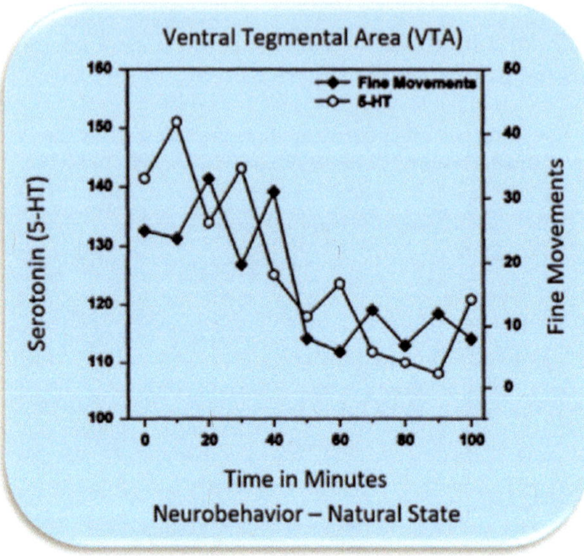

Figure 10.6 *Sensing Finer Motor Skills of the Hand and Mouth During Wellness. Natural Neurobehavior*: The *y* coordinate axis videotraces the instant synaptic release of the neurotransmitter, serotonin, (5-HT) (open circles) as imaged with the BRODERICK PROBE® at ventral tegmental brain reward A_{10} cell bodies, VTA, in real time while the subject is using finer motor skills of the hand and mouth (closed circles). The *x* axis denotes tempo at points in time (first hour-exploratory) and subsequent *post-saline* injection (second hour-habituation). *Note that the time sensitive juxtaposition of these algorithmic data showing brain sensing of neurotransmission with fine motor behavior is discretely different than that seen in VTA derived walking.*

10.4.3 Sub-Summary

Figure 10.5 *Sensing Walking During Wellness-Temporal Synchrony with Rhythmicity is shown.*

Figure 10.6 Sensing *Finer Motor Skills During Wellness-Temporal Synchrony with Rhythmicity is shown.*

Clinical applications: When temporal synchrony is present in the A_{10} ventral tegmental reward center, depression comorbid with substance abuse, for one example, may be avoided.

10.5 Algorithms for Sensing Mobility in Healthy Subjects Are Decoded: Wellness—Temporal Synchrony: Part 1

10.5.1 Decoding the Algorithm for Walking: The Brain Reward Terminals—Ventrolateral (vl) Nucleus Accumbens (NAcc)

Figure 10.7 demonstrates the release of 5-HT in **vlNAcc** (an A_{10} DA terminal field) as the subject walks.

Sensing Mobility for Walking—Signaled from the Brain Reward Nerve Terminals During Wellness

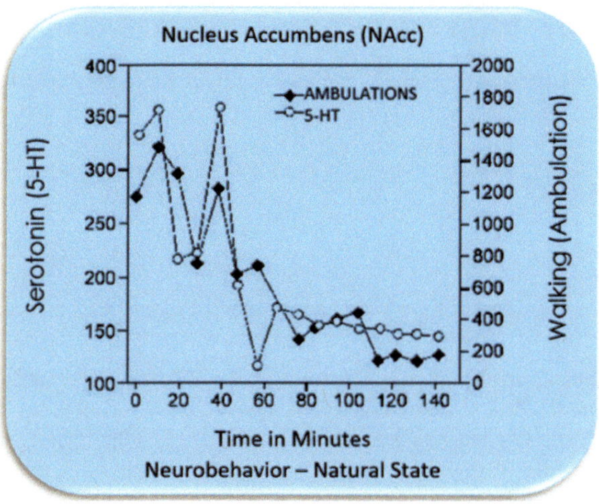

Figure 10.7 *Sensing Walking During Wellness. Natural Neurobehavior*: The *y* coordinate axis videotraces the instant synaptic release of the neurotransmitter, serotonin, (5-HT) (open circles) as imaged with the BRODERICK PROBE® at accumbens brain reward A_{10} terminals, NAcc in real time while the subject is walking (closed circles). The *x* axis denotes tempo at points in time (first hour-exploratory) and subsequent *post-saline* injection (second hour-habituation). ***Note the simultaneous and synchronous rhythm in the videotrace.***

10.5.2 Decoding the Algorithm for Hand and Mouth Skills: The Reward Center—Accumbens Nerve Terminals

Figure 10.8 demonstrates the release of 5-HT in **vlNAcc** (an A_{10} DA terminal field) as the subject performs finer motor responses of the hand and mouth.

Sensing Mobility for Motor Skills of the Hand and Mouth— Signaled from the Brain Reward Nerve Terminals During Wellness-NAcc

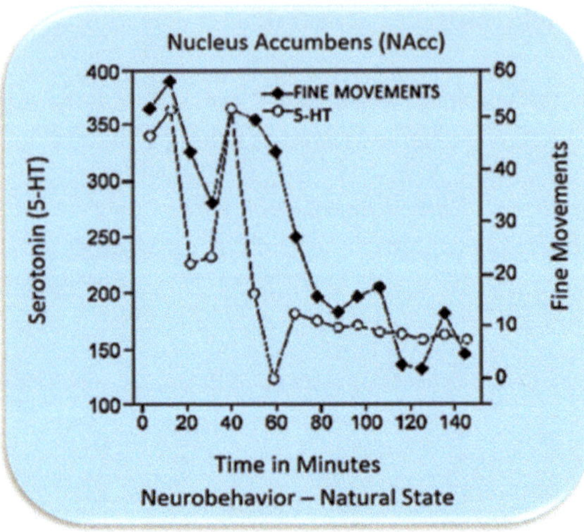

Figure 10.8 Sensing *Finer Motor Skills During Wellness: Natural Neurobehavior*: The *y* coordinate axis characterizes the instant synaptic release of the neurotransmitter, serotonin, (5-HT) (open circles) as imaged with the BRODERICK PROBE® at A_{10} terminals nucleus accumbens (NAcc) in real time while the subject conduces finer delicate movements of the hand and the mouth (closed circles). The *x* axis denotes tempo at points in time (first hour-exploratory) and subsequent *post-saline* injection (second hour-habituation). ***Note the simultaneous and synchronous rhythm in the videotrace.***

10.5.3 Sub-Summary

Figure 10.7 *Sensing Walking During Wellness-Temporal Synchrony with Rhythmicity is shown.*

Figure 10.8 Sensing *Finer Motor Skills During Wellness-Temporal Synchrony with Rhythmicity is shown.*

Clinical applications: When Temporal Synchrony is present in the A_{10} terminal reward center, depression, for one example, may be avoided.

10.6 Algorithms for Sensing Mobility in Healthy Subjects Are Decoded: Wellness—Temporal Synchrony: Part 2

10.6.1 Decoding the Algorithm for Walking: The Basal Ganglia-Striatal Nerve Terminals

Figure 10.9 demonstrates 5-HT release at A_9 terminal field, **DStr**, while the subject walks.

Sensing Mobility for Walking—Signaled from the Striatal Nerve Terminals During Wellness

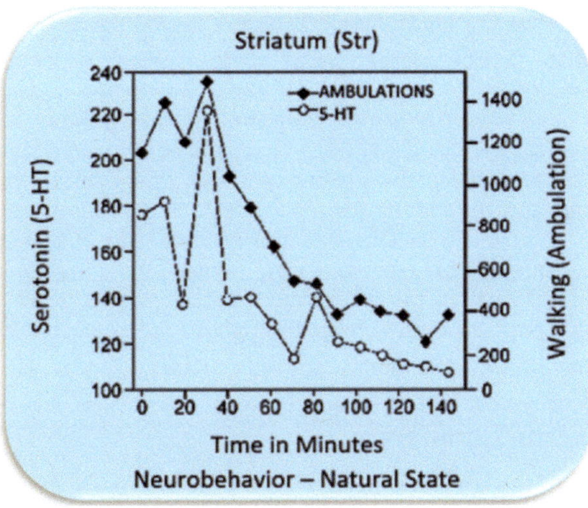

Figure 10.9 *Sensing Walking During Wellness. Natural Neurobehavior*: The *y* coordinate axis videotraces the instant synaptic release of the neurotransmitter, serotonin, (5-HT) (open circles) as imaged with the BRODERICK PROBE® at striatal basal ganglia A_9 terminals, DStr in real time while the subject is walking (closed circles). The *x* axis denotes tempo at points in time (first hour-exploratory) and subsequent *post-saline* injection (second hour-habituation). *Note the simultaneous and synchronous rhythm in the videotrace.*

Figure 10.10 demonstrates 5-HT release at A_9 terminal field, **DStr**, while the subject conduces finer delicate movements of the hand and the mouth.

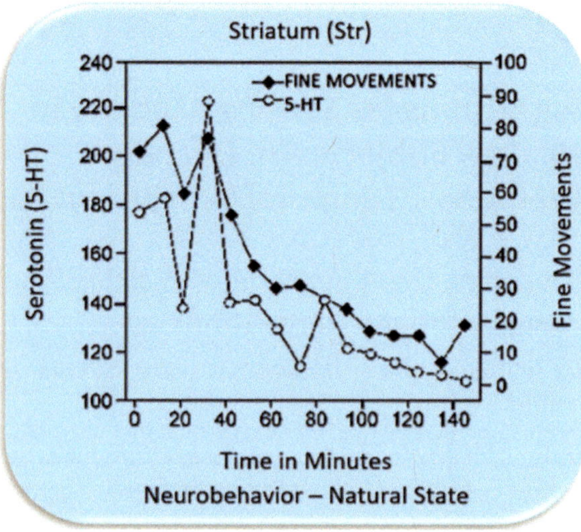

Figure 10.10 Sensing *Finer Motor Skills During Wellness: Natural Neurobehavior*: The *y* coordinate axis characterizes the instant synaptic release of the neurotransmitter, serotonin, (5-HT) (open circles) as imaged with the BRODERICK PROBE® at striatal basal ganglia A_9 terminals, DStr in real time while the subject conduces finer delicate movements of the hand and the mouth (closed circles). The x axis denotes tempo at points in time (first hour-exploratory) and subsequent *post-saline* injection (second hour-habituation). *Note the simultaneous and synchronous rhythm in the videotrace. Note the skillful certainty of the nanoprobe. Serotonin directs hand and mouth motor coordination while the reverse is clear for waling.*

10.6.2 Sub-Summary

Figure 10.9 *Sensing Walking During Wellness-Temporal Synchrony with Rhythmicity is shown.*

Figure 10.10 Sensing *Finer Motor Skills During Wellness-Temporal Synchrony with Rhythmicity is shown.*

Clinical applications: When Temporal Synchrony is present in the basal ganglia, Parkinson disease may be avoided.

10.7 Algorithms for Sensing Mobility in Unhealthy Subjects Are Decoded: Temporal Asynchrony

NMI studies were conducted on freely moving and behaving non-prescient mammals. Concurrently, infrared photocell studies of open-field behavior in a novel environment were performed. Thus, each neurochemical data point had a corresponding, correlative behavioral data point. In the control, natural and physiological condition, basal 5-HT release was detected at synaptic environments at

- A_{10} terminal field, Ventrolateral Nucleus Accumbens (vlNAcc)
- A_{10} somatodendrites, Ventral Tegmental Area (VTA)

In each case in the following studies, the murine model, free to think, albeit anthropomorphically, and behave unrestrained, is administered a drug. In this case, the drug is cocaine.

10.7.1 Decoding the Algorithm for Walking: The Brain Reward Nerve Terminals: NAcc—Psychostimulant

Figure 10.11 illustrates instant release of the neurotransmitter, serotonin (5-HT) at A_{10} cell bodies, **NAcc**, while the subject walks under psycho-stimulation.

Figure 10.12 demonstrates 5-HT release at A_9 terminal field, **NAcc**, while the subject conduces finer delicate movements of the hands and the mouth.

10.7.2 Sub-Summary

Figure 10.11 *Sensing Walking During Psycho-Stimulation: Temporal Synchrony is lost-Rhythmicity remains intact.*

Figure 10.12 Sensing *Finer Motor Skills During Psycho-Stimulation-Temporal Synchrony is lost-Rhythmicity remains intact.*

Clinical applications: When Temporal Asynchrony is present in the **NAcc** under psycho-stimulation and juxtaposition of signals do not mimic the natural cable cell body network, axonal decisions may be impaired. The outcome may be undesirable, that is, other neural pathways are blocked downstream blocking attempts to walk and creating motoric imbalance.

Sensing Mobility for Walking—Signaled from the NAcc During Cocaine (Drug) Psycho-Stimulation

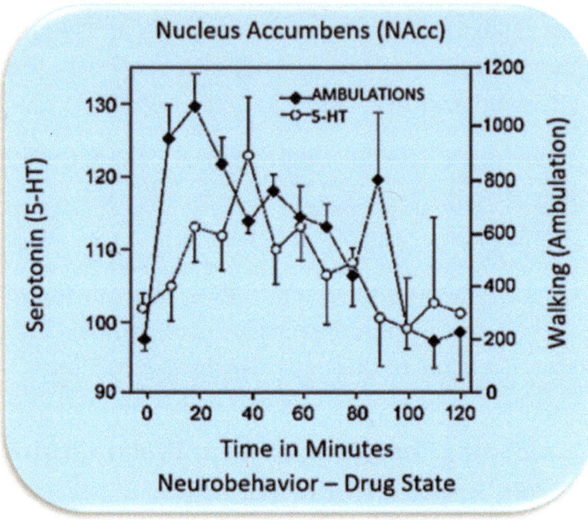

Figure 10.11 *Sensing Walking During Psycho-Stimulation: Cocaine (Drug) Neurobehavior*. The *y* coordinate axis video-traces the instant synaptic release of the neurotransmitter, serotonin, (5-HT) (open circles) as imaged with the BRODERICK PROBE® at accumbens neurons, **NAcc**, in real time while the subject is walking (closed circles). The *x* axis denotes tempo at points in time (first hour-exploratory) and subsequent *post-saline* injection (second hour-habituation). *The algorithmic data shows some temporal rhythm; the algorithm shows temporal asynchrony after drug.*

Note: LIVE electrochemical signals inspired cerebrally in **NAcc** while walking under the influence of a psychostimulant are temporally asynchronous. Thus, a single dose of cocaine will change algorithms for NAcc sensing mobility for walking. Temporal synchrony abates while neural rhythm is somewhat maintained. The data are most important for psychostimulant differential diagnosis used by artificial intelligence for drug discovery in multiple fields of psychiatry and neurology.

Sensing Mobility for Finer Motor Skills of the Hands and Mouth—Signaled from the NAcc During Cocaine Psycho-Stimulation

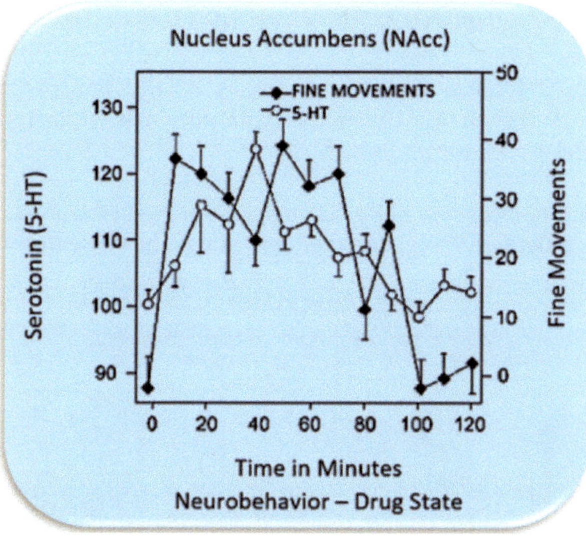

Figure 10.12 Sensing *Finer Motor Skills During Psycho-Stimulation: Cocaine (Drug) Neurobehavior*: The *y* coordinate axis characterizes the instant synaptic release of the neurotransmitter, serotonin, (5-HT) (open circles) as imaged with the BRODERICK PROBE® at Nucleus Accumbens A_{10} cell bodies, (NAcc) in real time while the subject conduces finer delicate movements of the hand and the mouth (closed circles). The *x* axis denotes tempo at points in time (first hour-exploratory) and subsequent *post-saline* injection (second hour-habituation). *The algorithmic data shows some temporal rhythm; the algorithm shows temporal asynchrony in the fine motor movements of the hand and mouth after drug.*

Note: LIVE electrochemical signals inspired cerebrally in **NAcc** while performing ***fine motor movements of the hand and mouth*** under the influence of a psychostimulant are temporally asynchronous. Thus, a single dose of cocaine will change algorithms for NAcc sensing mobility for ***fine motor movements of the hand and mouth.*** Temporal synchrony abates while neural rhythm is somewhat maintained. The data are most important for psychostimulant differential diagnosis used by artificial intelligence for drug discovery in multiple fields of psychiatry and neurology.

10.8 Algorithms for Sensing Mobility in Unhealthy Subjects Are Decoded: Temporal Asynchrony VTA-Psychostimulant

Figure 10.13 illustrates instant release of the neurotransmitter, serotonin (5-HT) at A_{10} cell body cable network while the subject walks under psycho-stimulation.

Sensing Mobility for Walking—Signaled from the Cell Body Neurons during Cocaine (Drug) Psycho-Stimulation: The Cable Network

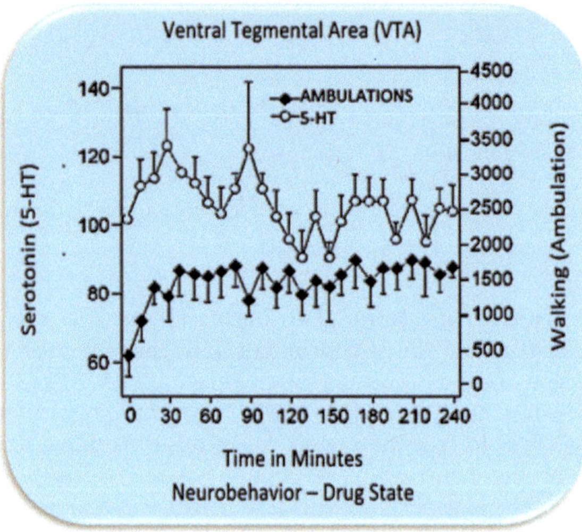

Figure 10.13 (*Continued*) *Sensing Walking During Psycho-Stimulation:* Cocaine conduces neurochemistry with behavior: The *y* coordinate axis video-traces the instant synaptic release of the neurotransmitter, serotonin, (5-HT) (open circles) as imaged with the BRODERICK PROBE® at ventral tegmental neurons, **VTA**, in real time while the subject is walking (closed circles) under the influence of a drug, a psychostimulant. The *x* axis denotes tempo at points in time (first hour-exploratory) and subsequent *post-saline* injection (second hour-habituation). *The algorithm shows signals for walking that plateau and signals for the release of 5-HT that do not plateau. The two parts of the equation shows that the algorithm is not rhythmic. The algorithm displays temporal asynchrony.*

Note: LIVE electrochemical signals inspired cerebrally while signally walking derived online from VTA during psycho-stimulation are not rhythmic and are temporally asynchronous. ***The response is from a single dose of cocaine.***

Figure 10.14 illustrates instant release of the neurotransmitter, serotonin (5-HT) at A_{10} cell body cable network while the subject uses finer motor skills of the hands and mouth under psycho-stimulation.

Sensing Mobility for Hands and Mouth Motor Coordination—Signaled from the Cell Body Cable Network During Cocaine (Drug) Psycho-Stimulation

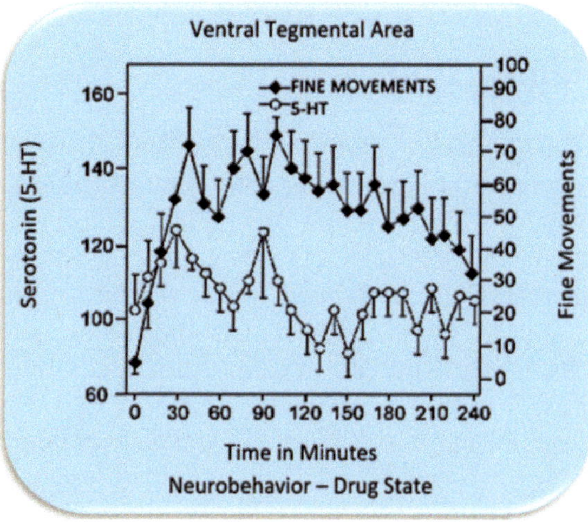

Figure 10.14 Sensing *Finer Motor Skills During Psycho-Stimulation: Cocaine (Drug) Neurobehavior*: The *y* mathematical axis characterizes the instant synaptic release of the neurotransmitter, serotonin, (5-HT) (open circles) as imaged with the BRODERICK PROBE® at ventral tegmental A_{10} cell bodies, VTA, in real time while the subject conduces finer delicate movements of the hand and the mouth (closed circles). The *x* axis denotes tempo at points in time (first hour-exploratory) and subsequent *post-saline* injection (second hour-habituation). *The algorithm for data from imaging finer motor skills of the hand and mouth commensurate and online with 5-HT release, LIVE, shows rhythmicity and yet the algorithm shows temporal asynchrony.*

Note: LIVE electrochemical signals inspired cerebrally while signally walking derived online from VTA during psycho-stimulation are not rhythmic and are temporally asynchronous. ***The response is from a single dose of cocaine.***

10.8.1 Sub-Summary

Figure 10.13 *Sensing Walking During Psycho-Stimulation: Temporal Synchrony is lost-Rhythmicity is lost.*

Figure 10.14 Sensing *Finer Motor Skills During Psycho-Stimulation- Temporal Synchrony is lost-Rhythmicity is not lost.*

Clinical applications: When Temporal Asynchrony is present in the **VTA** under psycho-stimulation, the cable cell body network as well as axonal decisions may well, indeed, be impaired. The outcome may be most likely undesirable at several levels of the central and peripheral nervous system.

10.9 Human Studies

The patients undergo partial cortical resection with intra-operative *in vivo* studies using NMI and BRODERICK PROBE® biosensors before the resection procedures. These studies begin with electroencephalographic (EEG) monitoring wherein the site of cortical resection is defined with subdural grid epilepsy electrodes, placed on the surface of the brain with epilepsy strip electrodes placed deeper in the temporal lobe (i.e. hippocampus).

In Figure 10.15, a photo of an epilepsy patient's brain is depicted. Note the ***Great Vein***, called the ***Vein of Galen***, is clear.

Exposed Temporal Lobe of the Epilepsy Brain during Intraoperative Surgery

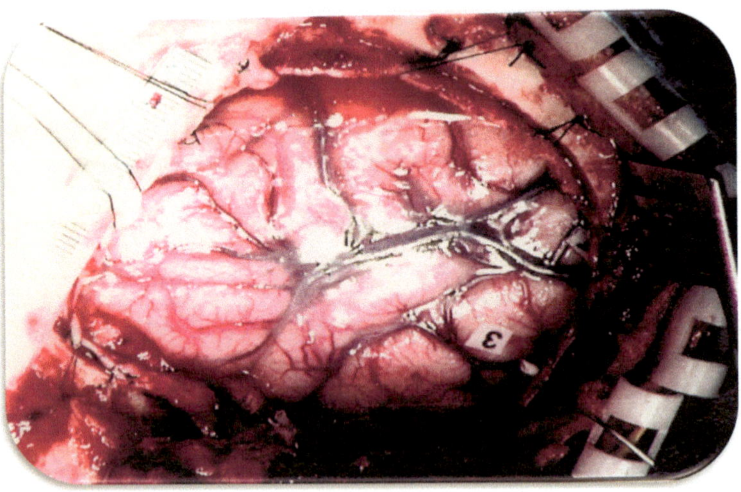

Figure 10.15 The tiny probe is fixed safely in place in the anterior cortex of the epilepsy patient.

LIVE *In Vivo* Data Online During Intraoperative Surgery in the Epilepsy Patient

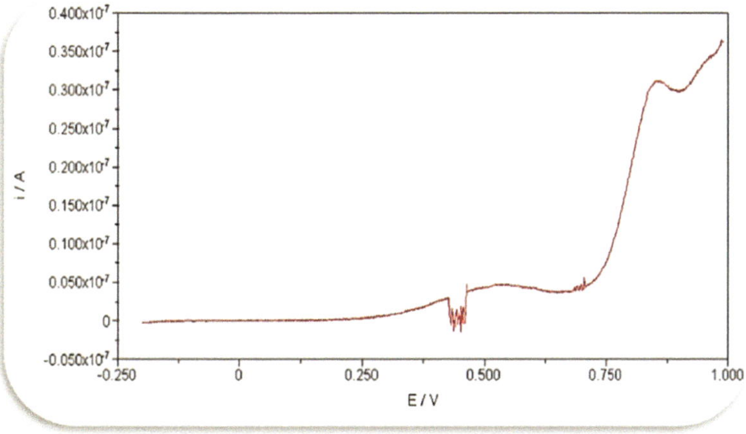

Figure 10.16 Intra-Operative NMI *in vivo;* the image is derived from the human epilepsy patient using the BRODERICK PROBE®. Linear circuit: On the *x*-axis: oxidation potential in millivolts (mV). On the *y*-axis: current in μamps. Starting at oxidation potential, 0.450 V, HVA, L-TP, DYN, & SRIF are imaged. L-TP and Peptides are present in higher concentration than the monoamines or indoleamines. Biogenic amines are below detectable limits.

Neuromolecular Imaging (NMI) with the BRODERICK PROBE®: *In Vitro* Data

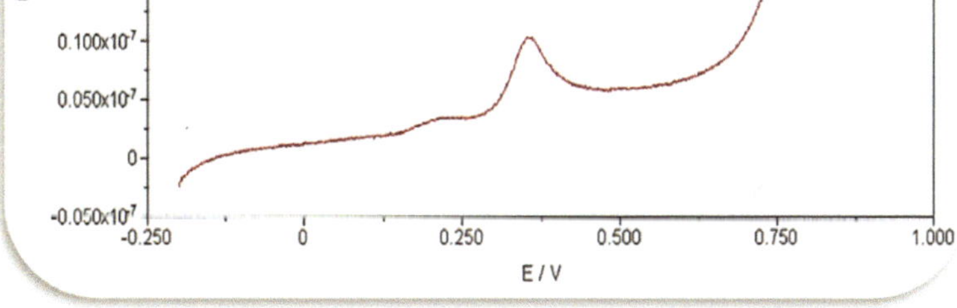

Figure 10.17 NMI and BRODERICK PROBE®: ***in vitro*** *calibration studies for neurotransmitters and neurochemicals imaged in epilepsy patient tissue in situ and epilepsy patients in vivo.* Linear circuit: On the *x*-axis: oxidation potential in millivolts (mV). On the *y*-axis: current in μamps times 10. Starting from right to left, DA, 5-HT, HVA, (*continued*)

Figure 10.17 (*continued*) L-TP, DYN, and SRIF are pipetted into phosphate buffer in the electrochemical call and neurotransmitters are subsequently studied in a series of scans. The higher concentration of HVA and SRIF are below detectable limit and the biogenic amines are predominant. The data confirm the patient data as L-TP and Peptides are present in higher concentration than the monoamines or indoleamines in the brain of the epilepsy patient.

Note: The abbreviations for neurotransmitters and neurochemicals are listed in Table 10.1.

Table 10.1 delineates molarity detection limits for biogenic amines and neuropeptides gleaned from pre-calibration and post-calibration *in vivo* and *in situ* empirical data from patients diagnosed with epilepsy. Also, *in vitro* data from calibration protocols in our laboratory provides further useful data toward the tabulation.

Table 10.1

MOLARITY-PRELIMINARY DATA

		Detection Limits - *In Vitro*	Detection Limits - *In Vivo*	Approximate Neurochemical Concentration
Catecholamine	DA/NE	2nM	20nM	~.2nM
Serotonin	5-HT	1nM	10nM	~.5nM
Homovanillic Acid	HVA	20nM	0.2*u*M	7nM
Tryptophan	L-TP	20nM	0.2*u*M	1*u*M
Dynorphin	DYN	4nM	40nM	2*u*M
Somatostatin	SRIF	3.7nM	37nM	4*u*M

10.10 Summary

10.10.1 Murine Studies

Endogenous serotonin (5-HT) release at the synapse surged and ebbed with rhythmic synchronicity at the nerve terminals, dorsal striatal A_9 terminals and ventral striatal A_{10} terminals in parallel with the rise and fall of motor coordinating from basal ganglia. Thus, temporal synchrony occurs between natural behaviors and brain neurochemistry in the physiologic state and creative algorithms are in progress to tack on artificial intelligence codes on the interfacial surface of the BRODERICK PROBE® sensor.

Indeed, the data illustrate a direct communication between the instant inventive imaging of 5-HT release in basal nuclei at both the A_9 terminal field, DStr as well as the A_{10} terminal field, vlNAcc. 5-HT release is imaged at the same magnitude and at the same time as each walking step is taken; no demonstration of a temporal gap is seen online. Thus, the communication between 5-HT release and naturally occurring behavior is temporally synchronous. Moreover, at both A_{10} and A_9 terminals, each increase in 5-HT release is simultaneous. Each increase in 5-HT release is concurrent with each increase in movement. Therefore, in the physiologic condition, the 5-HT response is positively correlated with walking as well as it is in concert with the finer hand and mouth delicate motor responses needed for a high quality of life.

In the experimental condition, the murine model is used in the same paradigm for comparison to the physiologic condition. Hence, subjects are treated with the psychostimulant, cocaine and creative algorithms ensue. The data demonstrate that 5-HT flowed and ebbed without synchronicity at the vlNAcc as the subject walked and used fine motor nuances; it is no surprise that cocaine causes dysfunctional movement. However, the contribution by the tiny nanoprobe lies in its preciseness. In front of our eyes, the online data shows a temporal gap between the synaptic release of serotonin and motor coordination. But, although the data are asynchronous, the curve defining the algorithm remains rhythmic. Significant first-time findings show a juxtaposition, a ***biomarker temporal lag*** that recognizes serotonin signals as the neurotransmission directs motion in real time. Therefore, signally in somatodendrites compared with nerve terminals, provided by NMI and the BRODERICK PROBE® shows that the two neuroanatomic sites present with distinctly different algorithms. The data allow personalized medicine in motor disorders to an extent not perceived previously.

Thus, it is likely that temporal asynchrony in the form of dysfunctional behavior from cocaine psycho-stimulation at distal presynaptic A_{10} circuits within 5-HT-DA interactions seems causal at least in part.

After cocaine administration, the creative algorithm derived from a non-physiologic condition, shows a dissident time sensitive correlation between endogenous 5-HT release at basal nucleus, A_{10} terminals, Nucleus Accumbent (vlNAcc) while walking or using the finer motor coordination of the hands and mouth. Indeed, a quite interesting juxtaposition, the ***biomarker temporal lag*** between the release of the neurotransmitter 5-HT and the behavioral component response as manifested in walking and finer movements like chewing, is exhibited.

Our laboratory was the first to report the exciting data from the BRODERICK PROBE® that showed that the psychostimulant exerted significant effects on the 5-HTergic synapses, somewhat competing with the conventionally sometimes overemphasized DA responses to cocaine. This first report that cocaine had a mechanistic effect on 5-HT release was met with skepticism. But, inherently intriguing then was the response of the microdialysis sector to attempt to prove our new exciting conceptual leap wrong. Lo and behold, the skeptics then had to show, after further study, that cocaine did act on 5-HT function in A_{10} terminals [1].

Interestingly, the psychostimulant effect of cocaine on 5-HT A_{10} somatodendrites in the ventral tegmental area (VTA) as this relates to directing motion from basal ganglia offers algorithms that appear non-rhythmic and non-synchronous; this response is shown for the first time; these algorithms are needed for medicinal therapeutics. *Given the caveat that the physiologic response between brain and behavior in VTA appears as a juxtaposed rhythm that is synchronous, rhythmic and simultaneous, the data that show that the reward terminal response to cocaine which is juxtaposed, rhythmic and non-synchronous is even more interesting. Walking is affected to a slightly greater degree than is the delicate movements of the hands and mouth.* Therefore, walking and finer motor control are somewhat spared from the stimulant psychosis of cocaine in neuronal terminals, NAcc. Cell bodies within the cable network of dendrites, VTA may be spared as well insofar as delicate finer motion is concerned after a single injection of cocaine. *The most interesting implications come to mind from these unique data. The nanoprobe images nuances neither perceived nor empirically previously shown. We recall that cocaine is an effective topical anesthetic in use in dental surgery worldwide; there are positive and negative effects of medicines routinely.*

10.10.2 Human Studies

In vivo live imaging of the epileptic neocortex during surgery revealed high L-TP and peptides compared to monoamines or indoleamines. Thus, L-TP and peptides may serve as *biomarkers* for pharmaco- and/or gene therapy for neurodegenerative processes.

Perhaps high L-TP and peptides is indicative of temporal asynchrony induced by the pathologic state of seizure activity. In this manner, temporal patterns may be used to create a dynamic data profile in the clinical setting. Moreover, the dynorphin detected in epilepsy is also evidence of pathology. In conclusion, L-TP and neuropeptides may serve as neuronal biomarkers for pharmaco- and or gene therapy for neurodegenerative processes.

10.10.3 Dual Species Translational Evidence for Defective Brain Synchronous Rhythms Factoring in Disease

Although static neurotransmitter levels are currently the standard, dynamic data from NMI and BRODERICK PROBE® nanobiosensors provide a more complete portrait of the temporal harmony between neurotransmitter concentration and movement in the physiologic state, as well as the intensity of disease or drug-induced temporal cacophony on specific parts of microneuroanatomy in the pathologic state. Data, herein, describe rhythmic temporal synchrony and its converse, arrhythmic temporal asynchrony empirically. Thus, both heuristically and theoretically, these dual species conclusions further demonstrate the critical nature of the rhythmic brain translationally.

10.10.4 Conclusion

The tiny nanoprobe has uncovered something not previously conceived or discovered in any venue. The data from NMI and the BRODERICK PROBE® parses the subtle effects of a single dose of cocaine. The rhythm of delicate motor coordination of the hands and mouth is spared at the A_{10} nerve terminals after a single dose of cocaine.

As the dendrites form a cable network of sophisticated communication around the brain, signaling axons to make more decisions, our data show that cocaine traumatizes the dendrite and still, the nerve terminals survive after the acute dose at least in aspects of rhythmicity. After cocaine administration at a higher dose, sensing walking mobility show an affect in the ventral tegmental area, the dendritic cell bodies that form the cables soaring around the brain while finer motor skills of hand and mouth at the dendritic cell bodies and their axonal extensions remain rhythmically intact. The data see into strategies for specified protocols and highly advanced pharmacotherapies and physiotherapies for patients with compromised mobility and diseases of neurodegeneration. The BRODERICK PROBE® translated from the laboratory to the operating room is the first of its kind.

A consideration of the ultimate rhythm follows....

References

1. Bradberry CW, Nobiletti JB, Elsworth JD, Murphy B, Jatlow P, Roth RH (1993) Cocaine and cocaethylene: microdialysis comparison of brain drug levels and effects on dopamine and serotonin. *J Neurochem* 60(4): 1429–1435.
2. Broderick PA (1988) Distinguishing in vitro electrochemical signatures for norepinephrine and dopamine. *Neurosci Lett* 95: 275–280.
3. Broderick PA (1989) Cathodic Electrochemical Current Arrangement with Telemetric Application. *US Patent Application No. # 4, 883,057, November 28th, 1989*, Issued.

4. Broderick PA (1989) Characterizing stearate probes in vitro for the electrochemical detection of dopamine and serotonin. *Brain Res* 495:115–121.
5. Broderick PA (1991) In vivo voltammetric studies on release mechanisms for cocaine with gammabutyrolactone. *Pharmacol Biochem Behav* 40: 969–975.
6. Broderick PA (1993) In vivo electrochemical studies of gradient effects of (SC) cocaine on dopamine and serotonin release in dorsal striatum of conscious rats. *Pharmacol Biochem Behav* 46(4):973–984.
7. Broderick PA (1995) Microelectrodes and their use in cathodic electrochemical current arrangement with telemetric application. *US Patent Application No. # 5,433,710, August 22, 1995*, Issued.
8. Broderick PA (1999) Microelectrodes and their use in an electrochemical arrangement with telemetric application. *US Patent Application No. # 5,938,903, August 17th, 1999*, Issued.
9. Broderick PA (2008) Studies of oxidative stress mechanism using a morphine/ascorbate animal model and novel N-stearoyl cerebroside and laurate sensors. *J Neural Transm* 115(1): 7–17.
10. Broderick PA (2012) Real time signal processing at the neuromolecular level in the epilepsy and Parkinson's brain. IEEE Signal Processing in Medicine and Biology Symposium, Technical Program, Saturday, December 1, 2012 at the Grove School of Engineering, CCNY, in conjunction with the NYU Polytechnic Institute, New York.
11. Broderick PA (2013) Neuromolecular Imaging shows Temporal Synchrony Patterns between serotonin and movement within neuronal motor circuits in brain. *Brain Sciences* 3: 992–1012.
12. Broderick PA (2015) Noninvasive Electroactive Photonic Protein Sensor with Polymer Photovoltaic Optics for Memory Transduction using Organic and Inorganic Elements as Platforms. *US Provisional Application No. # 62/273,693, 2017.*
13. Broderick PA (2020) Identification, Diagnosis, and Treatment of Tau-Associated Pathologies Using Electrodes. *US Provisional Application No. 63/130,453.* December 24th.
14. Broderick PA (2021) Identification, Diagnosis, and Treatment of Neuropathologies, Neurotoxicities, Tumors, and Brain and Spinal Cord Injuries Using Electrodes with Microvoltammetry. *US Patent Serial No. 10,980,460, April 20th, 2021*, Issued.
15. Broderick PA (2021) Identification, Diagnosis, and Treatment of Tau-Associated Pathologies Using Electrodes. *US Provisional Application No 63/293/334, December 23rd.*

16. Broderick PA (2021) Noninvasive Electroactive Photonic Protein Sensor with Polymer Photovoltaic Optics for Memory Transduction using Organic and Inorganic Elements as Platforms. *US Non-Provisional Patent. International Application Patent # PCT/US16/68879. US Serial No. 11,197,615, B2, December 14th, 2021,* Issued.

17. Broderick PA, Doyle WK, Pacia SV, Kuzniecky RI, Devinsky O, Kolodny EH (2008) *A clinical trial of an advanced diagnostic biomedical device for epilepsy patients. J Long Term Eff Med Implants* 18: 50. *http://dx.doi.org/10.1615/JLongTermEffMedImplants.v18.i1.480.*

18. Broderick PA, Kolodny EH (2011) Biosensors for brain trauma and dual laser doppler flowmetry: Enoxaparin simultaneously reduces stroke-induced dopamine and blood flow while enhancing serotonin and blood flow in motor neurons of brain, *in vivo. Sensors* (Basel) 11, 138–161.

19. Broderick PA, Kornak EP, Eng F, Wechsler RW (1993) Real time detection of acute (IP) cocaine-enhanced dopamine and serotonin release in ventrolateral nucleus accumbens of the behaving Norway rat. *Pharmacol Biochem Behav* 46: 715–722.

20. Broderick PA, Pacia SV (2005) Imaging white matter signals in epilepsy patients: A unique sensor technology. In: Broderick PA, Rahni DN, Kolodny EH (ed.) *Bioimaging in Neurodegeneration,* Humana Press Inc., Springer, New Jersey p. 199–206.

21. Broderick PA, Pacia SV (2006) Identification, diagnosis, and treatment of neuropathologies, neurotoxicities, tumors, and brain and spinal cord injuries using microelectrodes with microvoltammetry. *US Patent Serial No #7,112,319 B2, Sept.26. 2006,* Issued.

22. Broderick PA, Pacia SV, Doyle WK, Devinsky O (2000) Monoamine neurotransmitters in resected hippocampal subparcellations from neocortical and mesial temporal lobe epilepsy patients: in situ microvoltammetric studies. *Brain Res* 878: 49–63.

23. Broderick PA, Phelix CF (1997) I. Serotonin (5-HT) within dopamine reward circuits signals open-field behavior II. Basis for 5-HT-DA interaction in cocaine dysfunctional behavior. *Neuroscience and Biobehavioral Reviews* 21(3): 227–260.

24. Bromfield EB, Cavazos JE, Sirven JI. An Introduction to Epilepsy [Internet]. 2006 [cited 16 February 2018]. Available from: https://www.ncbi.nlm.nih.gov/books/NBK2508/.

25. Center for Behavioral Health Statistics and Quality. 2016 National Survey on Drug Use and Health: Detailed Tables [Internet]. 2016.

[cited 16 February 2018]. Available from: https://www.samhsa.gov/data/sites/default/files/NSDUH-DetTabs-2016/NSDUH-DetTabs-2016.pdf.

26. Cunningham KA, Lakoski JM (1988) Electrophysiological effects of cocaine and procaine on dorsal raphe serotonin neurons. *Eur J Pharmacol* 148(3): 457–462.
27. Di Chiara G, Imperato A (1988) Drugs abused by humans preferentially increase synaptic dopamine concentrations in the mesolimbic system of freely moving rats. *Proc Natl Acad Sci* 85: 5274–5278.
28. Essman WD, Singh A, Lucki I (1994) Serotonergic properties of cocaine: effects on a 5-HT2 receptor-mediated behavior and on extracellular concentrations of serotonin and dopamine. *Pharmacol Biochem Behav* 49(1): 107–113.
29. Hernandez L, Guzman NA, Hoebel BG (1991) Bidirectional micro-dialysis in vivo shows differential dopaminergic potency of cocaine, procaine and lidocaine in the nucleus accumbens using capillary electrophoresis for calibration of drug outward diffusion. *Psychopharmacology* (Berl) 105(2):264–268.
30. Jacobs BL (1986) Single unit-activity of brain monoamine-containing neurons in freely moving animals. *Ann NY Acad Sci* 473: 70–79. doi: 10.1111/j.1749-6632.1986.tb23604.x.
31. Jacobs BL, Azmitia EC (1992) Structure and function of the brain-serotonin system. *Physiol Rev* 72:165–229.
32. Jacobs BL, Fornal CA (1991) Activity of brain serotonergic neurons in the behaving animal. *Pharmacol Rev* 43:563–578.
33. Koob GF, Bloom FE (1988)Cellular and molecular mechanisms of drug dependence. *Science* 242:715–723.
34. Pan ZZ, Williams JT (1989) Differential actions of cocaine and amphetamine on dorsal raphe neurons in vitro. *J Pharmacol Exp Ther* 251(1):56–62.
35. Parsons LH, Justice JB Jr (1993) Serotonin and dopamine sensitization in the nucleus accumbens, ventral tegmental area, and dorsal raphe nucleus following repeated cocaine administration. *J Neurochem* 61(5):1611–1619.
36. Volkow ND, Wang G-J, Telang F, Fowler JS, Logan J, Childress A-R, Jayne M, Ma Y, Wong C (2006) Cocaine Cues and Dopamine in Dorsal Striatum: Mechanism of Craving in Cocaine Addiction. *The Journal of Neuroscience* 26(24):6583–6588.

37. World Drug Report. Pre-Briefing to the Member States [Internet]. 2017 [16 February 2018]. Available: https://www.unodc.org/wdr2017/field/WDR_2017_presentation_lauch_version.pdf.

Chapter 11

Temporal Asynchrony in Epilepsy: The Ultimate Rhythmic Temporal Synchrony in the Epileptic Orgasm

Action potentials at the synapse in the epilepsy patient's brain are different from others; the firing threshold nests at a voltage lower than is natural state. Action potentials in the epilepsy patient are also temporally asynchronous with mobility as seizing generates action potentials in the epilepsy neuron that drop voltage even further!

Sudden neuronal depolarizations, speedy, repetitive, so many different types of phenomenal, unwanted, unconscious, subconscious and even conscious behavioral symptomatology directly derive from the neuronal electrochemical human and animal brain-such is epilepsy! With seizures, come powerful reflections of the stressed temporally hyper-asynchronous status epilepticus brain followed by nerve hyperpolarization then temporally synchronous with physiological natural motion, a seeming reversal of the previous asynchrony. What happens in-between is unfathomable. The neuron depolarizes at the most unfortunate times, anywhere, anyhow, anytime!

To put it bluntly, these electrical neurons, depolarizing, when they very well want to, "gets on the epilepsy person's nerves." The seizures interrupt life itself, right in the middle of anything or everything; the entire being is out of neuronal control. But, what about the aura, the warning, the focal ictal warning preceding the next seizure, i.e. the alert! The epilepsy brain can use the aura to its benefit to avoid danger. Canines use it to alert the

Neuroimaging: Sensing Biochemistry in the Brain
Patricia A. Broderick

ISBN 978-981-4613-49-1 (Hardcover), 978-0-429-02974-5 (eBook)
www.jennystanford.com

epilepsy person of pending danger. It may be that the epilepsy person needs to stay away from the stove at that moment before seizing, to avoid further physical danger perhaps coming as a severe fall during the seizing episode.

The Merriam Webster Dictionary calls the aura, "a subtle sensory stimulus, surrounding and coming from within the living organism itself." The author believes temporal synchrony is key to a healthy brain [14]. Thus, a rhythmic brain that is synchronous may well define a healthy brain [24]. But orgasmic rhythm, amidst pathophysiology and rhythmopathy in hippocampal theta oscillation, may cause memory loss in persons suffering with epilepsy [78].

11.1 Electrocatalyzing Dopamine for Brain Reward in the Epileptic Person

A rhythmically synchronous brain may well define a healthy brain. Yet, synchronous rhythmicity is performed *par excellence* amidst the pathophysiology of the asynchronous epilepsy brain. The ultimate rhythm happens in the temporal lobe. Biogenic amines and brain reward are decisive points of reference.

The epileptic aura, the warning, may well realize itself in a sexually sensual manner; an orgasmic state may indeed be reached such as what is self-reported by the Russian writer, Fyodor Dostoyevsky, who incorporated his ecstatic epileptic seizures into four different characters in his novels. It is said that Dostoyevsky most likely had his prolific writing ability, his hypergraphia, derived as part and parcel of his temporal lobe epilepsy. The temporal lobe is highly linked with the brain reward center; then, the temporal lobe is most likely one of the primary neuroanatomic sites of the sexually sensual experience as well.

Dostoyevsky describes his ecstatic auras to a friend,

> "For several moments I would experience such joy as would be inconceivable in ordinary life—such joy that no one else could have any notion of. I would feel the most complete harmony in myself and in the whole world and this feeling was so strong and sweet that for a few seconds of such bliss I would give ten or more years of my life, even my whole life perhaps" [71].

Figure 11.1 shows the human brain of the epilepsy person LIVE intraoperatively. The great cerebral vein, known as the *vein of Galen,* is clearly seen.

The Temporal Lobe of the Epilepsy Patient During Surgery

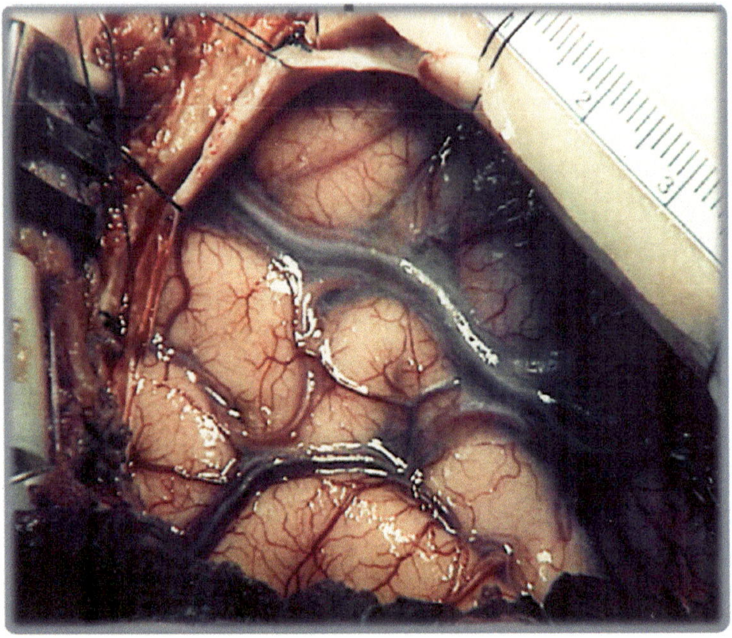

Figure 11.1 The human anterior temporal lobe of the epilepsy patient during surgery.

11.1.1 "*Coitus brevis epilepsia est*" ("Sex is a brief seizure")

"Epilepsy is the one disease that has been most frequently linked to sex. "*Coitus brevis epilepsia est*" ("Sex is a brief seizure") is an ancient proverb attributed to Galen, the famous physician of the Roman Empire. In the 18th and 19th centuries, some doctors, including Samuel-Auguste Tissot and Edward Sieveking, argued that excessive masturbation could cause epilepsy. At the time, castration and clitoridectomy (removal of the clitoris) were reportedly performed on people with severe epilepsy [6]."

The Federalist, Timothy Pickering, the third United States, Secretary of State, under Presidents George Washington and John

Adams, scripted in a letter to Loammi Baldwin, Jr., an associate, that "semen is a healthy bodily fluid and wasting such on masturbation, without *coitus*, is the pathway to epilepsy."[1]

Thus, it is apparent that the disease of epilepsy has not had its etiology, nor its diagnosis and/or treatment in perspective *a priori*. To report that castration and clitoridectomy are "therapy" for the person who is suffering from epilepsy, seems an atrocity and the derivation of such an atrocity may be an oxymoron. Indeed, the concept of sex and the concept of epilepsy have opposite meanings and yet these concepts do have at times been juxtaposed. Since sexuality and epilepsy are related, juxtaposing the two has an element of veracity. Yet, sex does not cause epilepsy; however, sexual behaviors are associated with epilepsy.

Epilepsy is fascinating and at the same time, epilepsy is frightening since spirituality and religion have been intertwined with epilepsy. The stigma remains. At one time, people suffering from the disease of epilepsy were not allowed to enter a public restaurant [33].

The philosopher, Aristotle, did not appear to assuage the prejudice. "Aristotle theorized that sleep was caused by the evaporations resulting from consuming food, which would subsequently rise and fall in the veins. He extended this hypothesis to the process occurring when one convulsed and thought this was the mechanism that caused epilepsy to affect levels of consciousness" [44].

The physician Hippocrates[1] brought the populace to its senses regarding the nature of epileptic seizures. Despite the fact that Hippocrates called epilepsy, "The Sacred Disease," the physician has made it clear to all that epileptic seizures are caused by abnormal activity in the brain. ***"It is thus with regard to the disease called Sacred: it appears to me to be nowise more divine nor more sacred than other diseases, but has a natural cause. Men regard its nature and cause as divine from ignorance and wonder, because it is not at all like to other diseases."***[2]

The invention of the electroencephalograph (EEG) [9], the advances in neurosurgery, the discoveries of antiepileptic drugs

---

[1]Onanism, or self-pollution, Timothy Pickering on Masturbation.

[2]On the Sacred Disease by Hippocrates, Written, 400 BCE, Translated by Francis Adams, *http://classics.mit.edu/Hippocrates/sacred.html*.

as well as advances in the depth of understanding pathophysiology *per se* made the twentieth century mark epilepsy as one of its greatest achievements. Among the most prestigious physicians connected with epilepsy, the work of Henry Gastaut, Wilder Penfield, and Herbert Jasper are notable. The most recent advances in the field of epilepsy include the development of advanced imaging techniques, the development of microsurgery, and the research on the connection between genetic factors and epileptic seizures [54].

In this Chapter, the author lends an explanatory note to the *modus operandi* that underlies the empirically derived *Temporal Synchrony Thesis* [14].

- *temporal synchrony in the physiological state*
- *temporal asynchrony in epilepsy*
- *the rhythmic temporal synchrony in the epileptic orgasm*
- *the brain's neuronal synchrony in the phenomenon of orgasm*
- *the intimately close link of orgasm to brain reward*
- *to the catecholamine neurotransmitter, dopamine*
- *to the indoleamine neurotransmitter, serotonin.*

11.1.2 The Temporal Lobe: Sexuality and Epilepsy

Sexual ictal manifestations predominate in women with temporal lobe epilepsy! This is a finding that supports once again that *sexual dimorphism* is alive and well in the human brain [69].

Too, Patricia has come across a series of relevant publications concerning sexuality in epilepsy patients that add more evidence relating sexuality to seizures. These studies further relate to nerve impulse changes at temporal lobe synapses in the human brain of the epilepsy patient.

The clinical neuropsychologist, Baird discusses the temporal lobe and sexuality and specifically, temporal lobe sexuality in epilepsy. In one of her reports, Baird shows that epilepsy patients perceived sexual changes, that is, improved sexual drive, after temporal lobe resection but not after extratemporal resection. In her study, fifty-eight temporal lobe and sixteen extra-temporal lobe patients were questioned on five areas of sexual functioning, *sexual drive, thoughts/fantasies, interest, masturbation, and activity.*

Each patient's perception of sexual changes relative to perceived levels of normal functioning was assessed. In patients who had undergone temporal lobe (64%) rather than extra temporal lobe (25%), ***positive sexual changes occurred*** and were significantly more frequent following resections in the right hemisphere. Females were more affected than males. Epilepsy resection surgery is associated with *sexual enhancement* [2].

An Artistic Depiction of the Seizing Epileptic Cortex

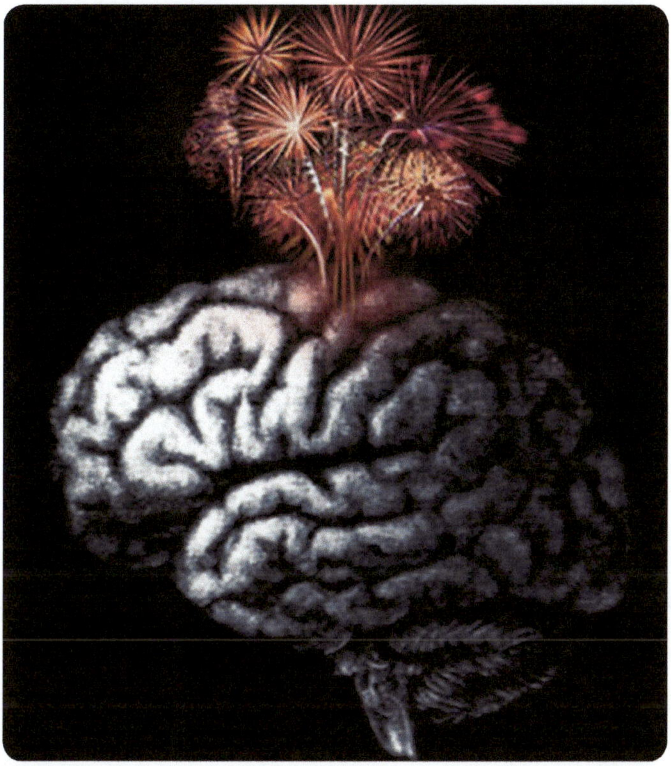

Figure 11.2 Artist, Andrew Baird's rendition of a seizing brain. Reprinted with permission from *Sex in the Brain* [5].

Baird and her colleagues also reported the role of the mesencephalon limbic structure, amygdala, in sex-related epileptic temporal lobe resections. Since the amygdala is known to enable brain reward, the aim was to demonstrate a change in the volume of the amygdala as correlated with changes in the sex drive of epilepsy patients postoperatively. In this study, forty-five patients with intractable temporal lobe epilepsy, who underwent surgical resection in the *Comprehensive Epilepsy Program at the Austin and Repatriation Medical Centre,* answered questions relating to sexual outcome after surgery. Volumetric analyses of both amygdalae were conducted on the patients' preoperative T (1)-weighted magnetic resonance imaging scans; 46 neurologically normal volunteers served as controls. The results show that ***enhanced sex drive*** was linked empirically to the neurons in the amygdala. Therefore, the amygdala, a temporal lobe neuronal site, neuromodulates sex drive in the epileptic brain [3]. Baird further reviews neurological control of human sexual behavior as specifically related to the mesial temporal lobe [4].

Figure 11.2 portrays the genitocortical brain during an epileptic seizure according to the author's knowledge of the epileptic brain and her knowledge of the brain *per se* [5].

11.1.3 Epilepsy Was Placed in Perspective

It is noteworthy that the physiology of epilepsy was placed in perspective by John Hughlings Jackson, the physician who is credited with the formation of a conceptual framework for clinical neurophysiology, the discipline that now underlies diagnostic neurology. Although Dr. Jackson was somewhat under the radar for some time, he became known for his acute recognition that focal epilepsy and focal necrosis can be viewed as reciprocal physiological processes [70].

The physician William Gowers assisted in placing epilepsy into its appropriate perspective in his unfinished work, the *Bible of Neurology. A particularly interesting one of his works is A Study of the So-Called Tendon-Reflex Phenomena. Med.-Chir. Trans. 62. Published by the Medical and Chirurgical Society. London: J.E. Adlard, 1879* [37].

Shorvon has published a systematic review of the field of contemporary therapy in epilepsy, by researchers and practitioners in neurological sciences from many countries and it is updated from its 1996 edition to incorporate the rise of molecular genetics, the integration into clinical practice of a range of novel anti-epileptic drugs and of more substantial and well-evidenced information about established medicaments, the increased attention to patient-centered issues, individuality, patient preference, and the individual clinical circumstances in which epilepsy manifests. The work brings a closer integration of the investigatory approaches into surgical therapy [77].

Epilepsy is a neuropathology, a neuropsychiatric disorder and a neurodegenerative disease. Epilepsy is a neurological disorder wherein the immune system, epigenetics, and structural changes in neurons are part of the equation. Epilepsy is considered a prionopathy, a proteopathy and is therefore one of the deleterious *proteinaceous unfolding* disorders for which *Tau* is now recognized LIVE using the BRODERICK PROBE® nanosensor [15–17].

11.1.4 The Neuroanatomy of the Temporal Lobe

The temporal lobe is one of the four lobes of the brain (along with the frontal lobe, parietal lobe, and occipital lobe). The temporal lobe largely occupies the *middle cranial fossa*, a shallow depression, the central part of which is comprised of the pituitary gland and two lateral parts accommodate the temporal lobe *via* providing communication signally. The three borderline bones are the sphenoid bone and the two temporal bones. ***Only primates have a temporal lobe.***

Gaillard and Bell [36] cite that "Venous drainage occurs *via* two routes:

- blood from the neocortex travels anteriorly to the superficial middle cerebral vein, then into the inferior anastomotic vein (*vein of Labbé*) which joins the transverse sinus
- blood from the mesial temporal lobe goes to the posterior choroidal vein which goes on to form the internal cerebral vein, eventually becoming the great cerebral vein (*vein of Galen*), continuing into the straight sinus"

The temporal lobe is divided into two main sections:

- **neocortex** (sometimes referred to simply as temporal lobe)
    - lateral and inferolateral surfaces
    - comprised of standard cerebral cortex

- **mesial temporal lobe** (sometimes referred to as the limbic lobe)
    - including the hippocampus, amygdala, parahippocampal gyrus

11.1.5 Neurons in the Temporal Lobe Provide Pleasure

Squirreling her jelly donut and a serendipitously found random issue of *Psychology Today* in her bag, Patricia rushed to the Lexington Avenue bus from her 63rd St and First Avenue home in Manhattan. She was traveling to her tenured teaching position in Cathedral High School, a roman catholic and private high school for girls, situated right in the heart of New York City. She began to read about the work of the scientist Dr. James Olds on her short trip to school and she found herself deeply engrossed in the subject of the neuroanatomy of the brain's pleasure centers. While reading, she was consummately fascinated; she realized that the neuroanatomical significance of the brain's ability to provide pleasure directly links to mental health.

11.1.6 Deep Brain Stimulation Begins: The Neuropsychopharmacological Revolution

Dr. Jose Rodriguez Delgado extended his animal experiments to twenty-five carefully chosen patients undergoing chronic treatment of refractory epilepsy and schizophrenia at a Rhode Island asylum and for whom there was no effective treatment available at the time. Delgado's ground-breaking paper describing the results was published in 1952 [28–30]. Therefore, Delgado, *Yale University*, was the first to electrically stimulate the brain. The medical doctor stimulated the ***caudate nucleus*** with a hand-held brain-computer interface, a transmitter. The caudate nucleus was chosen to be stimulated because the caudate then and it remains to this date, known to be involved in controlling voluntary movements. The caudate nucleus, also called the striatum, had not been separated into dorsal and ventral at that time. Was Delgado in the ventral striatum, also known now as the ***nucleus accumbens***?

"Delgado positioned himself between the burgeoning disapproval of mutilating surgical lobotomies and the belief that direct electrical stimulation of specific brain areas was scientifically and clinically superior to oral administration of drugs whose effects on the brain were inevitably mitigated by metabolism in the liver, obstruction by the blood-brain barrier and uncertain distribution throughout the brain. Delgado was not entirely alone in these beliefs [28, 29]."

Delgado's 1952 paper narrowly pre-empted the publication of somewhat similar research in humans by Dr. Robert Heath, Chairman of Neurology and Psychiatry at *Tulane University*.

Dr. Robert Heath, [39] *via* brain computer interfaces, implanted electrodes in the medial forebrain bundle of psychiatric and epilepsy patients and found the septal region to be the site most consistently implicated in the pleasure response. The septal region, emotionally charged, connects with efferents *via* the ***median forebrain bundle*** through the ***lateral hypothalamus*** *to the interpeduncular nuclei of the brain stem*.

Heath found that during orgasm, spike and large slow-wave electroencephalographic activity occurred in these regions of the medial forebrain bundle along with superimposed fast activity electrical stimulation that produced intense feelings of orgasmic pleasure. The intense orgasmic feelings could also be produced by stimulation of the neurotransmitter, ***acetylcholine***, observed in a psychiatric patient. Heath also found that the ***amygdala*** was involved in this intense electroencephalographic orgasmic reaction [39–41].

At about the same time, Peter Milner, who was trained as an electrical engineer, became intensely interested in electrical brain stimulation and brain reward *via* his wife, Brenda, a neuroscientist at McGill University in Canada. Peter began to work with Dr. James Olds and both scientists, Milner and Olds agreed with Heath that the ***septum*** was the pleasure center. It was amazing! When the septum of murine animals was electrically stimulated, the animals continued to electrically stimulate themselves incessantly even at the expense of their eating and drinking; the murine model showed us that they wished for septal stimulation even though their very sustenance was at stake [60].

The players who contributed to the field of neuro-psychopharmacology appeared to work closely together. The fabrication and use of a self-stimulating transistorized device in Heath's patients were inspired by the experiments of Olds et al. with small animals [60, 61]. Dr. Heath wrote, "The unit has provided an objective estimate of relative pleasurable or adversive effects of stimulation to different sites deep in the brain of man. The fact that patients most consistently choose to stimulate the septal region with this device concurs with their verbal reporting that a pleasurable response is most often elicited at this site. This technique has helped to demarcate the pleasure areas of the human brain (the medial forebrain bundle and interpeduncular nuclei of the mesencephalon), roughly corresponding to pleasure areas of brains of animal" [40].

But, it is so interesting that the Tulane group, under Robert Heath, had previously reported that septal stimulation induced a sexual motive state whereas stimulation of other pleasure sites (the vicinity of the medial forebrain bundle and interpeduncular nuclei), albeit a source of pleasure were not stimulators of sexual desire. However, the anatomical connections of *pleasure with sexual desire*, as discussed in later papers from Tulane, were described as within the region of the septum that includes other nuclear sites within the pathways of emotional expression, together with a demonstration of the functional interrelationship among sites within these pathways. *Thus, Heath defined the septal area to include the nucleus accumbens septi and the nucleus of the diagonal Band of Brocha [41], the hippocampi, the amygdalae, the orbital cortex, and specific sensory relay nuclei.*

Dr. Carl Wilhelm Sem-Jacobsen [76], in collaboration with experienced neurosurgeons in Oslo, conducted the very *first trials with deep brain stimulation* in patients with Parkinson's disease. Also for patients with Parkinson's disease, Drs. Alim Benabid and Pierre Pollak, the University Hospital of Grenoble, developed deep brain stimulation (DBS) in 1987. Drs. Benabid and DeLong won the 2014 Lasker-DeBakey Award for deep brain stimulation in Parkinson's patients. Interestingly, Olds worked with Heath who worked with Dr. Nathan Kline of the *Nathan Kline Institute*, formerly the *Rockland Psychiatric Institute* in New York. The late Dr. Kline is a Lasker Awardee, twice.

Dr. Kline honored Patricia with a scholarship from the *Rockland Psychiatric/Nathan Kline Institute* to support her doctoral studies that elucidate manic depression and the bipolar brain.[3]

11.1.7 The New Brain Sensing Revolution

Neuromolecular Imaging: The BRODERICK PROBE®

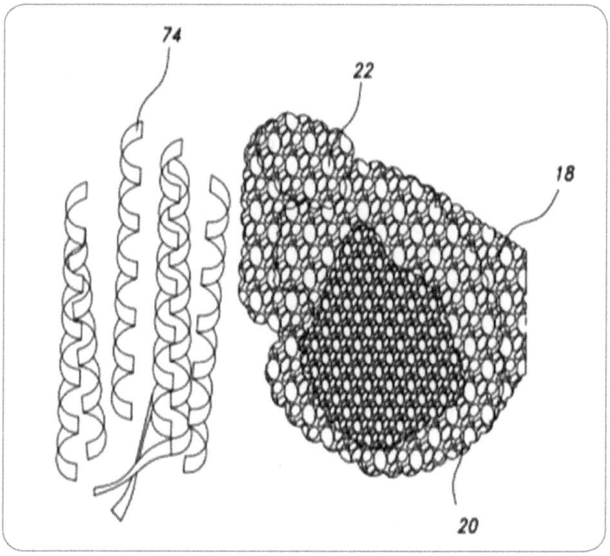

Figure 11.3 The rhodopsin fullerene BRODERICK PROBE®. This is the protein probe, a semiconductor and optical sensor for high-tech industries in enterprise virtual reality, food and aging, senior care and artificial intelligence, catalytic converters, switches, biomimetic production of pharmaceuticals and faster magnetic resonance and better diffusion tensor imaging technologies [16].

[3]Accepted by NY Medical College, her tuition comprised part of the *Nathan Kline Award*. A *1975 Plymouth Duster* was delivered to her home within a week of the receipt of her scholarship! Patricia served as a research technician at *Rockland Psychiatric Hospital* although Patricia was not privy to the Rockland Institute's electrical stimulation research. Also, accepted by *St. John's University College of the Arts and Sciences, Department of Pharmacology*, she worked under Dr. Vincent de Paul Lynch, Patricia's neuro-psychopharmacological studies on bipolar mental disorders showed that the medication, lithium, worked *via* the neurotransmitter, serotonin, through the amino acid, L-Tryptophan, as a synergist. Spectrofluorometric and atomic absorption techniques were used for analysis of brain and blood. *Neuropharmacology* (1982) Jul;21(7):671–679. doi: 10.1016/0028-3908(82)90010-7.

Using Neuromolecular Imaging, the BRODERICK PROBE® is patented preclinically and clinically as carbon and protein-based semiconductor nanoprobes with integrated interfacial surface algorithms for programming online unique neurobehavior. Biosensing recognizable neuromolecular elements in anion and cation ranges of neurotransmitters LIVE, while neurobehavior is imaged online, further advances material science and dramatically moves discoveries in neurotransmission forward. Peptides such as dynorphin and somatostatin and indeed, the Tau peptide is imaged LIVE by the BRODERICK PROBE® and difficult to image amino acids, such L-Tryptophan, precursor to serotonin, lie under the control of the nanoprobe [15–17].

Too, reading empirically derived signal processing-based algorithms on the BRODERICK PROBE® from multiple interconnected sites in brain, upper and lower limbs, these programmed sensing neuromolecular devices further equip the paralyzed person for agile mobility. Exquisite detail in information processing is essential and selective frequency of neurotransmission plus afferent or efferent signal dynamics from one neuroanatomic site to another is paramount to deciphering the significance of neurotransmission in concert with mobility, that is to say, neurobehavior is a cognitive look at motor coordination. Along with pure carbon polymers, the BRODERICK PROBE® fiber optical based programmable vector biosensing devices enable recognizable molecular elements for entropy in artificial intelligence (AI). The dementias, Alzheimer, Parkinson, epilepsy, stroke and brain reward circuits of orgasm and the like suit the tiny probes all the more and the work continues to progress,

> "Mental disorders are brain disorders and the brain is also the most important sexual organ where sexual desire, pleasure and orgasm, and also satisfaction are generated and where the genital response may be evoked or inhibited. Consequently, all disorders that affect the brain may also affect sexual functioning. Sexuality has a profound effect on general well-being, social functioning, and the quality of life of an individual. Sexual problems are commonly associated with feelings of failure, decreased self-esteem, fear of rejection, and difficulties in forming and maintaining relationships. Decreased mood or performance anxiety is often found in individuals presenting with sexual dysfunction.

So, there is a clear link between mental and sexual health. Every clinician should be aware of this important interaction and be able to use this knowledge in his/her daily practice. In an analogy to psychiatry, sexual medicine and therapy uses the bio-psycho-social model approach and moreover sexuality may be a paramount example on the interplay of somatic, psychological, and sociocultural factors" [51].

11.2 The Female Orgasmic Brain Reward Pleasure Centers

The study of sexuality is important clinically to treat any number of psychiatric and neurological disorders such as anorgasmia for one example. Understanding the normal physiology of orgasm might help address the problem of anorgasmia, the inability to achieve climax and the issues related to female sexual dysfunction, as drug makers have yet to come up with a Viagra for women. "It's a Woman's Prerogative to Change Her Mind"[4,5] at coitus as truth may deceive. Should she change her mind in the affirmative, a sublime and naturally achieved escape from accumulated sexual tension, sexual stimulation, arousal and internal sexual buildup may occur in a ride that is accompanied by ***neural relief***.

Female orgasm is a neuropsychological response. As the Masters and Johnson team investigated in the laboratory conditions, female orgasmic response is complemented by the contraction of some voluntary and involuntary musculature, such as vulva, vagina, uterus, pelvic floor muscles, some of body muscles (leg, abdomen and pelvic muscles) [56, 57]. Kinsey spent a lifetime understanding the human psyche in the male and female journeys to sexuality [45, 46].

Freud has mastered his legacy whether by chance or strategy. Freud's account of the development of femininity and female sexuality posited that a transfer of erotic zones in the young girl's life from the clitoris to the vagina should just take place as a

[4]A poetry book, a collection of deep essays regarding the female humanity: Elizabeth Macklin, *By Herself: Women Reclaim Poetry*, Last updated on May 16, 2022 22:50:22.

[5]The drugs, Addyi® and Vyleesi®, are FDA approved therapies to improve female sexual desire.

matter of routine. The process was interpreted by many analysts and commentators as a crucial part of a biological imperative to heterosexuality. Interestingly, Freud advised that healthy children achieved orgasm by stimulating the clitoris and he provides a caveat that vaginal orgasms stay in the realm of the mature woman. Should a woman show inability toward vaginal orgasm, the prediction was frigidity [35].

Masters and Johnson debunked Freud's conceptual leap to frigidity if vaginal orgasms could not be achieved and the ability to achieve clitoral orgasms received their rightful due. Indeed, women are known to attain an orgasmic state that may last for **43** seconds, a physiological state coined as *status orgasmus* [55, 56]. In some women who have developed an expanded sexual response (ESR), expanded orgasm, (EO), multiple orgasms and *status orgasmus* altered states of consciousness that can vary in duration as well as in number of minor orgasms they contain in the train of orgasmic patterns. *Status orgasmus* is the continuous form of blended orgasms and/or clitoral/vaginal orgasms that last from 1 minute to 10–15 minutes (or more).

Thus, a woman can have multiple orgasm whereas a man may be limited, that is, a man experiences orgasm but there is a refractory period after orgasm and ejaculation. Erection must wait as explained by the experts [56, 57, 72–75].

11.2.1 Vaginally Activated Orgasm (VAO)

There is general agreement that it is possible to have an orgasm by way of the direct simulation of the external clitoris. In contrast, the Jannini team discuss the possibility of achieving climax during penetration since this issue has met with controversy. Six scientists with different experimental evidence debate the existence of the vaginally activated orgasm (VAO).

Experts #1 and 2 reviewed data from the literature demonstrating the anatomical possibility for the VAO.

Expert #3 presents validating women's reports of pleasurable sexual responses and adaptive significance of the VAO.

Expert #4 induced echographic dynamic evidence to describe one single orgasm, obtained from stimulation of either the external or internal clitoris, during penetration.

Expert #5 reviewed his experiments showing the uniquely different sensory responses to clitoral, vaginal, and cervical stimulation.

Expert #6 presented findings on the psychological scenario behind VAO.

The team concluded, "The assumption that women may experience only the clitoral, external orgasm is not based on the best available scientific evidence" [43].

11.2.2 *"The Females' Psyche* Is Intriguing Neurologically, Psychologically and Neurochemically"

H. Umit Sayin and his colleagues [74] are known for their work on the *"Four Nerve Theory of female Orgasm."* Actually, there are six pathways and four nerves that are involved and hence, the theory is also called the "Four Nerve Six Pathway Theory of Female Orgasm." In Eastern literature centuries ago and in Western literature during the last decades, there are descriptions and discussions of intense female orgastic profiles. As mentioned by Masters and Johnson, in women, the orgasmic response can be enhanced and expanded. Also, it has been proposed that in a minority of women, orgasm reflexes can be triggered by means of the stimulation of glans clitoris, other clitoral structures, PC-muscle, cervix, other Deep Vaginal Erogenous Zones (DVZ: A-, O-, G-Spots, etc.), anus, nipples, and ear lobes.

In the ***Four Nerve-Six Pathway Theory of Female Orgasm***, the development of a single intense orgasm and/or ESR, six separate pathway systems may activate the female orgasms. These are pudental, pelvic, hypogastric, vagus and intercostal nerve-mediated reflex arches with oxytocin the neurotransmitter and prolactin, the hormone, in play. Therefore, four main nerve pathways and two oxytocin pathways mediate the formation of normal or single orgasms, as well as developed, sustained, intense and long-lasting orgasms in ESR. In the *Staircase Hypothesis of Attaining Different Orgasmic Reflexes*, ESR [73–75] can be learned and developed in a few women by training in a staircase pattern. Clitoral, vaginal, blended, intense blended orgasms and ESR can be explained by staircase orgasmic reflex hypothesis and four nerve-six pathway theory. Even though ESR can be attained by very few women today, after the discovery of the mechanisms of intense female orgasms and ESR, it will be possible for many women to experience these orgastic profiles. The contribution of training techniques for sexual competence is invaluable to therapy in the sexual realm; the contribution is a most important addition to the quality of relationships.

Therefore, pleasure and reaching sexual climax are learned. Sensuality, sexual brain, sexual body, sensitivity in the genitals and orgasm reflexes can also be developed and enhanced through partner involvement and Sayin advises that the partner should be trained as well. During prolonged orgasms, which may last for tens of minutes, some minority of women may achieve a different state of consciousness, ecstasy and blissful, oceanic, divine feeling after exercises and training. Dr. Sayin gives an example, "a woman who is sexually responsive (with clitoral and vaginal orgasm reflexes) can reach to 5 multiple orgasms in 30 minutes," [74]. Here, he says, "the most crucial thing is the capacity and approach of the partner." This is a prelude to the studies by Newberg et al., who suggest training as a positive approach to healthy relationships that include orgasms [59].

11.2.2.1 Epileptic orgasms are learned phenomenon

Just as orgasms without epilepsy are learned phenomena, so too epileptic orgasms and indeed epileptogenesis *per se* may well depend on learning. Plasticity is central to the ability of a neural system to learn and also to its ability to develop spontaneous seizures. Hsu et al. ask, "What is the connection between the two?" Since learning itself is known to be a destabilizing process at the algorithmic level, Hsu et al. discuss restraining factors on a spontaneously active "Hebbian learning" system. These authors have found that the ability to learn appears to confer an "intrinsic vulnerability to epileptogenesis on that system." Indeed, epilepsy and extrapolating to epileptic orgasms, if these are abnormal *learned responses* to certain repeated provocations, *unlearning* is a clear non-drug, non-surgery solution to epilepsy disorders and orgasms, at least in part [42]. Cognitive therapy to assist the patient in unlearning epileptogenesis is an option for therapy.

11.2.3 Reflex Seizures Entrain to Orgasms

Reflex seizures exhibit classical *absence* electroencephalographic patterns that may bio-mimic orgasmic sexual climax! In fact, there are stimuli for causation of epileptic orgasm that range from visual to auditory rhythmic inputs as does brain reward stimuli. Flashing lights and certain television shows induce reflex seizures that enable entrancement into sexual climax. Komisaruk discusses experiences which are similar to or are conceptual triggers for reflex seizures. These signals appear in the form of sensory input that performs non-genitally such as the signally that translates during aural seizures to orgasmic altered states of consciousness (ASC) [48].

Clinical case reports confirm that orgasms come from reflex seizure [26]. Another report discusses six cases of reflex-seizure-induced orgasm [62]. ***"In these studies, seizure-induced orgasms were indeed so pleasurable that the patient, a female, experienced orgastic pleasure from her reflex seizures to the degree that she did not desire sexual intercourse with her husband and such reflex-induced orgasms interfered with sexual intercourse during daily married life."***

A 41-year-old woman had nocturnal somatosensory seizures followed by orgasm. The interictal electroencephalogram (EEG) showed that the discrete left central parietal region exhibited paroxysmal activity that markedly increased during sleep. Eight electroclinical seizures were recorded during sleep polygraphic recordings. Orgasms occurred coinciding with generalization of paroxysmal activity. No abnormalities were found on serial neurologic examination or computed tomography scan. The attacks were completely controlled with antiepileptic drugs [25].

11.2.4 Orgasms Entrain to Seizures

Elevated synchrony, the ultimate synchrony in orgasm promotes conditions under which anomalous firing patterns are more likely to propagate across the brain and thereby produce the pathophysiology of epilepsy [72]. Cascades of neural firing promulgate because the *firing neurons* spread at geometrically increased numbers with associated frequencies corresponding to those of the rhythmic seizures and the result is disruption of function after entrainment [7]. Disrupted consciousness may be caused within the positive-feedback intensification contributing to trance [72] or to the sedative-hypnosis [68] responses due to the elevated neurotransmission *via* morphine and enkephalinergic opioids that are derived endogenously from the extended orgasm *per se*.

Empirical data show that seizures with orgasmic auras (warnings) originate in right inferolateral and temporal lobe structures which in other forms of epilepsy these ecstatic states precede lost consciousness [82]. In these ways, epilepsy-like dynamics may contribute to the phenomenology of orgasm as a 'little death'. Increased copulatory-phase theta activity may indicate the establishment of a network of enhanced pre-orgasmic functional connectivity with subsequent oscillating spindle activity [72] and/or central pattern generators in temporal synchrony [14]. These activation cascades may be more self-limiting in orgasm and more dysregulated in epilepsy.

Entrainment involves learning the partner's characteristics. Thus, orgasm may determine a person's mate, that is by associating features from climax-inducing partners with the rewarding nature of both copulation and orgasmic experience. Pfaus believes that although hormonal actions may set the tone for sexual activity by generating sexual excitement and arousal, it is each one's unique experience with sexual behavior and sexual reward that molds the strength of responses made toward sexual incentives. Pfaus believes that learning your partner's needs is the key to a healthy sexual brain [67].

When epilepsy triggers orgasm, it is possible that the seizures are partially caused by extended periods of rhythmic sexual stimulation, or *via* elevated excitatory neural activity during orgasm itself, or both. Considering the importance of neural synchrony in the rhythmic pathogenesis of seizures [80], these studies are consistent with rhythmic entrainment as being central to sexual experience and orgasm [72].

11.2.5 Rewiring the Brain for the Orgasm: Spinal Cord Injury Patients

Sometimes, a person is unable to feel genital stimulation. There may be multiple reasons for this. However, the brain might actually remap itself to allow such a person to reach orgasm. This is an amazing thought, at first glance, since generally, one may perceive of orgasm and sexual pleasure as being dependent on the stimulation of genitalia. But, this is not entirely the truth. The brain is plastic and can create new pathways to pleasure that don't involve our sex organs at all. In the case of trauma, remapping of the senses may occur that permits us to experience sexual and orgasmic sensations in other body parts In people who have suffered lower body paralysis, for example, the brain might actually rewire itself in order to allow a person to achieve orgasm through stimulation of other body parts, such as the skin of the arm or the nipples.

Indeed, the skin of spinal cord injury patients is a sensitive organ system and hypersensitivity is par for the course in these patients. Both sexes have reported such a phenomenon. The skin near the injury is hypersensitive to touch and even painful and intensely aversive to touch if one touches it by, for example, while handing a phone to the patient and skimming by the neck by mistake such as what happened to my quadriplegic sister, Josephine, for example.

But, when stimulated in the right way, by the right person, the patients report pleasure at the touch previously capable of causing unbearable pain. What is caused by the touch of the right person at the right time, is what the patients describe as orgasm in the skin. A woman with a complete spinal cord injury at her upper thoracic level, whose area of hypersensitivity was the neck and shoulder, claimed to have orgasms from stimulation of the skin of her neck. Komisaruk's laboratory measured the patient's heart rate and blood pressure and found that it increased markedly during self-application of a vibrator to her neck-shoulder junction and she described experiencing an orgasm accompanied by a "tingling" sensation in her vagina [79].

11.2.6 The Vagus Nerve Bypasses the Spinal Cord: *The Murine Model*

The sensory field of the nerves that are involved in orgasm were mapped:

> "the pelvic nerve receives sensory input from the vagina and cervix, the hypogastric nerve receives input from the uterus and cervix and the pudendal nerve receives input from the clitoral region and the vagus nerve receives input from the uterus and cervix" [49].

The pudendal nerve and the pelvic nerve enter the spinal cord at the sacral 2 ± 4 level whereas the hypogastric nerve ascends in the sympathetic chain and enters the spinal cord at thoracic 10 ± 12 levels. The hypogastric nerve could convey cervical stimulation-elicited activity to the brain in women with complete spinal cord injury below T-10, since the fibers of the hypogastric nerve enter the spinal cord at T-10 *and below*. The Komisaruk team hypothesized that women with complete spinal cord injury (SCI) above the level of entry of the hypogastric nerve (i.e. injury above T-10) would have no sexual response to vaginal or cervical self-stimulation.

Therefore, the team diagnosed spinal cord injury patients as having "complete" spinal cord injury "upper-level" injury group (absence of sensation and voluntary movement below T-9 (n = 6)) and a "lower-level" injury group (absence of sensation and voluntary movement starting below T-10 (n = 10)). Women without spinal cord injury comprised a control group (n = 5).

The finding from the Komisaruk, Whipple et al. team:

"The vagus conveys significant vaginocervical input directly to the medulla oblongata of the brain, completely bypassing the spinal cord."

The findings indicate the existence of multiple sensory pathways from the genital system to the brain. Perhaps in women with complete spinal cord injury with no input from the genitospinal nerves, the sensory vagus nerve may play a role in stimulating orgasm. There are nerve pathways from the sexual system to the brain that are functional in women with "complete" spinal cord injury. The vagus mediates orgasm [49, 84].

11.2.7 The Vagus Nerve Bypasses the Spinal Cord: *The Patient*

Using the murine animal model data to translate for patient care continued using magnetic resonance imaging (fMRI) to study spinal cord injury patients. The findings are of intense interest. The team found that women who were diagnosed with complete spinal cord injury (SCI) at T-10 or above do report vaginal–cervical perceptual awareness.

"***The vagus nerve provided the afferent pathway for the sexual response via the Nucleus Tractus Solitarii (NTS) region of the medulla oblongata to which the vagus nerves project.***" This nucleus is part of the brain reward circuit and is activated by vaginal–cervical self-stimulation in spinal cord injury women. Regional blood oxygen level-dependent (BOLD) signal intensity was imaged during cervical self-stimulation by the spinal cord injured person. The researchers used statistical parametric mapping analysis with brain motion artifact correction, much needed due to the large and bulky size of the head gear.

Psychiatric examination and MRI established the location and extent of spinal cord injury. In order to demarcate the NTS, a gustatory stimulus and several hand movements were used to activate the superior region of the NTS and the ***Nucleus Cuneatus*** that is adjacent to the inferior region of the NTS, respectively. Each of four women with interruption, or "complete" injury, of the spinal cord and one woman with significant but "incomplete" SCI, all at or above T10, showed activation of the inferior region of the NTS during cervical self-stimulation.

- Three women experienced orgasm during the cervical self-stimulation. The brain regions that showed activation during the orgasms included
- *hypothalamic*
- *paraventricular nucleus,*
- *medial amygdala,*
- *anterior cingulate,*

- *frontal, parietal, and insular cortices, and*
- *cerebellum.*

Therefore, the Komisaruk and Whipple team showed vaginal–cervical sensibility in women with complete spinal cord injury above the level of entry into spinal cord of the known genitospinal nerves! [49, 84].

11.3 The Neuroanatomy of Orgasmic Brain Reward: Epilepsy and the Electroencephalogram

Heath studied brain reward during the orgasm in the epilepsy and psychiatric patient and compared the psychiatric patient and the epilepsy patient in brain reward as therapy at that time. The neuroanatomy of orgasm in the brain was studied with electroencephalography (EEG). The photographs show the electroencephalogram during orgasm in the human [41].

11.3.1 Case Studies from Drs. Heath and Sem-Jacobsen

Heath implanted electrodes in several patients (preparatory to brain surgery for treating epilepsy) and enabled them to self-stimulate via a wearable control panel. One patient with an electrode implanted into the septum, "when asked why he pressed the septal button with such (high) frequency, the patient said the feeling was 'good' and made him feel as if he were building up to a sexual orgasm. He was unable to achieve the orgastic end point, however, and explained that his frequent, sometimes frantic, pushing of the septal button was an attempt to reach a 'climax' although at times this was frustrating and produced a 'nervous feeling" (p. 227) (from Komisaruk [47–49]).

Sem-Jacobsen (1968) applied electrical stimulation to the frontal cortex through a chronically-implanted electrode. One patient, saying he liked the stimulation and asking for more, responded with trembling, deep breathing, flushing, sudden relaxation, smiling and ejaculation. Another patient was described by Sem-Jacobsen as having "a nonsexual type of orgastic sensation. The patient liked it and wanted to be stimulated again, but when suddenly he became satisfied, he did not want any more electrical stimulation ... he reported feeling relaxed, pleasant ... it's like a sexual pleasure. No smell. No taste. I feel it in the whole body" (p. 172). Then, likely reflecting the contemporary scientific/cultural mores, Sem-Jacobsen stated, "Upon realizing that this was definitely a sexual response, no further stimulation was made" (p. 172) (from Komisaruk [47–49]).

11.3.1.1 Stereotaxic coordinates for the sexual response during orgasm in the epilepsy patient

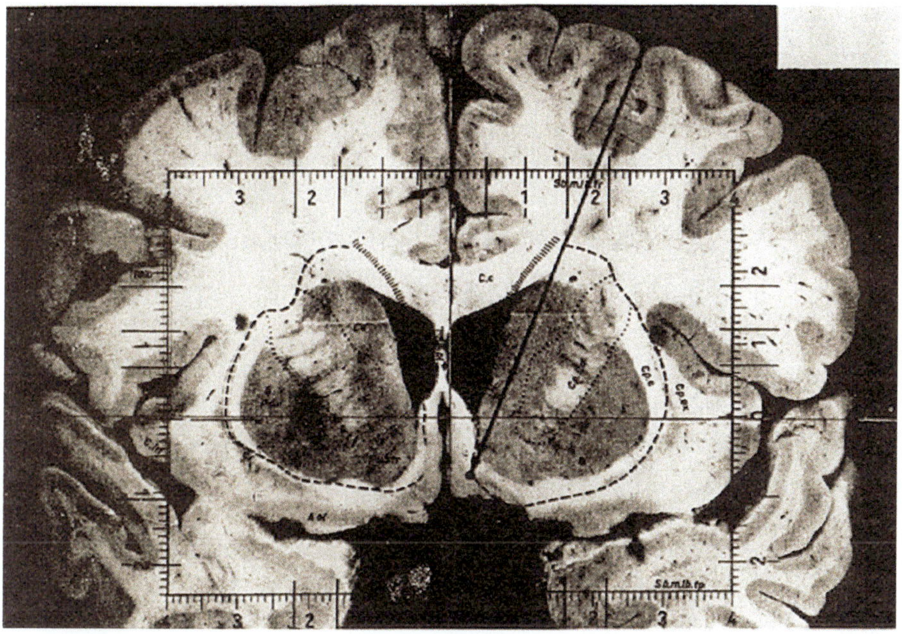

Figure 11.4 Electrode placements in the brain of the psychiatric and/or the epilepsy patient 11–19. (Diagrams are shown in *Introduction, Stereotaxis, an Atlas of the Human Brain,* (d. b). Schaltenbrand G., and Bailey, P. New York. Grune & Stratton, 1959.) Placements are shown in the left anterior septal region (part A) and right central nuclei of the ***thalamus*** (includes the centro-medial) (part B), brain sites from the septum was the neuroanatomical substrate of the most pronounced changes that was recorded during orgasm. Heath included other sites in the definition of the septum. Reprinted with permission from Grune and Stratton, New York, USA.

11.3.2 Combining the Features of Epileptic Seizures with Sexual Orgasm

A plethora of papers from the Komisaruk team shares much about orgasm whether the orgasm is in epilepsy or not [47].

Safron [72] abbreviates the work of Komisaruk.

"Firstly, Komisaruk combines the features of epileptic seizures with sexual orgasm, that is, that some forms of epilepsy are capable of generating orgasm-like pre-ictal auras. Second, they describe seizures as characterized by abnormally large degrees of neural synchrony and go on to speculate that elevated synchrony also accompanies 'the rhythmical and voluntary movement-generated timing of genital stimulation'. Finally, they observe that some aspects of orgasmic expression can be likened to the uncoordinated motor responses present during *grand mal* seizures and argue that widespread fMRI activations during orgasm suggest similar mass activations as observed in epilepsy."

11.3.3 The Penfield Homunculus: Mapping Epileptic Seizure Discharge

The Penfield Homunculus is a visual representation of the mapping of body space in the somatosensory cortex of the brain, with the size of the body representing the size of the area of cortex devoted to it, and hence the sensitivity of that region as well. From Penfield and Rasmussen (1950) [8].

11.3.4 The Penfield Homunculus in Men and the Genital Sensory Cortex in Women

Three body parts activated the sensory cortex in precise agreement with the "homunculus" map of Penfield and Rasmussen (1950), the finger representation lateral, the toe stimulation in the medial hemisphere wall paracentral lobule) and the nipple self-stimulation in the thoracic region of the homunculus. However, in response to nipple self-stimulation, the Komisaruk team also observed activation that partially overlapped the sites of activation produced by the clitoral, vaginal and cervical self-stimulation in the paracentral lobule.

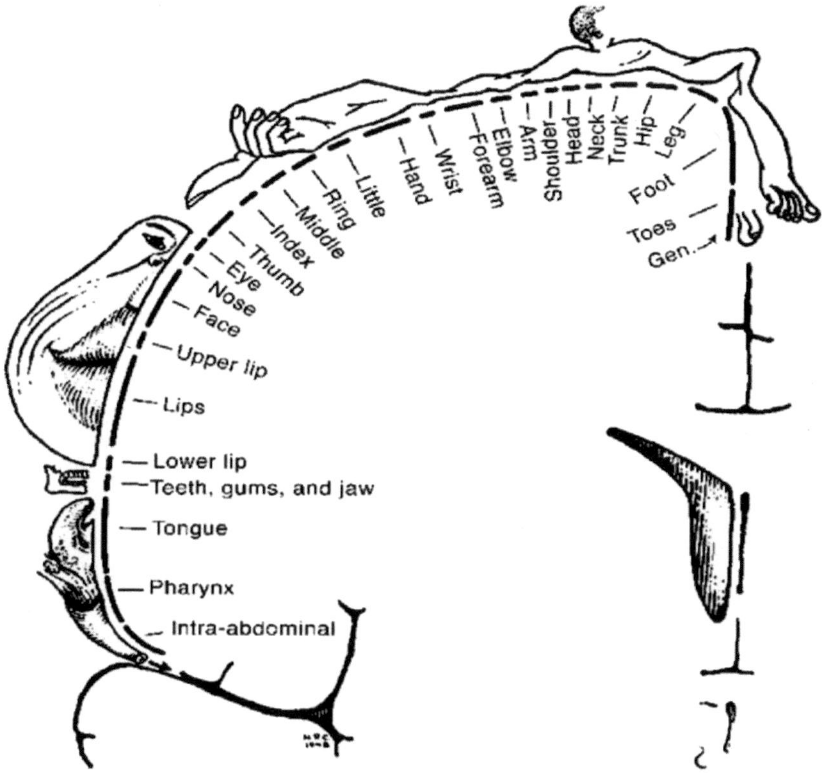

Figure 11.5 The sensory homunculus is a topographic representation of the sensory distribution of the body found in the cerebral cortex. The topograph has body parts illustrated along the surface of the postcentral gyrus of the parietal lobe. Over 400 craniotomies were performed to delineate the homunculus; the studies were performed under local anesthesia. The evidence was obtained from the effects of cortical stimulation (often in the patients' own words), the effects of cortical ablation, and the effects of epileptic seizure discharge. Separate chapters deal in detail with technique and various sensorimotor representations of the body in the cortex. *The Cerebral Cortex of Man: A Clinical Study of Localization of Function.* © 1950 Gale, a part of Cengage Learning, Inc. Reproduced by permission.

Mapping the sensory cortical fields of the clitoris, vagina, cervix, and nipple to underlying sexual response to contribute to sexual medicine and sexuality for mental health is inspiring and it is provided by Komisaruk et al. [50].

The nerve projections from the vagina, uterine cervix, and nipple to the sensory cortex in humans were the focus and functional magnetic resonance imaging (fMRI) was the technique utilized to study sensory cortical responses to clitoral, vaginal, cervical, and nipple self-stimulation. Points on the homunculus were the reference targets for responses to the thumb and the great toe (hallux) stimulation. Clitoral, vaginal, and cervical self-stimulation activated differentiable sensory cortical regions, all clustered in the medial cortex). Nipple self-stimulation by females activated the genital sensory cortex (as well as the thoracic) region of the homuncular map which was actually centered on the male.

Therefore, the genital sensory cortex, identified in the classical Penfield homunculus based on electrical stimulation of the brain only in men, was confirmed by the present study in women by applying clitoral, vaginal, and cervical self-stimulation, and observing their regional brain responses using functional magnetic resonance imaging (fMRI).

Vaginal, clitoral, and cervical regions of activation were different according to nerve input and output and behavioral correlates according to sex. The team noted that activation of the genital sensory cortex by nipple self-stimulation was unexpected, but suggests a neurological basis for women's reports of its erotogenic quality. "***The conclusion regarding nipple self-stimulation as erotic was a surprise to the males in the Komisaruk team but not a surprise to the females***."

11.3.4.1 Brain regions activated

The sensory, motor, reward, frontal cortical, and lower brainstem regions were activated such as the genital sensory cortex (paracentral lobule), secondary somatosensory cortex (operculum [SII] regions OP1 and OP4), precuneus, inferior parietal lobule, insula, hippocampus, amygdala, cerebellum, supplementary motor area, dorsal and ventral striatum (caudate, putamen, and nucleus accumbens), ventral tegmental area, hypothalamus, pons, anterior and posterior cingulate cortex, temporal pole, and prefrontal cortex.

Sensory Cortical Responses in Women in Relation to Penfield and Rasmussen's [classical sensory homunculus]: Finger Stimulus

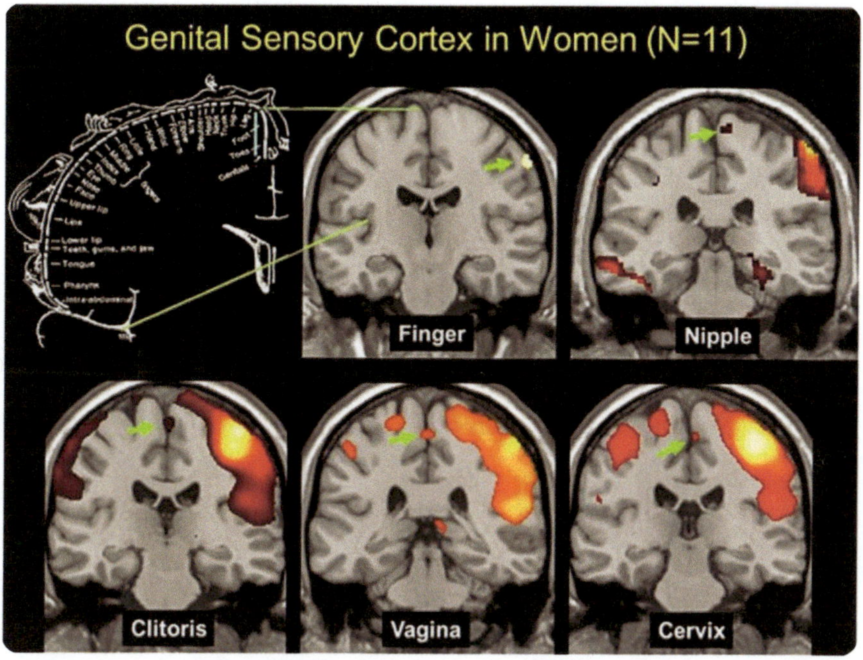

Figure 11.6 Sensory cortical responses (indicated by arrows) to finger stimulation and self-stimulation of clitoris, vagina, cervix, and nipple in relation to Penfield and Rasmussen's [classical sensory homunculus]. The female genital sensory responses are located in the same medial cortical region as that in the classical map, which was based on **men** [8]. *An intriguing exception is that in the study of Komisaruk, women's nipple self-stimulation activated their genital sensory region. The large activations in the lateral cortical regions were produced by the hand activity used for the self-stimulation.* Adapted with permission from Komisaruk B, Whipple B (2011) [48].

11.3.5 The Penfield Homunculus in Men and the Genital Sensory Cortex in Women: The Toe Stimulus

11.3.5.1 Brain regions activated

The activated brain regions included sensory, motor, reward, frontal cortical, and lower brainstem regions, such as the genital sensory cortex (paracentral lobule), secondary somatosensory cortex (operculum [SII] regions OP1 and OP4), precuneus, inferior parietal lobule, insula, hippocampus, amygdala, cerebellum, supplementary motor area, dorsal and ventral striatum (caudate, putamen, and nucleus accumbens), ventral tegmental area, hypothalamus, pons, anterior and posterior cingulate cortex, temporal pole, and prefrontal cortex. These brain regions showed differential patterns of activity during the post-orgasm recovery phase.

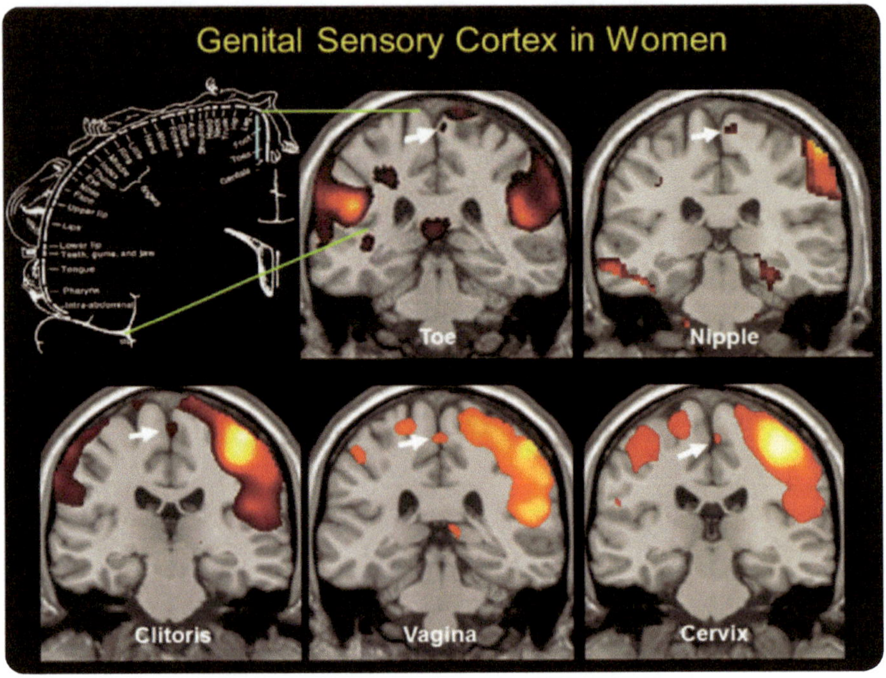

Figure 11.7 fMRI data showing (arrows) the region of the sensory cortex (paracentral lobule) in activated by stimulation of specific body regions, in relation to the classical Penfield sensory homunculus Note that self-stimulation of clitoris, vagina and cervix (group data from 11 women) **activate slightly different regions, but all in the "male genital" region of Penfield's map.** Hallux (great toe) stimulation by the investigator activated the foot region of the homunculus (individual data). Note that **nipple** self-stimulation in this individual also activated the ***medial paracentral lobule*** in the genital region. Adapted from Komisaruk BR et al. (2011) [50].

The sensory regions that activate the paracentral lobule are the cervix and vagina. The clitoris activates the paracentral lobule anteriorly to that activated by the cervix and vagina. See Figure 11.6.

The Paracentral Lobule in the Cortex of the Brain: Cortical Orgastic Reward

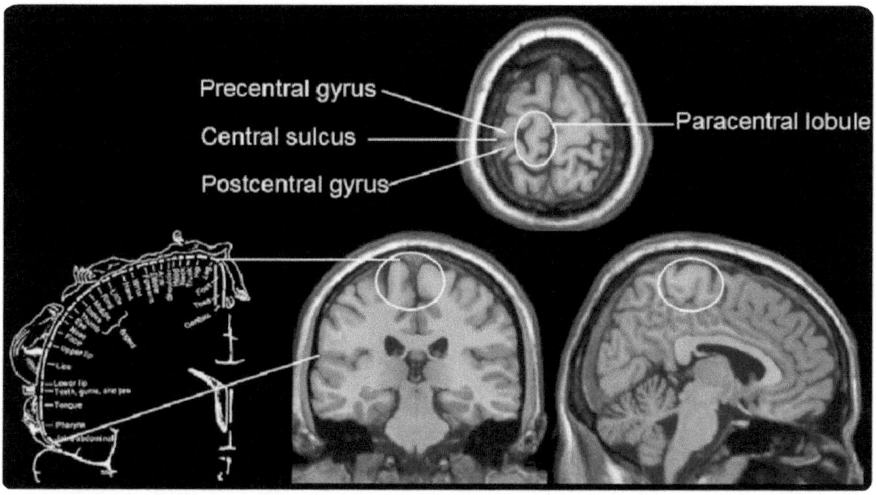

Figure 11.8 fMRI data showing the region of the paracentral lobule in the genital cortex in relation to adjacent cortical brain sites and the classical Penfield sensory homunculus [50].

11.3.6 Non-Genital Orgasms

Orgasms characteristically result from genital stimulation. However, there are many reports that other types of sensory stimulation, non-genital and these can also can generate orgasms. The Komisaruk team provides evidence from research and other sources of orgasms elicited by imagery, stimulation of nipple, lips, mouth, anus, rectum, prostate and other body regions, in phantom limbs, during sleep, in response to brain and ***spinal cord stimulation***, during childbirth, modulated by, and in response to, certain drugs, ***cocaine and marijuana*** for example and in relation to ***epileptic seizures***. A concept of generalized neurological orgasmic processes involves recruitment, peak excitement and resolution and genital orgasm is considered to be only one special case.

As one example, the Komisaruk authors have documented cases of women who claim they can experience orgasms just by imagery—***without any physical stimulation***. Their bodily reactions of doubling of heart rate, blood pressure, pupil diameter and pain threshold responses that are comparable in magnitude to when the same women induced such orgasms by genital self-stimulation, bear out their claim [48, 86].

11.3.7 Orgasmic Meditation

To facilitate a meditative state of connectedness and calm alertness between the two participants Newberg et al. [59] suggest a way for subjects to stimulate the clitoris and collect relevant data from fMRI. The technique is called *Orgasmic Meditation* (OM). Data was acquired on 20 pairs of subjects shortly following one of two states that were randomized in their order during the OM practice or during a neutral condition. The practice is performed while the female is lying down on pillows with the clitoris exposed. During the practice, the male performs digital stimulation of the ***clitoris*** for 15 min. Resting BOLD image acquisition was performed at completion of the practice to assess changes in functional connectivity associated with the performance of the practice. The results demonstrated significant changes in functional connectivity associated with the OM compared to the neutral condition.

For the ***entire group*** of males and females, there was altered connectivity following the OM practice involving the **left superior temporal lobe, the frontal lobe, anterior cingulate, and insula**.

In **female** subjects, there was altered connectivity involving the **cerebellum, thalamus, inferior frontal lobe posterior parietal lobe, angular gyrus, amygdala, middle temporal gyrus, and prefrontal cortex**.

In **male** subjects, there were **functional connectivity changes within the supramarginal gyrus, cerebellum, and orbitofrontal gyrus, parahippocampus, inferior temporal gyrus, and anterior cingulate**.

Overall, these findings suggest a complex pattern of functional connectivity changes occurring in both members of the couple pair that result from this unique meditation practice. The changes represent a hybrid of functional connectivity findings with some similarities to meditation-based practices and some with sexual stimulation and orgasm.

Some Notes on the sexuality studies discussed:

- *The parahippocampus to which the Newberg team is referring is most likely the parahippocampal cortex.*
- *The stimulation of both nipple and clitoris would be of interest to image.*
- *The stimulation of both nipple and other sexual organs would be of interest to image.*
- *The finger, toe and nipple stimuli are activated differentially from the cervix, vagina and clitoris insofar as the paracentral lobule activation is concerned.*
- *The finger, toe and nipple are shown as highly activated nonetheless.*
- *Although the cerebellum, known for stable balance, is currently in the mix of being imaged as a sexual organ, the cerebellar role in sexuality has been previously shown to be critical in this respect by Blakemore et al.* [10].

11.4 The BRODERICK PROBE® Sees the Orgasm through the Biogenic Amines

A seizing episode is not equal to one neuron firing. One seizing episode is not equivalent to one seizing neuron. To see a seizing episode on the encephalogram (EEG), thousands of neurons must fire to meet the threshold of detection on the EEG. Orgasm meets the threshold criteria. Can this phenomenon be studied *in situ*? We have some results on this question and provide our rationale for this yet preliminary outcome.

We imaged neuronal tissue that was in the physiological state of spiking or "*Firing,*" that is, tissue which was determined by the surgical team at NYU Tisch as ***within the epileptogenic zone*** of the patient and we compared such with epileptic neuronal tissue received from ***the same patient*** that was determined to be, by the surgical team at NYU Tisch Hospital, in the physiological state of non-spiking "*Non-Firing*" as the tissue was ***external to the epileptogenic zone***.

The neuronal tissue was received LIVE *en bloc* from the epilepsy patient during intraoperative surgery and the tissue was studied immediately upon receipt. We had subtyped the patient as mesial temporal lobe (MTLE) since MTLE presents when the neurotransmitter, norepinephrine is prevalent.

What was a very curiously interesting and scientifically sound outcome!

We were surprised at first glance! We thought that the possible signaling for orgasm from the spiking tissue would exceed that of the non-spiking tissue. Signaling in the spiking tissue was even below the norepinephrine detection limit! Thus, the firing tissue dissipated its catalyzed dopamine and norepinephrine. The non-spiking resected tissue that derived from the ***same patient*** *still signaled NE release that was typically imaged in the MTLE patient studies.*

But at second glance, let's look toward hippocampal brain reward to support the data.

Ventral tegmental stimulation elicits dopamine release in the hippocampus [83]. The evoked dopamine release from the ventral tegmentum to the hippocampus is catalyzed to norepinephrine in the prefrontal cortex [55]. Cell body, locus coeruleus spiking *behaves* in ***reverse*** to the prefrontal cortical release of the catecholamines, dopamine and norepinephrine [22, 23, 66]. The indoleamine, serotonin, is hypercritical in the coerulohippocampal brain reward circuit [83]. Indeed, this is the pathway that brings serotonin to the prefrontal cortex from the locus coeruleus and the raphe nuclei; imaging brain reward serotonin in the dopamine pathway at the position of our innovative nanoprobe is shown immunohistochemically [21, 65].

Thus, we predict that electroactive brain reward signals of orgasm may be released *via* catalyzed catecholamines and indoleamines in cell bodies, the somatodendrites, the neuronal firing sites *per se*.

11.4.1 The BRODERICK PROBE® Brings Dopamine to the Fore of Epileptogenesis

11.4.1.1 Enter Doyle and Pacia: A knock on the laboratory door

It was just another productive and pleasant day in the Broderick Laboratory in the year 1995 when Drs. Werner Doyle and Steven Pacia knocked on the laboratory door. Of course, the physicians had called first to share with me the dismay that had become their daily sentiments as they worked with microdialysis, intraoperatively, in epilepsy patients. You see the gentlemen doctors were discouraged with the microdialysis surgeries to which they were assigned by the late Dr. Dennis Spencer, Yale University, New Haven. *Both Werner and Steve were extremely concerned about the* ***clogging of microdialysis cannulas*** *that occurred routinely during the course of the intraoperative surgeries for resection.*

Patricia welcomed the doctors to her laboratory. Her team of medical and undergraduate students continued to perform their work as usual. Not only both gentlemen doctors wished to work with the BRODERICK PROBE® nanoprobe, but also they wished to observe and be taught about her work. Patricia allowed her guests to watch her students work on all aspects of her nanoprobe sensor to ask her students any questions without hesitation. Werner and Steve watched Patricia and her team performing two LIVE surgeries on cocaine and psychostimulant mechanisms in the brain. Patricia had *in vitro* calibration studies ongoing and at the same time, there were students manufacturing the nanoprobe. The laboratory is only four hundred square feet! Yet, the atmosphere, as per usual, was calm and respectful with a contentment that was immediately felt by Werner and Steve.

Werner was especially impressed with and quite vocal about how Patricia taught her nanotechnology, her surgical techniques and how she nimbly directed her laboratory. Werner was humble about his own work as a surgeon. Patricia was amazed that she was meeting with a neurosurgeon and reciprocally, Werner was amazed at the multifaceted original work going on in Patricia's laboratory. Werner commented:

"All I do is open the brain."

Oh, the news of the BRODERICK PROBE® nanoprobe reached Dr. Edwin Kolodny, Chairperson of Neurology at the New York University (NYU) Medical Center at the time. Dr. Kolodny enthusiastically offered Patricia an Adjunct Faculty position in the Department of Neurology, Langone, NYU Medical Center. Patricia accepted her new appointment but declined any salary for herself. What Patricia and her laboratory needed was funding for her nanoprobe for her work with the NYU Medical Center. The funding from the Department of Neurology came through with grants written by Patricia as adjunct faculty with permission granted by Dr. Kolodny on behalf of NYU. The professional relationship between Patricia and Ed Kolodny was of the highest caliber; grants were awarded by the *Leon Lowenstein Foundation* for Parkinson and the *FM Kirby Foundation* for stroke.

The news reached Dr Devinsky, Director of the *NYU Comprehensive Epilepsy Center,* Drs. Devinsky and Kuzniecky enthusiastically welcomed Patricia's work. Funding came from the PACE Foundation and other foundations for the diagnosis and treatment of epilepsy. As Patricia was adjunct faculty at NYU while her primary appointment was Medical Professor at the City University of New York Medical School, she was not allowed to be placed on any major *National Institute of Health* grant at NYU. It seems that Patricia had to suffer solitude in funding in this respect, Yet, everyone wanted her work and their names on her work. The *Comprehensive Epilepsy Center* funded scientists and physicians to learn the BRODERICK PROBE® nanotechnology but did not fund Patricia herself nor was her laboratory funded by the Center.

Patricia began the work nonetheless, putting all thoughts of herself aside as usual. Thus, cutting-edge work began. Patricia conceived and manufactured novel ways to image the epilepsy patient that were different from her technology expertise in the murine model. Her work was abundantly novel. In the white matter of the cortex of the epilepsy patient, she and her team delineated each and every layer of the six, including the *Band of Baillarger.* The tiny probe sought each neurotransmitter profile with ease. *Publications and presentations are knowledge worldwide!* The team headed by Patricia put the biogenic amines at the forefront of the diagnosis and treatment of epilepsy [18, 20, 31, 31].

The clinical trial is described in Broderick et al. [19]. The publication was preceded by Patricia's speech at the Polytechnic Institute, April 2009.

Patricia worked for and achieved ***NYU Institutional Review Board*** approval for the intraoperative use of the BRODERICK PROBE® nanoprobe. Steve attended the meetings. The document was under the purview of Dr. Kuzniecky who signed as *Principal Investigator*. There were intense interviews, documents and meetings galore with the physician administrators of NYU Medical Center. ***The process took ten years toward completion.*** The BRODERICK PROBE® was approved for intraoperative surgical studies for the epilepsy patient. The BRODERICK PROBE® won IRB approval!

11.4.1.2 Enter Bekker, Haile, Li and Wenning: Another knock on the laboratory door

This time, the knock was from the members of the ***Anesthesiology Department*** at NYU Medical Center. Dr. Alex Bekker was a significant force in the department and he too was intensely interested in placing the BRODERICK PROBE® into his and the department's work on "cognitive outcome in patients after surgically-induced post-operative hypoxic hypoxia and ischemic-hypoxia." Dr. Bekker requested a luncheon visit with me and the members of his team, Drs. Haile, Li, Leslie, and Wenning. It was a lively meeting in which Dr. Bekker was deciding on the electrochemist of his choice. After many meetings and discussion, Alex chose "moi" as his electrochemist of choice. Years of lively and diligent work commenced [24, 38] and the relationship was blended into comradery, collegiality and long-term friendships.

Vivid and fond memories come to Patricia's mind, that is, when Dr. Bekker observed that Patricia was working on the sixth floor in the hallway of Tisch Hospital to proceed with her work for epilepsy patients, he gave her an office with the anesthesiologists and a locker in which to keep her nanotech equipment safely. There is brain reward in Patricia's prefrontal cortex as these memories come across so blithely as Alex also invited her to dinners with the department of anesthesiology multiple times and Alex himself mentored Patricia on the "know how" of the NYU operating room.

11.4.2 A Life-Threatening Accident

There was a life-threatening accident between the time spent with the *Neurology Department, the Comprehensive Epilepsy Center* and the time spent with the Anesthesiology department in NYU Medical Center.

The accident occurred in Boston at the *Hynes Convention Center* at the ***American Epilepsy Society,*** *December 4, 2009.* Patricia was to present her break-through intraoperative research in the epilepsy patient. Accompanied by the Vice President of the *Broderick Brain Foundation*, Don Odom Jr., founder of Dystenium, neither expected the horror that happened. Fractured limbs were only part of the physically damaged Patricia as she exited the Convention Center with Don. Instead of dining out before the important presentation, the two ended up in the Emergency Department.

But Patricia had a presentation. What was going to happen? Too, Patricia was at the end of her manuscript on therapies for stroke using her nanoprobe. Who was to present the data? How would the manuscript be submitted?

Albeit badly damaged physically, Patricia still dictated the last part of her manuscript to Don right from the Westin Hotel in Boston! Indeed, Patricia could not walk and her eyes and teeth were badly and visibly damaged. The pain was so intense that it was not scrutable. The two worked together as Patricia dictated the finishing paragraphs to her colleague, Don. Dr. Kolodny, the co-suthor and Patricia talked. It just was not easy.

Nonetheless, Don presented "***neurotransmitter subtyping***" for diagnosing epilepsy patients! From the hotel bedside, Patricia met with investors on schedule despite the trauma. Patricia's paper on Parkinson disease and bromocriptine therapy was completed, submitted, accepted by the journal, *Pharmaceuticals* and published.[6]

The details of the publication in *Pharmaceuticals* follow in the footnote.

---

[6]Broderick, PA, Kolodny, EH, (2009). Real-time imaging of biomarkers in the Parkinson's brain: Pharmaceutical therapy with bromocriptine. *Pharmaceuticals*, 2(3): 236–249 (2009).

11.4.3 The Genes that Code the Proteins for Dopamine for Epilepsy

Bozzi and Borrelli [11–13] propose that altered D1 and D2 receptor signaling might contribute to epileptogenesis *via* the activation of the neuronal cell death cascades, activated by the c-AMP dependent protein kinase/extracellular-related protein kinase (PKA/ERK) and mTOR (a target of rapamycin and a member of the phosophatidylinositol 3-kinase-related family of protein kinases) an intracellular signaling pathway regulating the cell cycle. The beneficial effects of selective dopamine agonists at the ***DA2 receptor*** that blocks seizures need more attention (https://www.ncbi.nlm.nih.gov/pmc/articles/PMC3774988).

Also, in addition, the works of Starr, Dailey et al. and Epps and Weinshenker are noteworthy [27, 34, 81].

11.4.4 A Word about Glutamate

The inhibitory action on the excitatory amino acid, glutamate, produced by the inhibitory neurotransmitter, gamma-aminobutyric acid (GABA), may indeed itself sensitize a small number of excited neurons, actually causing the brain of the patient to seize despite the presence of inhibitory GABA therapy. One reason for the seizure caused by too much inhibition may be the GABA-related opposing effects in the GABA interneuron. Given the caveat that the Glutamate/GABA biochemistry may go awry in favor of glutamate and not GABA, there are other concerns regarding the efficacy of GABA at least as a primary pharmacotherapy.

Indeed, there is a tremendous number of reports that deal with empirical evidence that GABA may in fact not be the primary neurotransmitter around which epilepsy pharmacotherapeutics should revolve. Even the SV2A protein anti-epilepsy work may be leading to somewhat negative results as reported in work on binding sites with the anti-epilepsy medication, levetiracetam (Keppra®) [51]. It seems that Keppra® was able to bind to the SV2A vesicle protein but the binding was unproductive in that it was unable to enhance GABA levels in healthy human subjects. The studies were done with baseline GABA ***Magnetic Resonance***

Spectroscopy. It is noteworthy that Keppra® was successful in relieving different levels of seizure intensity in animal models and in fact, it was following these successful animal studies (see review [53]) that the SV2A protein binding site was discovered. Lynch et al. [53] showed that the photo affinity stereo selective labeling binding site for Keppra® is the SV2A protein.

The SV2A synaptic vesicle protein is thought to alleviate the debilitating epileptic episodes by modulating exocytosis at the level that precedes the final event of vesicle fusion, that is, at the very nexus of synaptic release of epileptically significant neurotransmitters including GABA. Also important is the role of SV2A proteins to maintain vesicle pools of amino acid neurotransmitters that remain available at any moment in time for electrolyte cations, for example, calcium, to perform exocytosis presynaptically by way of action potential voltage signally a seizure at the ictal level.

Medications based on glutamate/GABA concepts have met with some success clinically. The principle underlying glutamate/GABA neuronal circuits has always struck this author as intriguing because excessive inhibition by GABA, may actually cause the few action potentials driving GABA in the epilepsy brain to generate to seizure state despite the inhibitory predominance in neurotransmission. Also, consider the short half/life of glutamate metabolized to GABA and its counterpart, GABA synthesized to glutamate. This fleeting temporal relationship makes the study of glutamate/GABA even more interesting. It is this complexity that leads scientists and physicians to look into the precursor molecule, glutamine to understand the glutamate/GABA circuit more clearly [1].

11.5 Commentary

A breakdown of the connectivity among critical sites of the reward circuit shows that the following aspects are missing in the ***orgasm*** studies presented and thus, these aspects would need to be explored.

- The habenula function is missing. After all, the thalamus is shown as crucial to orgasm since 1954 and the habenula is connected to the thalamus.
- The pineal gland function is missing. After all, one's circadian rhythm may impinge on the essence of orgasm whether epileptic or not.
- The function of the female ejaculation is missing. After all, it is the female prostatic fluid of fragrance, the embodiment of desire. It is what all humans seek and few achieve.

11.6 NMI and the Focal Epilepsies

Wieser [85] categorized five types of psychomotor seizures but the major subtypes currently postulated as major consists of two, mesial and neocortical temporal lobe epilepsy (TLE). With data from electroencephalogram (EEG) and electrocorticography (EcoG) [63, 64], NMI and the BRODERICK PROBE® nanobiosensors library of formulas works to enable seizure-free outcomes and refine the epileptogenic zone by delineating new definitions of neurochemical subtypes for mesial and neocortical TLE, named MTLE and NTLE.

Let's continue then, with our clinical cases ...

References

1. Bak LK, Shousboe A, Waagepetersen (2006) The glutamate/GABA-glutamine cycle: aspects of transport, neurotransmitter homeostasis and ammonia transfer. *J Neurochem* 98:641–653.
2. Baird AD, Wilson JC, Bladin PF, Saling MM, Reutens DC (2003) Sexual outcome after epilepsy surgery. *Epilepsy Behav* 4(3):268–278. doi: 10.1016/s1525-5050(03)00085-4.
3. Baird AD, Wilson JC, Bladin PF, Saling MM, Reutens DC (2004) The amygdala and sexual drive: insights from temporal lobe epilepsy surgery. *Ann Neurol* 55(1):87–96. doi: 10.1002/ana.10997.
4. Baird AD, Wilson JC, Bladin PF, Saling MM, Reutens DC (2007) Neurological control of human sexual behaviour: insights from lesion studies. *J. Neurol Neurosurg Psychiatr* 78(10):1042–1049. Published online 2006 Dec 22. doi: 10.1136/jnnp.2006.107193.

5. Baird A (2020) *Sex in the Brain: How Seizures, Strokes, Dementia, Tumors, and Trauma Can Change Your Sex Life*. Columbia University Press.
6. Baird A (2020) The truth about sexual seizures-orgasms and epilepsy. *Psychology Today* January 11.
7. Bartolomei F, Naccache L (2011) The global workspace (GW) theory of consciousness. *Behav Neurol* 24:67–74. doi: 10.3233/BEN-2011-0313.
8. Becker RF (1953) *The Cerebral Cortex of Man. By Wilder Penfield and Theodore Rasmussen. The Macmillan Company,* New York, N.Y. 1950. 248 pp.; *Am J Phys Anthropol* 11(3):441–444. doi: https://doi.org/10.1002/ajpa.1330110318.
9. Berger H (1929) Uber das elektrenkephalogramm des menschen [Article in German] Berger H. *Archiv f Psychiatrie.* 87:527–570.
10. Blakemore SJ, Wolpert DM, Frith CD (1999) The cerebellum contributes to somatosensory cortical activity during self-produced tactile stimulation. *Neuroimage* 10(4):448–459. doi: 10.1006/nimg.1999.0478.
11. Bozzi Y, Borelli E (2006) Dopamine in neurotoxicity and neuroprotection: what do D2 receptors have to do with it? *Trends Neurosci* 29(3):167–174.
12. Bozzi Y, Borrelli E (2013) The role of dopamine signaling in epileptogenesis. *Front Cell Neurosci* 7:157.
13. Bozzi Y, Dunleavy M, Henshall DC (2011) Cell signaling underlying epileptic behavior. *Front Behav Neurosci* 5:45.
14. Broderick PA. (2013) Neuromolecular imaging shows temporal synchrony patterns between serotonin and movement within neuronal motor circuits in the brain. *Brain Sci* 3(2):992–1012. doi: https://doi.org/10.3390/brainsci3020992.
15. Broderick PA (2020) Identification, diagnosis, and treatment of tau-associated pathologies using electrodes. *US Provisional Application No.* 63/130,453. December 24.
16. Broderick PA (2021) Tau peptide signals are seen in Parkinson subjects by Broderick Probe® sensors *MOJ Clin Med Case Rep* 11(2):46–50. doi: 10.15406/mojcr.2021.11.00380.
17. Broderick, PA (2022) Identification, diagnosis, and treatment of tau-associated pathologies using electrodes. *US Provisional Application No.* 63/293,334, December 23.

18. Broderick PA, Doyle WK, Pacia SV, Devinsky O (1997) The neurochemistry of the temporal neocortex in resected tissue from epilepsy patients with cortical dysplasia pathology versus patients with primarily a mesial temporal lobe pathology. In: Session entitled: *Treatment of Epilepsy in the 21st Century*, Participatory Roundtable. (D. Treiman and D. Naritoku, Chairs), American Epilepsy Society, Boston, MA.

19. Broderick PA, Doyle WK, Pacia SV, Kuzniecky RI, Devinsky O, Kolodny EH (2009) A clinical trial of an advanced diagnostic biomedical device for epilepsy patients. *J. Long-Term Eff Med Implants* 18, June 1.

20. Broderick PA, Jean-Baptiste P, Rigg J., Doyle WK, Pacia SV, Devinsky O (1997) Monoamine neurotransmitters in human neocortical temporal lobe epilepsy (NTLE): microvoltammetric studies. *Epilepsia* 38(Suppl. 8):112; Presentation at American Epilepsy Society, Boston, MA.

21. Broderick PA, Phelix CF (1997) I. Serotonin (5-HT) within dopamine reward circuits signals open-field behavior. II. Basis for 5-HT–DA interaction in cocaine dysfunctional behavior. *Neurosci Biobehav Rev* 3:227–260.

22. Broderick PA, Piercey MF (1990) Regulation of nerve impulse frequency and transmitter release by serotonergic autoreceptor agonists. Presynaptic receptors and the question of autoregulation of neurotransmitter release. *Ann NY Acad Sci* 604(1):596–597. doi: https://doi.org/10.1111/j.1749-6632.1990.tb32035.x.

23. Broderick PA, Piercey MF (1999) 5-HT1A agonists uncouple noradrenergic somatodendritic impulse flow and terminal release. *Brain Res Bull* 27(5):693–696. doi: 10.1016/0361-9230(91)90047-n.

24. Broderick PA, Wenning L (2018) Neuromolecular Imaging and BRODERICK PROBE® nanobiosensors reveal a temporal synchrony in brain rhythms in neural transmission online with movement designs during natural physiology: Temporal asynchrony is imaged online in the same subject during pathology. *Med Res Arch* 6(7): J1–18. doi: https://doi.org/10.18103/mra.v6i8.1765.

25. Carpizo R, Berciano J (1988) Orgasmic epilepsy. *Epilepsia* 29(5): 635–639. doi: https://doi.org/10.1111/j.1528-1157.1988.tb03774.x.

26. Chaukimath SP, Patil SS (2015) Orgasm induced seizures: a rare phenomenon. *Ann Med Health Sci Res* https://www.ncbi.nlm.nih.gov/pmc/articles/PMC4804666/ 5(6):483–484. doi: https: doi: 10.4103/2141-9248.177993.

27. Dailey JW, Jobe PC (1985) Anticonvulsant drugs and the genetically epilepsy-prone rat. *Fed Proc* 44(10):2640–2644.

28. Delgado JMR (1952) Permanent implantation of multi-lead electrodes in the brain. *Yale J Biol Med* 24:351–358.

29. Delgado JMR, Hamlin H, Chapman P (1952a) Technique of intracranial electrode placement for recording and stimulation, and its possible value in psychiatric patients. *Conf Neurol* 12:315–319.

30. Delgado JMR (1955) Evaluation of Permanent Implantation of Electrodes within the Brain. *Electroencephalogr Clin Neurophysiol* 7:637–644.

31. Doyle WK, Pacia SV, Devinsky O, Jean Baptiste P, Rigg, JL, Broderick PA (1997) Preliminary results defining the epileptic substrate in patients with refractory seizures using *in situ* microvoltammetry. *The Congress of Neurological Surgeons,* New Orleans, LA.

32. Doyle WK, Pacia SV, Devinsky O, Jean-Baptiste P, Green, S, Broderick, PA (1998) Norepinephrine (NE) serotonin (5-HT), and dopamine (DA), in refractory human temporal lobe epilepsy (TLE) with *in situ* microvoltammetry. *Am Assoc Neurosurgeons,* Philadelphia, PA.

33. Epilepsy Foundation of America (1992) *The Legal Rights of Persons with Epilepsy: An Overview of Legal Issues and Laws.* 6th ed. Maryland, USA.

34. Epps SA, Weinshenker D (2013) Rhythm and blues: animal models of epilepsy and depression comorbidity. *Biochem Pharmacol* 85(2):135–146. doi: 10.1016/j.bcp.2012.08.016.

35. Freud S. (1961) Female sexuality. In: Strachey J, ed., *The Standard Edition of the Complete Psychological Works of Sigmund Freud.* Vol. 21. Hogarth; London, pp. 225–243.

36. Gaillard F, Bell D (2020) Temporal lobe. Reference article, *Radiopaedia.org* (accessed on April 7, 2022). doi: https://doi.org/10.53347/rID-10093.

37. Gowers, W (1879) A study of the so-called tendon-reflex phenomena. *Med.-Chir. Trans.* 62. Published by the Medical and Chirurgical Society. J.E. Adlard, London.

38. Haile M, Broderick PA, Li YS, Blanck T, Quartermain D, Bekker A (2011) Nimodipine reverses the elevation of synaptic striatal dopamine and serotonin during *in vivo* hypoxia. *Open Neurosci J* 5: 1–7. 1874-0820/11 2011 Bentham Open.

39. Heath RG and the Tulane Department of Psychiatry and Neurology. *Studies in Schizophrenia.* Harvard Univ. Press, Cambridge, Massachusetts, 1954. (The initial report of findings in 26 patients

prepared with depth electrodes was made in 1952 at a meeting in New Orleans, Louisiana.)

40. Heath RG, John SB, Fontana C J (1968) The pleasure response: Studies by stereotaxic technics in patients. In: Kline, N., and Laska, E., eds., *Computers and Electronic Devices in Psychiatry,* pp. 178–189. Grune & Stratton, New York.

41. Heath RG (1972) Pleasure and brain activity in man: deep and surface electroencephalograms during orgasm. *J Nerv Ment Dis* 154(1):3–18.

42. Hsu D, Chen W, Hsu M, Beggs JM (2008) An open hypothesis: is epilepsy learned, and can it be unlearned? *Epilepsy Behav* 13(3): 511–522. doi: 10.1016/j.yebeh.2008.05.007.

43. Jannini EA, Rubio-Casillas A, Whipple B, Buisson O, Komisaruk BR, Brody S (2012) Female orgasm(s): one, two, several. *J Sex Med* 9: 956–965.

44. Kaculini CM, Tate-Looney AJ, Seifi A (2021) The history of epilepsy: from ancient mystery to modern misconception. *Cureus* 13(3): e13953. Mar 17. doi: 10.7759/cureus.13953.

45. Kinsey AC, Pomeroy WB, Martin CE (1948) *Sexual Behavior in the Human Male.* W.B. Saunders, Philadelphia.

46. Kinsey AC, Pomeroy WB, Martin CE Gebhard P (1953) *Sexual Behavior in the Human Female.* W.B. Saunders, Philadelphia.

47. Komisaruk, BR, Beyer-Flores C, Whipple B (2006) *The Science of Orgasm.* Johns Hopkins University Press.

48. Komisaruk B, Whipple B (2011) Non-genital orgasms. *Sex Relationship Ther* 26(4):356–372. doi: 10.1080/14681994.2011.649252.

49. Komisaruk B, Whipple B, Crawford A, Grimes S, Liu W-C, Kalnin A, Mosier K (2004) Brain activation during vaginocervical self-stimulation and orgasm in women with complete spinal cord injury: fMRI evidence of mediation by the vagus nerves. *Brain Res* 1024, 77–88. doi: 10.1016/j.brainres.2004.07.029.

50. Komisaruk BR, Wise N, Frangos E, Liu, W-C, Allen K, Brody S (2011) Women's clitoris, vagina and cervix mapped on the sensory cortex: fMRI evidence. *J Sex Med* 8:2822–2830.

51. Krüger THC, Giraldi A, Tenbergen G (2020). The neurobiology of sexual responses and its clinical relevance. In: Lew-Starowicz M, Giraldi A, and Krüger THC, eds., *Psychiatry and Sexual Medicine,* pp. 74–82. doi: https://doi.org/10.1007/978-3-030-52298-8_7.

52. Kuzniecky R, Pan J, Burns A, Devinsky O, Hetherington (2008) Levetiracetam has no acute effects on brain gamma-aminobutyric acid levels. *Epilepsy Behav* 2:242–244.

53. Lynch BA, Lambeng N, Nocka K, Kensal-Hammes P, Bajjalieh SM, Matagne A, Fuks B (2004) SV2A: more than just a new target for AEDS—the synaptic vesicle protein SV2A is the binding site for the antiepileptic drug levetiracetam. *Proc Natl Acad Sci USA* 101: 9861–9866.

54. Magiorkinis E, Diamantis A, Sidiropoulou K, Panteliadis C (2014) Highlights in the history of epilepsy: the last 200 years. *Epilepsy Res Treat* 582039.

55. Masana M, Bortolozzi A, Artigas F (2011) Selective enhancement of mesocortical dopaminergic transmission by noradrenergic drugs: therapeutic opportunities in schizophrenia. *Int J Neuropsychopharmacol* 14(1):53–68. doi: 10.1017/S1461145710000908. Epub 2010 Aug 12.

56. Masters WH, Johnson VE (1966) *Human Sexual Response*, Bantam, 1981, 1st ed. ISBN 978-0-553-20429-2.

57. Masters WH, Johnson VE, Kolodny RC (1988). *Masters and Johnson on Sex and Human Loving*. Little, Brown and Company. ISBN 978-0316501606.

58. Merrill MA Clough RW, Dailey JW, Jobe PC, Browning RA (2007) Localization of the serotonergic terminal fields modulating seizures in the genetically epilepsy-prone rat. *Epilepsy Res* 76(2–3):93–102.

59. Newberg AB, Nancy A. Wintering, C Hriso, F Vedaei, M Stoner, R Ross (2021) Alterations in functional connectivity measured by functional magnetic resonance imaging and the relationship with heart rate variability in subjects after performing orgasmic meditation: an exploratory study. *Front Psychol* 11, November 2021. doi: https://doi.org/10.3389/fpsyg.2021.708973.

60. Olds J, Milner P (1954) Positive reinforcement produced by electrical stimulation of septal area and other regions of rat brain. *J Comp Physiol Psychol 47*:4HH27.

61. Olds J, Olds M E (1964) The mechanisms of voluntary behavior. In: Heath, RG, ed., *The Role of Pleasure in Behavior,* pp. 23–53. Hoeber, Harper & Row, New York.

62. Ozkara C, Ozdemir S, Yilmaz A, Uzan M, Yeni N, Ozmen M (2006) Orgasm-induced seizures: a study of six patients. *Epilepsia* 47(12): 2193–2197. doi: 10.1111/j.1528-1167.2006.00648.x.

63. Pacia SV, Devinsky O, Perrine K, Ravdin L, Luciano D, Vazquez B, Doyle WK (1996) Clinical features of neocortical temporal lobe epilepsy. *Ann Neurol* 40:724–730.

64. Pacia SV, Ebersole JS (1997) Intracranial EEG substrates of scalp ictal patterns from temporal lobe foci. *Epilepsia* 38:642–654.

65. Phelix CF, Broderick PA (1995) Light microscopic immunocytochemical evidence of converging serotonin and dopamine terminals in ventrolateral nucleus accumbens. *Brain Res Bull* 37:37–40.

66. Piercey MF, Broderick PA, Hoffman WE, Vogelsang, GD (1990) neurotransmitter release and impulse frequency studies of dopaminergic autoreceptor agonists. Presynaptic receptors and the question of autoregulation of neurotransmitter release. *Ann NY Acad Sci* 604 (Issue 1):xi–xii, 1–650. doi: https://doi.org/10.1111/j.1749-6632.1990.tb32035.x.

67. Pfaus J, Kippin T, Centeno S (2001) Conditioning and sexual behavior: a review. *Horm Behav* 40:291–321.

68. Reinoso-Barbero F, de Andrés I. (1995) Effects of opioid microinjections in the nucleus of the solitary tract on the sleep-wakefulness cycle states in cats. *Anesthesiology* 82(1):144–152.

69. Rémillard GM, Andermann F, Testa GF, Gloor P, Aubé M, Martin JB, Feindel W, Guberman A, Simpson C. (1983) Sexual ictal manifestations predominate in women with temporal lobe epilepsy: a finding suggesting sexual dimorphism in the human brain. *Neurology* 33(3):323–330. doi: 10.1212/wnl.33.3.323. PMID: 6681877.

70. Reynolds EH (2005) The John Hughlings Jackson 1935 Centenary Congress Medal. *J Neurol Neurosurg Psychiatry* 76:858–859.

71. Robertson C, Johnson J (2000) Fyodor Dostoevsky. The Arts Council of England; London Arts Board.

72. Safron A (2016) What is orgasm? A model of sexual trance and climax via rhythmic entrainment. *Socioaffect Neurosci Psychol* 6. doi: https://doi.org/10.3402/snp.v6.31763.

73. Sayin HU (2011) Altered states of consciousness occurring during expanded sexual response in the human female: preliminary definitions. *NeuroQuantology* 9(4). *doi*: 10.14704/nq.2011.9.4.486.

74. Sayin HU (2012) Doors of female orgasmic consciousness: new theories on the peak experience and mechanisms of female orgasm and expanded sexual response. *NeuroQuantology* 10(4): 692–714.

75. Sayin HU (2019) Five cases with expanded sexual response (ESR): case report. *Ann Clin Case Stud* 1: Article 1005, ISSN 2688-1241.

76. Sem-Jacobsen CW (1968) *Depth-Electrographic Stimulation of the Human Brain and Behavior*. Charles C Thomas, Springfield, IL.

77. Shorvon SD (2011) The causes of epilepsy: changing concepts of etiology of epilepsy over the past 150 years. *Epilepsia* 52(6): 1033–1044. doi: 10.1111/j.1528-1167.2011.03051.x.

78. Shuman T, Amendolara B, and Golshani, P (2017) Theta hemopathy as a cause of cognitive disability in TLE. *Epilepsy Curr* 17(2):107–111. doi: 10.5698/1535-7511.17.2.107.

79. Sipski M, Komisaruk B, Whipple B, Alexander CJ (1993) Physiologic responses associated with orgasm in the spinal cord injured female. *Arch Phys Med Rehabil* 74:1270.

80. Sobayo T, Mogul DJ (2016) Should stimulation parameters be individualized to stop seizures: Evidence in support of this approach. *Epilepsia* 57(1):131–140. doi: https: doi: 10.1111/epi.13259.

81. Starr M. S. (1996) The role of dopamine in epilepsy. *Synapse* 22:159–194. doi:10.1002/(SICI)1098-2396(199602)22:2<159: AID-SYN8>3.0.CO;2-C.

82. Voskuil PHA (2013) Epilepsy in Dostoevsky's novels. *Front Neurol Neurosci* 31:195–214. doi: http://dx.doi.org/10.1159/000343236.

83. Weitemier AZ, McHugh TJ (2017) Noradrenergic modulation of evoked dopamine release and pH shift in the mouse dorsal hippocampus and ventral striatum. *Brain Res* 1657:74–86. doi: https://doi.org/10.1016/j.brainres.2016.12.002.

84. Whipple B, Komisaruk BR (1997) Sexuality and women with complete spinal cord injury. *Spinal Cord* 35:136–138. doi: https://doi.org/10.1038/sj.sc.310037.

85. Wieser HG (1983) Depth recorded limbic seizures and psychopathology. *Neurosci Biobehav Rev* 7:427–440.

86. Wise NJ, Frangos E, Komisaruk, BR https://pubmed.ncbi.nlm.nih.gov/27791966/ - affiliation-2 (2016) Activation of sensory cortex by imagined genital stimulation: an fMRI analysis. *Socioaffect Neurosci Psychol* 6:31481. doi: 10.3402/snp.v6.314.

Chapter 12

Nanosensing Subtypes for Temporal Lobe Epilepsy in Patients

Novel imaging paradigms for mesial and neocortical subtypes of temporal lobe epilepsy (TLE) are considered. With the true meaning of electrocatalysis unearthed throughout the monograph, we look at the biogenic amines to further assist us in the diagnosis and treatment of epilepsy. The foci are the catecholamines, indoleamines and cofactors that are known to be highly implicated in the modus operandi of temporal lobe epilepsy (TLE). Thus, Neuromolecular Imaging (NMI) with BRODERICK PROBE® *biosensors subtype the temporal lobe epilepsy (TLE) syndrome electrochemically and in situ.*

12.1 Clinical Case 1

12.1.1 The Hippocampus as It Relates to the Limbic System

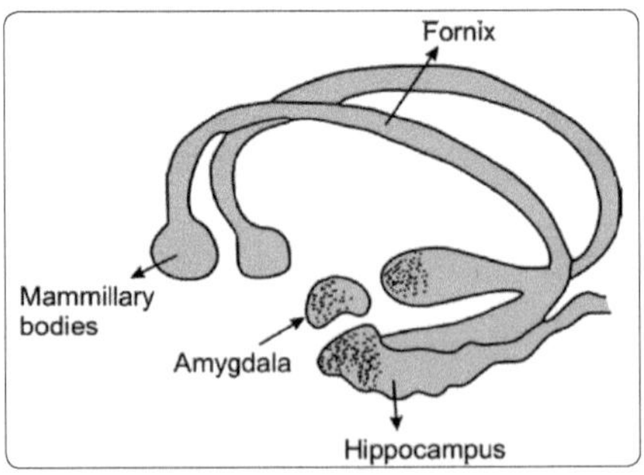

Figure 12.1 Shows the hippocampus as it relates to the limbic system, the amygdala, the fornix and the mammillary bodies. Note that (*continued*)

Neuroimaging: Sensing Biochemistry in the Brain
Patricia A. Broderick

ISBN 978-981-4613-49-1 (Hardcover), 978-0-429-02974-5 (eBook)
www.jennystanford.com

Figure 12.1 (*Continued*) the hippocampus is a C-shaped structure having amygdaloid body at its anterior end, culminating deep into mammillary bodies. Figure courtesy of AITBS publishers India [2].

12.1.2 About the Hippocampus

The hippocampus and the adjacent amygdala form the central axis of the Limbic System, which is known as "the feeling brain." The brain structure is known to act as a critical means to *event memory* and *associational recollection*, for just two examples Hippocampal phosphorylation of Tau can present in Alzheimer's disease [3] Yet, the phosphorylation itself in the hippocampus of the Alzheimer as well as the Parkinson and the epilepsy patient, may not be the culprit. It may be the *degree* of Tau phosphorylation in the hippocampus presenting the danger [12–14]. The hippocampus is divided into the hippocampus proper and the dentate gyrus. The key anatomic structures of the hippocampus are the dentate gyrus, cornu ammonis pyramidal layers, subiculum, entorhinal cortex, alveus, and finally, the fimbria that forms the crux of the fornix. The cornu ammonis, meaning Ammon's Horn, is comprised of CA1 to CA4.

In the polysynaptic pathway that takes care of semantic memory, the alveus of the hippocampus is crucial as the hippocampus receives afferent connections from parietal, temporal, and occipital areas *via* entorhinal cortex and then sends efferent projections to dentate gyrus → CA3 → CA1 → subiculum → *alveus* → fimbria → fornix → mammillothalamic tract → anterior thalamus → posterior cingulated → retrosplenial cortex [2].

Yes, the factors contributing to epileptic seizures are multifaceted, but *the excitation perforant path* within neuronal circuits in the hippocampus is certainly one critical factor. The present studies have addressed the neurochemistry of the hippocampal sites centered around and in the perforant path which is the major input to the hippocampus.

12.1.3 The Perforant Path of Excitation: The Firing Neuron in its Never-Ending Loop

The axons of the perforant path of excitation arise principally in layers II and III of the entorhinal cortex (EC), with minor contributions from the deeper layers IV and V. Axons from layers II/IV project to the granule cells of the dentate gyrus (DG) and pyramidal cells of the CA3 region, while those from layers III/V project to the pyramidal cells of the CA1 and the subiculum [2].

What happens when the neurons seize, that is, fire? The seizing/firing neurons follow what is known as the perforant path of excitation and this expression tells us how powerful the neuronal seizing/firing neuron actually is. The excitation perforant path acts in a trisynaptic loop. What does this mean for epilepsy patients? The excitatory current, classically the theta rhythm signals, start in the entorhinal cortex which sends projections to the dentate gyrus, to Ammon's horn, cornu ammonis, CA 3-CA 1 and from CA 1 to subiculum.

The subiculum sends the excitatory seizing/firing signals back to the entorhinal cortex, thus, automatically signaling neurons to fire and the seizure occurs again and the loop, the seizing circuit, proceeds again and again.

The neurons continually seize; we can make an analogy. The perforant path acts in *simili modo* to a relay tournament but in the relay tournament, there is a STOP at the finish line. In the epilepsy person with seizure, the neurons will not stop seizing; the signal is on a feedback circuit! Such is the perforant path of excitation in the hippocampus in the temporal lobe of the epilepsy person!

When the neurons will not cease seizing, the syndrome is called "status epilepticus." There are pharmaceuticals galore to try to stop the seizure, that is, the relay of non-stop neuronal seizing/firing. But patients continue to be refractory to the effects of the pharmaceuticals. This is where resection surgery is appropriate.

Nanosensing the perforant path brain sites that peruse and enter the hippocampus in temporal lobe epilepsy patients enables mesial TLE (MTLE) to be distinguished from neocortical temporal lobe epilepsy (NTLE) *via* electrochemical redox profiles using the BRODERICK PROBE® series of biosensors. Thus, the neurochemistry of the epileptic brain is subtyped for diagnosis and treatment.

Figure 12.2 shows a photograph of the hippocampus of an epilepsy Patient.

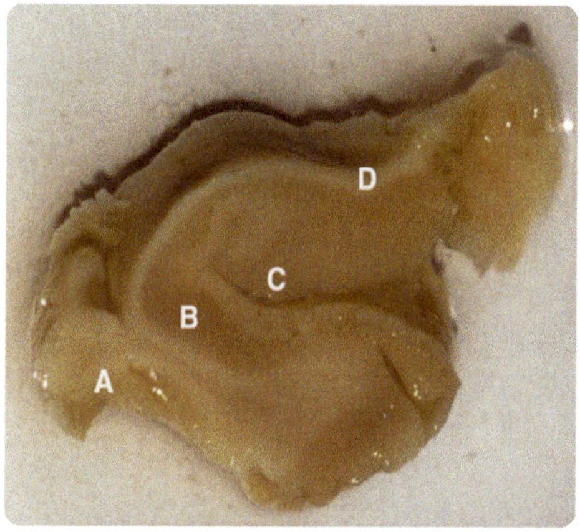

Figure 12.2 A longitudinal section of the hippocampus resected *en bloc* just prior to NMI. *Key*: A, Subiculum; B, Pyramidal cell layer, C, Granular cell layer of the Dentate Gyrus (DG), D, Alveus.

12.2 Subtyping the Hippocampus of Epilepsy Patients with NMI

12.2.1 Patient Classification

Thirteen consecutive patients who had temporal lobectomies for intractable seizures were studied. Four patients, 2 males and 2 females ages 36.75 ± 8.4 years were classified as having NTLE based on the lack of hippocampal (HPC) atrophy on MRI and tissue examination. The classification was based on the patient's demonstration of seizure onset in temporal neocortex during chronic intracranial electroencephalographic (EEG) studies with lateral temporal subdural grid electrodes as well as multiple baso-mesial temporal sub-dural strip electrodes. Patients were classified as having MTLE if pathologic examination of the resected temporal lobe revealed severe HPC neuronal loss and gliosis and if examination of the neocortex revealed no other etiology for the patient's epilepsy. Nine patients (4 males and 5 females, ages 31.4 ± 3.9 years) were classified as MTLE based on these features [15, 19].

Prior to implantation of intracranial electrodes, i.e., subdural grid and strip electrodes (AdTec Inc., Racine, Wisconsin), all patients had presurgical evaluations that included audiovisual and EEG monitoring of interictal and ictal activity, MRI with 3-mm coronal cuts through the temporal lobes, a neuropsychological evaluation and angiogram-Wada testing. Patients were referred for intracranial EEG monitoring after a review of the presurgical evaluation by a multi-disciplinary epilepsy team of neurologists, neurosurgeons and neuropsychologists. Functional mapping of cortex was performed in patients with dominant hemispheric seizure foci, using a Grass Instrument S-12 stimulator.

12.2.2 Surgery and Pathology

Following intracranial EEG seizure localization and functional mapping of language for dominant temporal lobe patients, all patients had anterior and mesial temporal lobe resections with complete resection of the amygdala and HPC and variable extent of the lateral resection, depending on the location of the ictal onset and its proximity to language function. The limits of the tissue resection were established by chronic and intra-operative EcoG. The limits of the tissue resection included the epileptogenic zone in all patients. The amygdala and the HPC were removed in NTLE as well as in MTLE despite the lack of involvement of the mesio-basal temporal region at ictal onset in NTLE patients. This was done because of the rapid spread of seizures into mesial temporal regions in many patients with neocortical foci because of data from animal studies revealing the HPC to be a pacemaker or amplifier of seizures beginning in the temporal lobe. Additionally, the rationale included a demonstrated memory dysfunction on presurgical testing of the NTLE patients. The test is called the WADA screen.

12.2.3 Tissue Preparation

Each specimen was placed in Ringer's Lactate (pH = 6.27) (McGaw Inc. Irvine, CA, USA) and studied with *in situ* neuromolecular imaging. Both room and specimen temperatures were controlled at 20–22°C. Specimen size was 2 × 15 × 20 mm^3. Visualization of the cytoarchetectonic features in HPC was performed with the OptiVISOR® (Optical Glass Binocular Magnifier, Donegan Optical Company, Lenexa, KA, 66219, USA). The subfields of the Pyrimidal Layer were recognizable by using density differences characteristic between CA 1 as compared with CA 2 as previously described by Afifi et al. [1] Moreover, since the CA 1 subfield of the Pyramidal Layer is adjacent to the Prosubiculum, this subparcellation was used as an initial marker for CA 1 subfield recognition.

12.2.4 Study Protocol

In separate experiments, each resected HPC specimen was studied within a Faradaic chamber in which a Ag/AgCl reference and stainless steel auxiliary electrode were placed in contact with the Subiculum. A BRODERICK PROBE® nanobiosensor was then placed 1 mm dorsoventrally into each HPC subparcellation, i.e.,

(a) Granular Cells of the Dentate Gyrus;
(b) Polymorphic Layer of the Dentate Gyrus;
(c) Pyramidal Layer, selecting either the CA 1 or the CA 2 subfields.

Recordings were taken from each hippocampal substrate. Gray matter was the focus of these hippocampal studies. Oxygen was bubbled through a Ringer's Lactate medium to diminish the possible effects of anoxia; the electrochemical signals for dopamine, norepinephrine and serotonin have been studied and are easily recognizable by NMI and the BRODERICK PROBE®. After the triple microelectrode assembly was put in place for scanning, potential (E_{app}) in mV was applied to the nanobiosensor *via* a modified potentiostat developed in the Broderick research laboratory: parameters-scan rate of 10 mV with an initial E_{app}

of 0.2 V and an ending E_{app} of 0.45 V. Sensitivity was set at 5 nV, time constant was set at 1 tau. The potentiostat detector was electrically connected to a Minigard Surge Suppressor (Jefferson Electric Magnetik, NY, USA) which was then connected to an electrical ground isolation.

12.3 NMI Protocols for Studies in Epilepsy Patients

12.3.1 General Description

BRODERICK PROBE® nanobiosensors do not sense membrane potentials. Instead, indicator nanobiosensors/microelectrodes pass small but finite currents, as neurotransmitters close to the microelectrode surface, undergo oxidation and/or reduction. Detection of electrochemical signals from neuroanatomic brain sites is termed Faradaic because the amount of the oxidative and/or reductive species detected at the surface of the indicator microelectrode/nanobiosensor surface, is calculated by a derivation of Faraday's Law.

12.3.2 Recordings

NMI signals were recorded on a strip chart recorder (Sci. Parts and Services, Inc., Teaneck, N.J.). One recording took place in 60 seconds. Charging (background) current was recorded and eliminated in the first 25 seconds of each recording. Charging current is a current pulse which flows through the capacitance double layer (C_{dl}) of the BRODERICK PROBE® indicator nanobiosensor that allows potential to accumulate at its surface. This is a necessary requirement to get to the point where Faradaic electron transfer can begin. Since charging current is proportional to electrode surface area, our nanobiosensor has the advantage of decreasing charging current effects. Our biosensor is a low resistance sensor which provides more advantages of decreased charging current and amplified Faradaic current. Moreover, the use of a beginning applied potential (E_{app}) of −0.2 V, additionally minimizes effects of charging current. In between scans or recordings, although any time frame can be selected, we used the parameter of a two min cell activation/deposition [4–8].

12.3.3 NMI: Electrical Circuit

The prototype potentiostat, advanced by the author and reported in her patent portfolio is equipped with a circuit known as semiderivative, used interchangeably with the term semidifferential. This circuit uses a linear scanning methodology as its basis. Semiderivative algorithm of voltammetric data means that the signals are recorded mathematically as the first half derivative of the linear analog signal. A semiderivative circuit combines an additional series of resistors and capacitors, called a "ladder network" with the traditional linear scanning technology. The addition of the "ladder network" to the original linear scanning circuit allowed for sharper and more clearly defined peaks than those previously obtained from the linear scanning methodology. It was in the Broderick lab that the semiderivative circuit for use with NMI biosensors was first discovered and described as a reduction electrical circuit [9–11].

12.3.4 Signatures for the Biogenic Amines

Signatures for each neurotransmitter and/or neurochemical for the biogenic amines are experimentally derived.

12.3.5 How Does the Nanoprobe Interpret the Signatures that Are Imaged?

Chemically, dopamine, (DA), norepinephrine (NE) and serotonin (5-HT) neurotransmitters have amine groups that are protonated at neutral pH and because of this, the molecules exist as cations, whereas the metabolites of the biogenic amines are deprotonated at neutral pH and thus exist as anions. Thus, the metabolites, such as the metabolites of DA, 3,4-dihydroxyphenylacetic acid, (DOPAC), 3,4-dihyroxyphenylglycol (DHPG-DOPEG) and homovanillic acid (HVA) do not interfere with the detection of DA at the same half-wave potential (the potential at which the maximum current is generated) that is characteristic for DA. Metabolites of NE, such as 3-methoxy-4-hydroxyphenyl glycol (MHPG), 3-methoxy-4-hydroxyphenylethyl alcohol (MOPET), 3-methoxy-4-hydroxyphenylethylamine (metanephrine) and 3-methoxy-4-hydroxyethanolamine (normetanephrine) do not interfere with the detection of NE at the same half-wave potential that is characteristic for NE.

In Vitro Scans for Imaging the Biogenic Amines Using the Nanoprobe

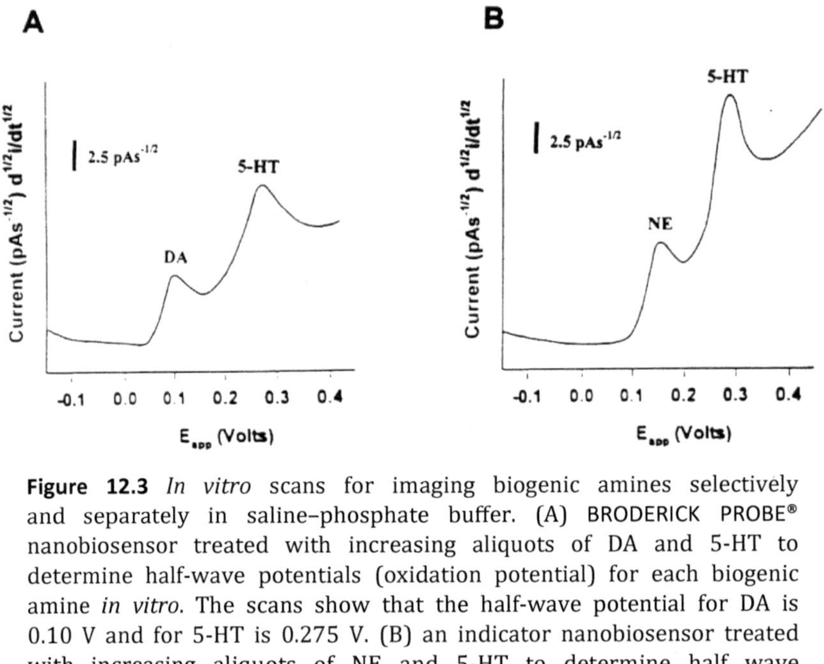

Figure 12.3 *In vitro* scans for imaging biogenic amines selectively and separately in saline–phosphate buffer. (A) BRODERICK PROBE® nanobiosensor treated with increasing aliquots of DA and 5-HT to determine half-wave potentials (oxidation potential) for each biogenic amine *in vitro*. The scans show that the half-wave potential for DA is 0.10 V and for 5-HT is 0.275 V. (B) an indicator nanobiosensor treated with increasing aliquots of NE and 5-HT to determine half wave potentials for each biogenic amine *in vitro*. A representative recording shows that the half-wave potential for NE is 0.15 V and for 5-HT is 0.28 V. Data are chosen from the third preconcentration. The *x*-axis denotes the applied potential to the indicator nanosensor in mV and the *y*-axis denotes the current derived from each biogenic amine. *Key*: DA, dopamine; NE, norepinephrine; 5-HT, serotonin; AA, neurotransmitters ascorbic acid cofactor.

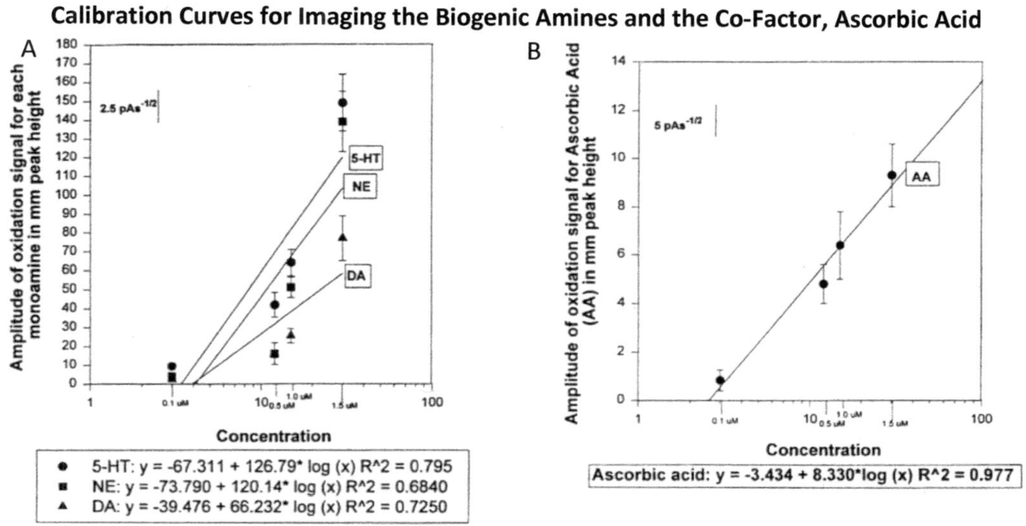

Figure 12.4 *In vitro* studies of biogenic amines and AA in saline–phosphate buffer. (A) *x*-axis represents increasing concentration of each monoamine, DA, NE and 5-HT (B), the x-axis represents increasing concentration of AA. Detection of AA required the presence of higher concentrations than did the monoamines. In both (A) and (B), the *y*-axis represents increasing current. Current was derived from the oxidation of each monoamine and AA as each neurochemical diffuses to the surface of the indicator nanosensor. Concentration of each monoamine and AA was directly proportional to the current derived from each monoamine and AA according to the Cottrell Equation. Concentrations were calculated by measuring the amplitude of the electrochemical signal at the half-wave potential (the potential at which maximum current was generated). Data were derived from the third preconcentration of biosensors (N = 57). Lipid Amplification Number (LAN), i.e., the operational stability of adsorption on the surface of the polymeric carbon biosensor after exposing each sensor to each biogenic amine and AA in saline-phosphate buffer. The response for the third time vis-à-vis the first time was measured and the delta number was calculated. LAN values for DA, NE and 5-HT were 5, 4 and 6.5, respectively. AA exhibits LAN to a lesser degree than does the biogenic amines (See Chapter 6 of this monograph).

Moreover, 5-HT was detected without the interference at the same half-wave potential (E_{app}) from either its metabolite, 5-hydroxyindoleacetic acid (5-HIAA) or uric acid (a constituent of brain with similar electroactive properties to those of 5-HT. Factors, e.g., the significantly lower sensitivity of the nanobiosensors to anions, the charge and diffusion characteristics of each catecholamine or indoleamine *vis-a-vis* its metabolites, generally preclude such interference.

12.3.6 Nanoprobe Selectively Images DA in the Presence of AA: Electrocatalysis Reviewed

Ascorbic acid (AA), the deprotonated form of vitamin C is found in high concentration in the mammalian brain at much higher levels than those of the various catecholamines and their metabolites. AA appears to act as a brain neuromodulator. AA is a cofactor in the enzymatic conversion of DA to NE. But, not only has AA been studied in terms of its inherent involvement in neuronal processes, we have tested AA in terms of evaluating methods for determining neurotransmitters in its presence.

This reaction mechanism involves the initial electro-oxidation of DA to an o-quinone, followed by the oxidation of AA by the resulting o-quinones; some of which may regenerate the DA signal. DA may then be available for oxidation at the electrode surface. The distinct electroactive signaling process for the selective image of DA as it unfolds from AA is published by the Broderick laboratory for the first time; the data are published in detail in Chapter 5 of this monograph.

12.3.7 NMI Selectively Images DA and NE *In Vitro* after Exposure to Human HPC Tissue

The neurotransmitters DA and NE, as well as the related compound, epinephrine, undergo a two-electron oxidation characteristic of catechols but subsequently are involved in a follow-up reaction which has been classified as an internal Michael cycloaddition. In this reaction, a pH sensitive intramolecular redox reaction in which the ethylamino side chain adds to the o-quinone ring in the 6-position to form a dihydroxyindoline compound which at least in one reaction, undergoes oxidation.

The catecholamines, DA, NE and epinephrine are structured similarly. The difference in structure between the DA, NE and epinephrine resides in the number of hydroxy groups on the benzene ring, with DA exhibiting two and NE and epinephrine exhibiting three. However, NE and epinephrine are further distinguished by a difference in the nitrogen group wherein NE exhibits an amine and epinephrine exhibits an amide on the alkyl side chain. Since epinephrine has been found to be essentially non-existent in human epileptogenic tissue, the two primary catecholamines of interests in the present studies, are DA and NE. DA was oxidized first; it is a smaller molecule than is NE although charges are similar. Thus, these catecholamines, DA and NE, exhibit close half-wave potentials.

Selective Detection of NE in the Presence of DA in Human HPC Tissue *In Situ*

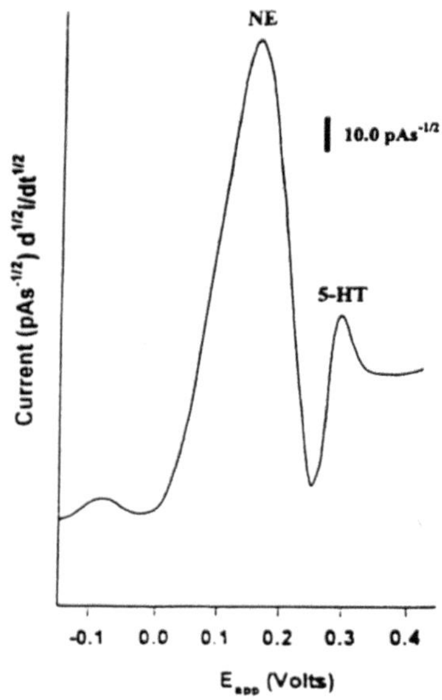

Figure 12.5 Selective detection of NE in the presence of DA in **human** HPC tissue *in situ*: A representative microvoltammogram for NE and 5-HT *in situ is* shown by an indicator microelectrode which was pretreated with DA. NE exhibited its characteristic half-wave potential at 0.15 V. The first waveform at minus 0.08 V represents the signature for NE after pretreatment with DA. 5-HT maintained its characteristic half-wave potential at 0.29 V within the same temporal and spatial resolution.

Table 12.1 Biogenic Amine (Monoamine) Release in Separate Hippocampal Substrates in NTLE patients

Patient no.	Hippocampus					
	Granular cells of the DG		Polymorphic layer of DG		Pyramidal layer	
10	DA	5-HT	DA	5-HT	DA	5-HT
	17.0	1.5	42.5	1.9	25.8	1.3
11	AA	5-HT	DA	5-HT	AA	5-HT
	32.6	3.3	71. 8	5.1	30.5	2.0
12	NE	5-HT	NE	5-HT	NE	5-HT
	102.5	3.6	115.5	4.7	13.2	4.0
13	DA	5-HT	NE	5-HT	DA	5-HT
	16.7	3.2	7.3	2.5	43.6	1.5

[a]Values for monoamines dopamine (DA), norepinephrine (NE) and serotonin (5-HT) were calibrated in nM concentrations. Values for ascorbic acid (AA) were calibrated in mM concentrations. Neurochemical signatures for NTLE patients consisted of the presence of DA, 5-HT and AA, and a relative deficiency of NE in each HPC subparcellaton with the exception of patient 12 and Polymorphic Layer of DG in patient 13.

Images from NTLE Patient 13 and MTLE Patient 8 of the Granular Cells of DG

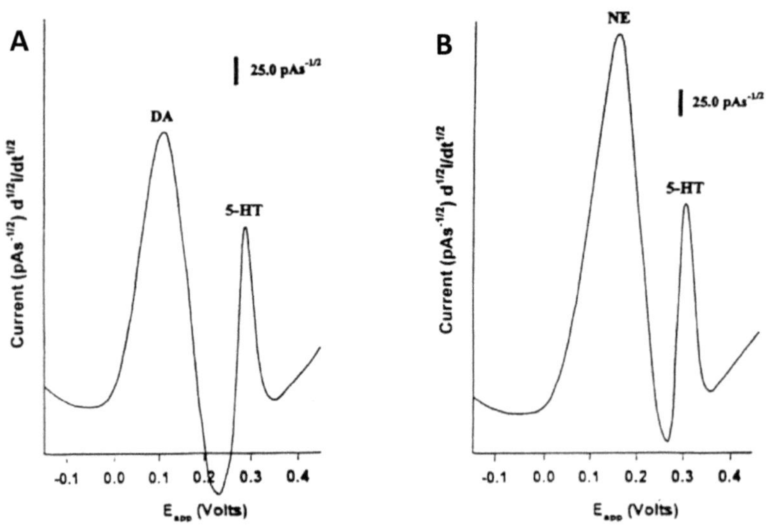

Figure 12.6 Images from NTLE patient 13 and MTLE patient 8 of the Granular Cells of DG. NTLE patient 13 exhibited DA and 5-HT signals on the recording at half-wave potentials of 0.11 V for DA and 0.29 V for 5 HT. MTLE patient 8 exhibited NE and 5-HTsignals on the recording at half-wave potentials of 0.16 V for NE and 0.3 V for 5-HT.

Recordings from NTLE Patient 13 and MTLE patient 8 of the Pyramidal Layer

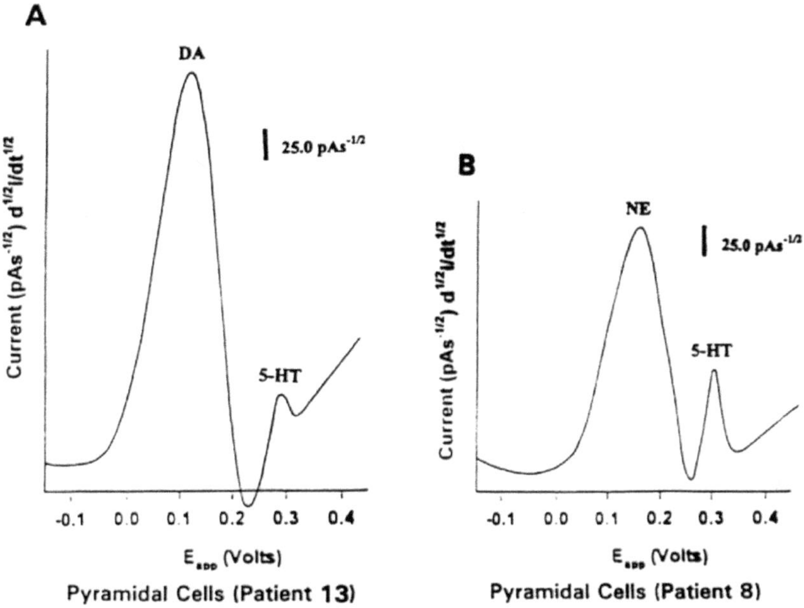

Figure 12.7 Recordings from NTLE patient 13 and MTLE patient 8 of the Pyramidal Layer. NTLE patient 13 exhibited DA and 5-HT signals on the recording at half-wave potentials of 0.11 V for DA and 0.29 V for 5-HT. MTLE patient 8 exhibited NE and 5-HT signals on the recording at half-wave potentials of 0.155 V for NE and 0.3 V for 5-HT. DA vs. NE signals differed significantly from each other by 50 mV *in vitro* and *in situ*.

Table 12.2 Validation for Biogenic Amine Signals

Biogenic amine	Validation results
NE	• Decreased electrochemical signal after administration of the anxiolytic diazepam • Increased signal after the direct receptor acting, a2 adrenergic antagonist, yohimbine • Decreased signal after administration of the direct receptor acting a2 agonist

Biogenic amine	Validation results
5-HT	• Increased signal after administration of the 5-HT precursor, L-tryptophan • Decreased electrochemical signal after the administration of the direct receptor acting $5\text{-}HT_{1A}$ agonists, buspirone and ipsapirone • Decreased electrochemical signal after the anxiolytic diazepam
DA	• Inhibition of DA release after administration of the impulse flow blocker, γ butyrolactone (γBL) • Increased electrochemical signal for DA in CPU after administration of the direct receptor acting DAD_2 antagonist haloperidol • Increased electrochemical signal for DA in NAcc after the administration of the DAD2, DAD3, autoreceptor antagonist, (+)-AJ76

12.3.8 Synopsis

NTLE: Ascorbic acid in

a. Granular cells of the Dentate Gyrus;
b. Polymorphic Layer of the dentate gyrus;
c. Pyramidal Layer (CA1 or CA2 subfields).

MTLE: Predominant were the monoamine neurotransmitters, norepinephrine in

a. Granular Cells of the Dentate Gyrus;
b. Polymorphic Layer of the Dentate gyrus;
c. Pyramidal Cells CA1 or CA 2.

12.3.9 Discussion

Resected hippocampal (HPC) tissue from specimens in the NTLE subtype population served as control for these studies because MRI evaluation did not show sclerotic pathology in HPC in this group. HPC specimens from NTLE patients exhibited dopamine and serotonin or ascorbic acid as compared with norepinephrine or serotonin in 8 of 12 HPC subparcellations in the NTLE group. Despite the limited number of NTLE specimens, dopamine was found to occur significantly more frequently in the Pyramidal Layer specimens as compared with MTLE.

Pyramidal Layer: Dopamine concentrations were significantly higher NTLE vs. MTLE specimens in all three subparcellations. These data are consistent with results from the neocortical tissue from the same patient population. Furthermore, these data are consistent with a study by Mori et al., in 1987 [17] of monoamines in resected HPC of four complex partial seizure, TLE patients. Given the caveat that Mori et al. [17] did not study HPC subparcellations and did not attempt to distinguish subtypes of TLE, their results show an interesting parallel, i.e., dopamine was present in three and absent in one of four TLE patients.

Our data recorded in TLE patients showed that significant amounts of dopamine are consistent with an expected increased dopamine in epilepsy as hypothesized by others. This hypothesis consisted of a postulated dopamine increase by glutamate further enhanced in the epilepsy brain and yet further enhanced after the subsequent down regulation of dopamine receptors. A receptor mechanism may fit since DAD 1 and DAD 2 receptors are present in human HPC and binding studies for DAD2 receptors have been found in the granular cells of the dentate gyrus. CA 1 subfield. Therefore, significant differences in the biogenic amine signatures from the hippocampal subparcellations in patients with MTLE and NTLE have been observed.

The discovery by our nanoprobe is especially important from the standpoint of biomarkers for the diagnosis and therapy of Alzheimer's disease as well as epilepsy. Ohm [18] has pointed out that the dentate gyrus, mentioned earlier to play a crucial role in associative memory, delays the formation of plaques, tangles and neuronal death until the late stages of these neurodegenerative diseases. Our NMI data that enable neurochemical profiles in the dentate gyrus, *in situ,* may be crucial for the dentate gyrus in particular to be a LIVE biomarker for cerebral dysfunctional memory caused by Alzheimer's disease as well as Parkinson disease and the epilepsies.

12.4 Clinical Case 2

12.4.1 The Neocortex

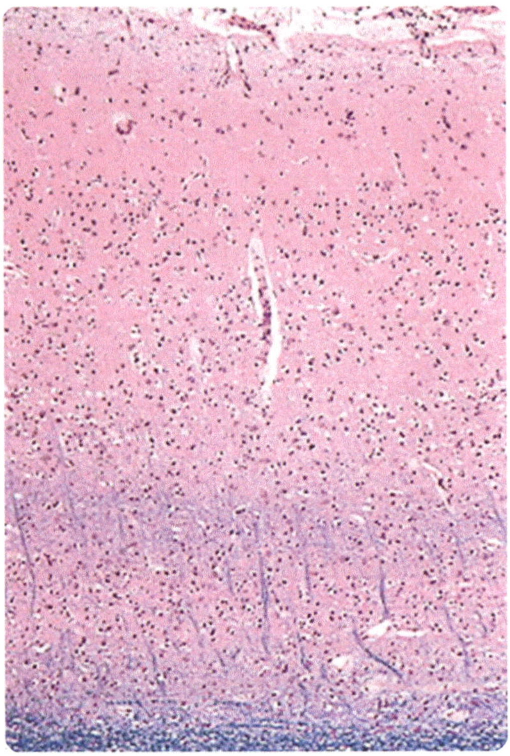

Figure 12.8 Micrograph of the visual cortex (predominantly pink). The blue, horizontal band in the lower half of the image are the bands of Baillarger/the line of Gennari. Subcortical white matter (predominantly blue) is seen at the very bottom of the image. Luxol fast blue stain, abbreviated LFB stain or simply LFB, is a commonly used stain to observe myelin under light microscopy, created by Heinrich Klüver and Elizabeth Barrera in 1953. LFB is commonly used to detect_demyelination in the central nervous system (CNS), but cannot discern myelination in the peripheral nervous system. https://en.wikipedia.org/wiki/Luxol_fast_blue_stain.

12.4.2 Subtyping Neocortex of Epilepsy Patients with NMI and the Broderick Nanoprobe

The Temporal Neocortex of the Epilepsy Patient

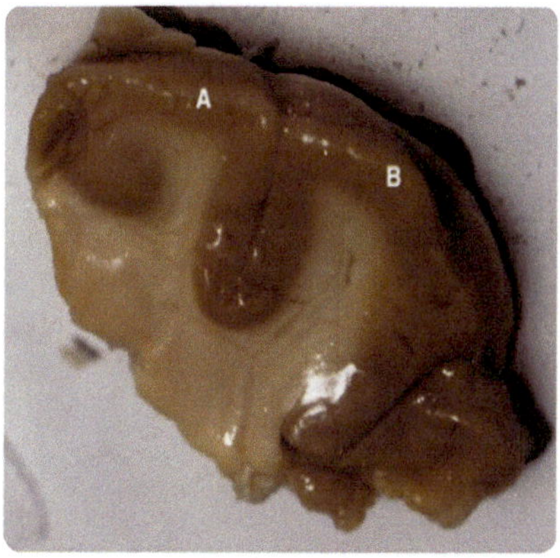

Figure 12.9 A section of the anterolateral temporal neocortex through the inferior and middle temporal gyrus. A, inferior temporal gyrus; B, middle temporal gyrus.

12.4.3 Patient Classification

Fourteen patients who had temporal lobe surgery for intractable seizures were studied. Patients underwent surgery within the same time period. Patients were classified as having mesial temporal lobe epilepsy if pathologic examination of the resected temporal lobe revealed severe hippocampal neuronal loss and gliosis and if examination of the neocortex revealed no other etiology for the patient's epilepsy. Nine patients were classified as mesial temporal lobe epilepsy (MTLE) based on these features. Five patients were classified as having neocortical temporal lobe epilepsy (NTLE) based on the lack of hippocampal atrophy on magnetic resonance imaging (MRI) and demonstration of seizure onset in temporal neocortex during chronic intracranial EEG study with lateral temporal subdural grid electrodes and multiple baso-mesial temporal subdural strip electrodes [15, 19].

12.4.4 Study Protocol

Resected temporal lobe tissue from fourteen epilepsy patients was examined in a medium of Ringer's lactate buffer solution. Samples were taken from the antero-lateral temporal neocortex. In a Faradaic chamber, a reference and an auxiliary electrode were placed in contact with the surface of the specimen and a stearate indicator nanobiosensor was inserted dorsoventrally, 2 mm into the brain tissue, in situ. With the triple microelectrode assembly in place, potentials were applied and scanned at a rate of 10 mV/s from an initial voltage of 0.2 V up to a voltage of 0.9 V. The electrochemical signals were recorded on a strip chart recorder.

12.4.5 Pathology and Results

Pathology for patients 1 through 9 is hippocampal disease; pathology for patients 10 through 14 is neocortical disease. Values for monoamines norepinephrine, serotonin, and dopamine are calibrated in nM concentrations. Values for ascorbic acid were calibrated in mM concentrations. Neurochemical signatures for mesial temporal lobe epilepsy patients consisted of the presence of norepinephrine and serotonin and the absence of dopamine in each neocortical specimen with the exception of patient #6, who exhibited dopamine and not NE. Neurochemical signatures for NTLE patients consisted of the presence of DA, 5-HT, AA in each neocortical specimen with the exception of patient, #14 who exhibited NE and not DA.

12.4.6 NMI Data from the Neocortex

Figures 12.10A,B show semiderivative voltammograms from MTLE patient 8 and NTLE patient 14, when the indicator microelectrode was placed in anterolateral temporal neocortex.

Serotonin was detected in the temporal neocortex of all fourteen patients, with a mean concentration of 2.2 nM±0.32 for the MTLE patients and 2.58 nM±0.59 for the NTLE epilepsy group. Ascorbic acid was found in the temporal neocortex of one of the NTLE patients at a concentration of 0.1 uM, but in none of the MTLE patients.

Table 12.3 Monoamine concentrations in the neocortex (Band of Baillarger, layer IV) in MTLE and NTLE patients[a]

Patient	MTLE		Patient	NTLE	
1	N.E.	5-HT	10	DA	5-HT
	48.3	1.5		17.8	2.0
2	N.E.	5-HT	11	DA	5-HT
	8.6	1.7		29.2	2.1
3	N.E.	5-HT	12	AA	5-HT
	5.6	1.3		0.11	4.9
4	N.E.	5-HT	13	N.E.	5-HT
	4.3	1.3		44.8	1.7
5	N.E.	5-HT	14	DA	5-HT
	17.2	1.9		3.0	2.2
6	DA	5-HT			
	17.2	2.0			
7	N.E.	5-HT			
	25.8	3.5			
8	N.E.	5-HT			
	8.5	3.2			
9	N.E.	5-HT			
	11.6	3.6			

[a]Pathology for patients 1 through 9 is hippocampal disease; pathology for patients 10 through 14 is neocortical disease. Values for monoamines [norepinephrine (NE), serotonin (5-HT), dopamine (DA)] are calibrated in nM concentrations. Neurochemical signatures for MTLE patients consisted of the presence of NE and 5-HT, and the absence of DA in each neocortical specimen, with the exception of patient 6 who exhibited DA and not NE. Neurochemical signatures for NTLE patients consisted of the presence of DA, 5HT and AA and the absence of NE in each neocortical specimen, with the exception of the patient 14 who exhibited NE and not DA.

Semiderivative Voltammograms from MTLE Patient 8, and NTLE Patient 14

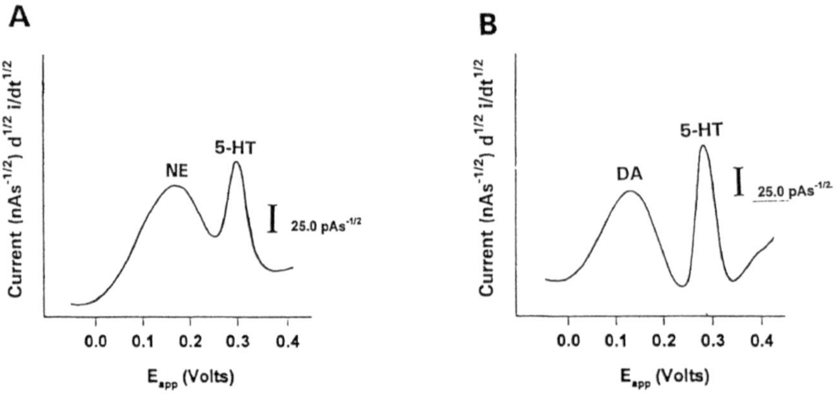

Figure 12.10 A and B show semiderivative voltammograms from MTLE patient 8, and NTLE patient 14, when the indicator microelectrode was placed in anterolateral temporal neocortex. The x-axis denotes the applied potential to the indicator nanobiosensor in mV and the y-axis denotes the current derived from each monoamine. The electrochemical signals are plotted and the characteristic oxidative potentials for each neurotransmitter was identified. MTLE patient 8 exhibited the NE and 5-HT signature on the recording at peak oxidation potentials of 0.17 V for NE and 0.305 V for 5-HT. NTLE patient 14 exhibited DA and 5-HT signals on the recording at peak oxidation potentials of 0.115 V for DA and 0.295 V for 5-HT. Data are drawn from the original recording, recorded in situ, i.e., actual raw data are presented.

12.4.7 Discussion

The most striking finding of this study was the marked norepinephrine depletion seen in the temporal neocortex of NTLE patients compared with that of the MTLE patients.

In experimental models, norepinephrine may differentially enhance or inhibit GABAergic cells. However, in numerous experimental models of epilepsy, norepinephrine depletion has been shown to enhance the frequency, intensity, and spread of seizures. A consistent depletion in neocortical norepinephrine has not been previously demonstrated in human epilepsy. One study by Goldstein et al. [16], using EcoG, showed increased concentrations of norepinephrine in spiking cortex compared with nonspiking cortex. This increased concentration was hypothesized to be a compensatory, inhibitory role of norepinephrine. However, since this study used whole tissue homogenates of large cortical regions, it is difficult to compare these data to the restricted neocortical samples examined in this example. Subtypes of temporal lobe epilepsy were also not considered. In another study of 20 patients with TLE by Pintor and colleagues [20], no difference in norepinephrine concentrations were found between spiking and nonspiking regions.

Although these previous studies are contributive, mechanisms are conflicting. NMI with the BRODREICK PROBE® that senses biochemical pathways in brain neuronal circuits provide a plausible rationale to subtype NTLE and MTLE *via* electrochemistry to image neurotransmitter profiles.

Subtyping the amino acid. L-Tryptophan in hippocampus and neocortex in epilepsy patients is next. L-Tryptophan is an amino acid and is precursor to biogenic amine, serotonin.

References

1. Afifi AK, Bergman RA (1980) *Basic Neuroscience: A Structural and Functional Approach*. Urban and Schwarzenberg Publishers, Baltimore-Munich.
2. Anand KS, Dhikav V (2012) Hippocampus in health and disease: an overview. *Ann Indian Acad Neurol* Oct-Dec; 15(4):239–246. doi: 10.4103/0972-2327.104323.
3. Avila J, Gómez de Barreda E, Engel T, Lucas JJ, Hernández F (2010) Tau phosphorylation in hippocampus results in toxic gain-of-function. *Biochem Soc Trans* 38:977–980.
4. Broderick PA (1987) Striatal neurochemistry of dynorphin-(1-13): in vivo electrochemical semidifferential analyses. *Neuropeptides* 10:369–386.
5. Broderick PA (1988) Distinguishing in vitro electrochemical signatures for norepinephrine and dopamine. *Neurosci Lett* 95:275–280.
6. Broderick PA, (1989) Characterizing stearate probes in vitro for the electrochemical detection of dopamine and serotonin. *Brain Res* 495:115–121.

7. Broderick PA (1990) State-of-the-Art microelectrodes for in vivo voltammetry. *Electroanalysis* 2, 241–251.
8. Broderick PA (1991) In vivo voltammetric studies on release mechanisms for cocaine with gamma-butyrolactone. *Pharmacol Biochem Behav* 40:969–975.
9. Broderick PA (1995) Microelectrodes and their use in cathodic electrochemical circuits. U.S. Patent # 5,433,710.
10. Broderick PA (1996) Microelectrodes and their use in cathodic electrochemical arrangement with telemetric application. European patent # 90914306.7.
11. Broderick PA (1999) Microelectrodes and their use in cathodic electrochemical arrangement with telemetric application. U.S. Patent # 5,938,903.
12. Broderick PA (2020) Identification, diagnosis, and treatment of tau-associated pathologies using electrodes. *US Provisional Application No.* 63/130,453. December 24.
13. Broderick PA (2021) Tau peptide signals are seen in Parkinson subjects by Broderick Probe® sensors *MOJ Clin Med Case Rep* 11(2):46–50. doi: 10.15406/mojcr.2021.11.00380.
14. Broderick, PA (2022) Identification, diagnosis, and treatment of tau-associated pathologies using electrodes. *US Provisional Application No.* 63/293,334, December 23.
15. Broderick PA, Pacia S V, Doyle WK, Devinsky O (2000) Monoamine neurotransmitters in resected hippocampal subparcellations from neocortical and mesial temporal lobe epilepsy patients: In situ microvoltammetric studies. *Brain Res* 878:48–63.
16. Goldstein DS, Nadi NS, Stull R, Wyler AR, Porter RJ (1988) Levels of catechols in epileptogenic and nonepileptogenic regions of the human brain. *J Neurochem* 50:225–229.
17. Mori A, Hiramatsu M, Namba S, Nishimoto A, Ohmoto T, Mayanagi Y, Asakura T (1987) Decreased dopamine level in the epileptic focus. *Res Comm Chem Path Pharmacol* 56:157–163.
18. Ohm TG (2007) The dentate gyrus in Alzheimer's disease. *Prog Brain Res* 163:723–740. doi: 10.1016/S0079-6123(07)63039-8.
19. Pacia SV, Doyle WK, Broderick PA (2001) Biogenic amines in resected human neocortex in patients with neocortical and mesial temporal lobe epilepsy: identification with in situ microvoltammetry. *Brain Res* 899, 106–111.

20. Pintor M, Mefford IN, Hutter I, Pocotte SL, Wyler AR, Nadir NS (1990) Levels of biogenic amines their metabolites, and tyrosine hydroxylase activity in the human epileptic temporal cortex. *Synapse* 5:152–156.

Chapter 13

The Broderick Nanoprobe: A Theragnostic Imaging Device

Engineered carbon BRODERICK PROBE® *polymers represent a cutting-edge, versatile tool for clinical work since the tiny polymers enable a patient diagnosis while imaging the molecule that identifies the diagnosis. Thus, the nanoprobe is a theragnostic sensor that advances electrochemical nanotechnology. Applications are published within and across disciplines such as biomedical engineering, neuroscience and mental health.*

Theragnostic medicine includes personalized medicine, pharmacogenomics, and molecular imaging [53]. The terms, theragnostics and theranostics are existentially the same. Indeed, theragnostics is considered the basis of modern clinical medicine [25].

13.1 Broderick Carbon Polymer Nanoprobes Are Theragnostics

13.1.1 About the Imaging Agent of the Theragnostic Nanoprobe

A sleek nanoprobe, the BRODERICK PROBE®, is smaller than a human hair and importantly, the tiny probe is named after the author's father, Paddy Broderick, who immigrated to America from County Clare, Ireland. The nanoprobe is designed to

Neuroimaging: Sensing Biochemistry in the Brain
Patricia A. Broderick

ISBN 978-981-4613-49-1 (Hardcover), 978-0-429-02974-5 (eBook)
www.jennystanford.com

diagnose, treat neurodegenerative disorders and in addition to find strategies, primarily, but not limited to the amelioration of the dreaded consequences of amyloidosis of the brain, spinal cord and skin [4, 5, 6, 12].

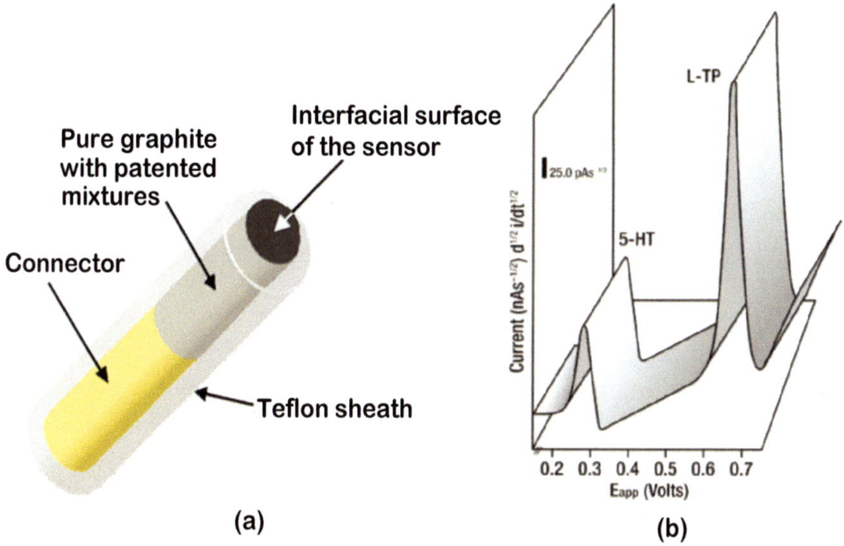

Figure 13.1 The Broderick nanoprobe: The BRODERICK PROBE® nanoprobe is a miniature theragnostic scientific and medical device. Patient studies were approved by the *NYU Institutional Review Board protocol.* The nanoprobe is a sensor comprised of the imaging agent, the nanoprobe and the therapeutic agents are comprised of an indoleamine, serotonin and its precursor, the amino acid, L-Tryptophan. In this case, the patient presented with epileptic seizures. The LIVE specimens are hippocampal and neocortical resected tissues from the epilepsy patient *en bloc* and LIVE. (a) The imaging agent of the theragnostic nanoprobe. (b) The therapeutic agent(s) of the theragnostic nanoprobe. On the x-axis, the oxidation potential in millivolts; on the *y*-axis, current. Note that oxidation potential at which serotonin (5-HT) generates its maximum current is at a peak oxidation potential of +0.290 V (experimentally derived (standard error is ±0.015 V). Note also that the oxidation potential at which L-Tryptophan (L-TP) generates its maximum current is at a peak oxidation potential of +0.675 V (experimentally derived (standard error is ±0.025 V). Also note that oxidation potentials/half wave potentials are operationally selective for each member of the Broderick patented formulary of nanobiosensors. Here, the nanobiosensor, was comprised in part of a saturated fatty acid, laurate.

The nanoprobe is available globally for manufacture and market; its *start-up* presale currency value is currently considered at about USD 100,000 per partial share of one share of equity held by *Eazysense Nanotechnologies Incorporated,* New York, NY, USA.

The nanoprobe is comprised of biologically compatible materials that enable online neurotransmitter signals, derived from redox reactions eventuating in the brain and spinal cord. for example, during the natural state, the diseased state and the medically treated state. The tiny sensor visualizes and tracks signals *in the same subject before and after therapy in real time.* This is an unusual feat in neuromolecular neuroscience. The signals derive from the brain as the tiny probe links with cerebral neurons and electroactive signals flow smoothly and LIVE as the subject walks!

This exciting nanotechnology images cerebral white matter to be imaged online with ease thus surpassing other bulkier attempts at imaging white matter such magnetic resonance imaging with diffusion tensor imaging. The data are shown herein. Thus, the nanoprobe library of sensors may well advance clinical therapies for leukodystrophies, for example.

The BRODERICK PROBE® nanotechnology is a non-invasive nanotechnology as data are collected during motor coordination in murine animals. Moreover, this nanotechnology is successful intraoperatively during epilepsy surgery in patients. In one of its embodiments, the nanoprobe senses, images the brain without opening the brain. Gliosis is not produced by the nanoprobe as empirically assessed by pathologists upon post-resection examination in the epilepsy patient. Thus, the BRODERICK PROBE® shows clinical evidence to be successful in the patient without producing scar tissue. This nanoprobe biosensor is shown empirically by trained immunologists to avoid bacterial infection in both short- and long-term studies.

The innovative sensing device is demonstratively patented [7–9, 13–16, 19, 22]; recent issue dates are 2020 and 2021. Another recent *United States Patent and Trademark Office* application by Dr Broderick shows the value of the nanoprobe as a topical therapeutic for skin health [22]. The longevity, reproducibility and viability of this LIVE imaging technique are established.

13.1.2 About the Therapeutic Agent(s) in the Theragnostic Nanoprobe

13.1.2.1 The precursor control of neurotransmitter synthesis

Importantly, the rates at which certain neurons produce and release their neurotransmitters can be affected by precursor availability. Therefore, precursor availability changes the plasma composition of precursors that occur after *ingestion* of the precursors in purified form or as *constituents of foods.* Thus, levorotary tryptophan administration or a plasma ratio of the tryptophan to other large neutral amino acids (*valine, isoleucine, leucine, tyrosine, phenylalanine and methionine*) tend to increase brain tryptophan levels, increase the substrate *saturation* of tryptophan hydroxylase, and accelerate the synthesis and release of serotonin [72]. But, tryptophan increases brain serotonin *via* unsaturated enzymatic tryptophan hydroxylase further enhancing serotonin [24]. Tryptophan manufactures serotonin proportionally without contingent benefits *via* nerve impulse frequency, supporting the work of Fernstrom [57]. Dr. Zigmond's group in Pennsylvania have come up with dramatic data very early on that show the neurotoxin, 5,7-dihydroxytryptamine, cannot stop tryptophan hydroxylase in its enzymatic quest to manufacture serotonin [61].

Elucidating precursor control of neurotransmitters is not a minor issue as its importance lies in the clinical use of levorotary tryptophan as a therapeutic theragnostic. It is, indeed, as the author will point out, the basis for the use of the therapeutic theragnostic, L-Tryptophan, within the inventive art of the BRODERICK PROBE® sensor.

The isomer of tryptophan, of interest herein is its levorotary plane, signified by the upper-case letter, "L". The origin of the levorotary tryptophan relies in its ability to rotate light to the left in a polarized field. When a molecule is determined to be levorotary or dextrorotary, its activity is a consequence of and/or a correlation with its direction in polarized light.

The amino acid, L-Tryptophan, is an essential amino acid; it is not synthesized in the human body, nor is it synthesized in many amniotes. The amino acid must pass the blood–brain barrier and be transported to the brain. Thus, this amino acid, in fact, *must* be ingested. L-Tryptophan is precursor to the neuromodulatory neurotransmitter, serotonin. The Cleveland Clinic describes serotonin quite simply as,

> "Serotonin is a chemical that carries messages between nerve cells in the brain and throughout your body. Serotonin plays a key role in such body functions as mood, sleep, digestion, nausea, wound healing, bone health, blood clotting and sexual desire. Serotonin levels that are too low or too high can cause physical and psychological health problems"[1]

13.1.2.2 The biosynthesis of serotonin (5-HT)

In amniotes, serotonin, (5-HT), neurons are present only in, for example, the median raphe nuclei as this cell body neuroanatomic substrate is a rich source of this neurotransmitter; the raphe is located in the brain stem. In anamniotes. 5-HT cell bodies, that is, dendrites, are also located in pretectal areas and basal forebrain. Tryptophan hydroxylase converts levorotary tryptophan to serotonin. There are different isoforms for tryptophan hydroxylase 2 (TPH2) that show a Km for its substrate, some of which are in the range of *in vivo* brain levels of tryptophan, (L-TP). Thus, the rate of 5-HT synthesis in cells expressing TPH2 is drastically affected by changes in the availability of its precursor. In particular, this is an effect of a precursor that likely is unseen in serotonergic cells expressing other TPH isoforms [34].

Moreover, the rate of serotonin, (5-HT) synthesis is believed to be reflected in the release of serotonin (5-HT), often quantified as the concentration of its catabolite 5-hydroxyindole acetic acid (5-HIAA), or the 5-HIAA/5-HT ratio or turnover. Thus, changes in tryptophan availability may have direct effects on serotonergic tone. Therefore, a person's dietary habits are important. To say that nutrition is important is an understatement. Moreover, stress, mobility and immune system regulation further affect plasma levorotary tryptophan release and concentrations.

---

[1]https://my.clevelandclinic.org/health/articles/22572-serotonin.

Upstream effects on brain tryptophan availability and raphe serotoninergic activity become a necessary consequence. Such tryptophan (L-TP) related changes in serotonergic activity directly affect behavior as well as do endocrine physiological states through serotonin (5-HT) projections to telencephalic and hypothalamic areas. Mood and affect are greatly influenced by the biosynthesis of serotonin [Broderick and Lynch 20]. Altered tryptophan hydroxylase activity has been demonstrated in the genetically epilepsy prone rodent [62] and a serotonergic positive influence in epilepsy is reported [43].

13.1.2.3 Dietary uptake of L-tryptophan to the brain: Manufacturing serotonin

Think about having a meal that comes from a normal *protein* source that usually contains something like 0.5–1% of the amino acid, L-Tryptophan. This will result in a relatively small increase in the tryptophan amino acid but a larger elevation of plasma concentrations of the other large neutral amino acids will occur. The outcome is a reduction in the plasma tryptophan/large neutral amino acid ratio and thereby the tryptophan influx to the brain is actually diminished.

On the other hand, dietary *carbohydrates* increase brain levels of L-tryptophan. This is due to elevated insulin. Interestingly, insulin promotes the uptake of large neutral amino acids except in skeletal muscles and thus, the outcome after a carbohydrate meal is to enhance the ratio of plasma tryptophan to large neutral amino acids. Thus, the ultimate and intriguing result of carbohydrate ingestion is an increase in the amino acid, L-Tryptophan that is transported into the brain, thereby allowing more serotonin to be manufactured in the brain.

Finally, the transport of L-Tryptophan to the brain is promoted by the common *large neutral amino acid transporter protein* that resides within the blood–brain barrier. It is amazingly noteworthy that ingesting carbohydrates will increase levorotary tryptophan and subsequently, brain serotonin, more so than does ingestion of levorotary tryptophan [34]!

13.1.2.4 The kynurenine pathway

Most interestingly, the consensus relates to the established data that a small fraction of the L-tryptophan pool is utilized for serotonin biosynthesis. The reports read that in mammals, the majority of L-tryptophan (L-TP) enters the kynurenine pathway and the amino acid is converted to other bioactive substances other than serotonin, such as kynurenic acid and quinolinic acid.

The first stage of this pathway is comprised of catalysis by the hepatic enzyme tryptophan 2,3-dioxygenase (TDO) and the extrahepatic enzyme indoleamine 2,3-dioxygenase (IDO), enzymes that are induced by glucocorticoids and pro-inflammatory cytokines, respectively. Thus, chronic stress and infections can shunt available tryptophan toward the kynurenic pathway and thereby decrease brain serotonin synthesis while simultaneously increasing the production of other tryptophan-based bioactive substances [49].

Moreover, brain tryptophan manufactures serotonin with only a 5 % contribution in this route as it is well known that a majority of tryptophan follows the kynurenine pathway (about 95%). Thus, one can think of relatively small changes in the activity of this pathway as greatly impacting tryptophan transport to the brain. Critical observations have pointed out that a decreased levorotary tryptophan passing the blood–brain barrier and entering into the brain, whether to manufacture serotonin or catabolize to the kynurenines, is dependent on a myriad of factors that we explain herein.

However, that the kynurenine pathway is one significant catabolic route in the mechanism of action for levorotary tryptophan is unquestionable. Ingested and/or administered amino acid, tryptophan, and its consequences on transport from the periphery to the brain derived from stress and/or inflammation/infection-induced neuroimmune activation of the kynurenic pathway, in particular, is also so noted. Of course, stress or inflammation/infection may underly, for example, mental illness comorbid with immune dysregulation and disorders in the neuroendocrine stress axis [17, 34, 60].

13.2 L-Tryptophan Diversifies

Let us look at the *L-Tryptophan/Serotonin/Indoleamine Pathway* for clues to these novel findings for the diagnosis and treatment of patients with epileptic seizures using the BRODERICK PROBE® as a theragnostic sensing device.

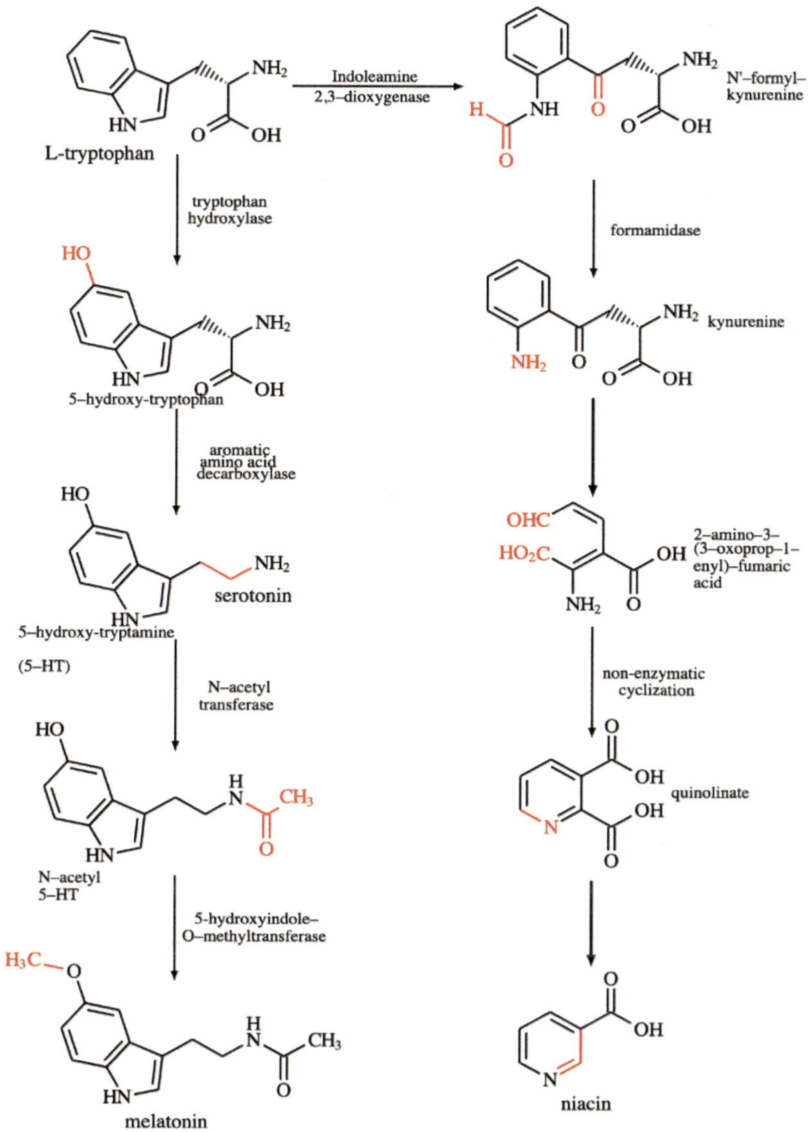

Figure 13.2 The biochemical biogenic amine, indoleamine pathway for the precursor to serotonin, which is L-tryptophan, L-TP, the levorotary tryptophan of interest here. There are alternative routes for synthesis of serotonin that involve kynurenine and kynurenine-related compounds. *Note*: The intriguing metabolism of *L-tryptophan* that proceeds into *serotonin* and *melatonin* (left) and *niacin* (right). The transformed functional groups after each chemical reaction are (*continued*)

Figure 13.2 (*continued*) highlighted in red. Amino acids biosynthesize protein, a requirement for living. But, many mammalian animals and humans do not enable synthesize of L-TP. It is diet that brings the amino acid, L-TP, to be called, essential and this levorotary tryptophan, L-TP, plays important structural or functional roles. In fact, tryptophan and tyrosine residues play a part in "anchoring" membrane proteins within the cell membrane. Further, *auxin* is a class of phyto-hormones synthesized from levorotary tryptophan and auxins act as plant growth regulators [41, 52]. The precursor role of L-Tryptophan is critical for health. https://en.wikipedia.org/wiki/Tryptophan#:~:text=Tryptophan%20(symbol%20Trp%20or%20W,polar%20aromatic%20beta%20carbon%20substituent.

Note: It is the shikimate pathway that represents the seven-step metabolic pathway used by bacteria, archaea, fungi, algae, some protozoans, and plants for the biosynthesis of folates and aromatic amino cids such as tryptophan, phenylalanine, and tyrosine. This shikimate pathway is not found in animal cells.

13.2.1 The L-Tryptophan/Serotonin/Kynurenine Pathway

Figure 13.2 delineates the serotonin indoleamine pathway. it is noteworthy that *tryptophan dioxygenase* or *indoleamine 2,3, dioxygenase* can reliably predict much different outcomes. L-TP does not always produce enzymatic synthesis to the indoleamine neurotransmitter, serotonin (5-HT). Indeed, the alternate biochemical enzymatic induction is to the synthesis of kynurenine and related kynurenine products such as quinolinic acid/quinolinate, a toxic kynurenine metabolite of tryptophan, *via* indoleamine 2,3 dioxygenase.

13.3 The Indoleamine Pair as Therapeutic Theragnostic

The interpretation of our clinical epilepsy work using the BRODERICK PROBE® nanoprobe is continued. We sensed the neurochemistry of the nanoprobe to classify patients into MTLE and NTLE as the classification relates to the indoleamines and the nanoprobe's utility as a theragnostic. Its utility as a theragnostic is practical and it is intriguing. simultaneously. It is inexorably critical to take a much closer look at the contribution of the indoleamine neurotransmitter, serotonin and its fascinating precursor, levorotary tryptophan as this molecular pair relates to the everyday nutritional life of epilepsy patients.

We sensed resected tissue from five epilepsy patients who were classified as *Neocortical Temporal Lobe Epilepsy* (NTLE) and nine epilepsy patients who were classified as *Mesial Temporal Lobe Epilepsy* (MTLE); the classification was based on magnetic resonance imaging (MRI) and intracranial EEG evaluations [71]. The BRODERICK PROBE® was placed (2 microns) within the neocortical neuroanatomic substrate, Layer IV, the Band of Baillarger, derived from **14** epilepsy patients *en bloc* and in separate studies, the placement of the nanoprobe sensor was placed within hippocampal tissue also derived from the same group of epilepsy patients. The hippocampal (HPC) subparcellations that were studied in **12 of 14** epilepsy patients were (1) granular cells of the dentate gyrus (DG) (gray matter) (2) polymorphic layer of the Dentate Gyrus, the alveus (white matter) and (3) pyramidal layer (gray matter). The neocortical layer, Band of Baillarger, Layer IV (white matter) in temporal neocortex was studied in the entire group of 14 patients.

The purpose of the studies was to oxidize the neurotransmitter, serotonin and its precursor, L-Tryptophan, *via* a reduction circuit specific for each neurochemical compound. The signature for serotonin (5-HT) and for L-Tryptophan (L-TP) was determined *via* the redox potential for each neurochemical; these digital values were experimentally derived by assaying each neurochemical *in vitro* in Ringer's Lactate and PO_4 Buffer.

13.4 Therapeutic Analytes for the Theragnostic Broderick Nanoprobe

13.4.1 Hippocampal and Neocortical Values in MTLE and NTLE Patients

Tables 13.1 and 13.2 show release concentrations, reflecting synaptic release of 5-HT and L-TP, directly related to current exhibited *via* Broderick nanoprobe signals, according to the Cottrell Equation, to which the underlying electrochemical principles of Neuromolecular Imaging (NMI) nanotechnology subscribe. Provocatively, in neocortical layer IV, NTLE patients had significantly lower L-TP concentrations than MTLE patients. The hippocampal polymorphic layer, the alveus, responded in similar fashion as did neocortical Layer IV as the imaging theragnostic nanoprobe images white matter distinguished from grey matter readily and reliably.

Table 13.1 5-HT release and L-TP release in concentration in hippocampal neuronal substrates

	Granular cells of DG		Polymorphic layer		Pyramidal cell layer	
	5-HT (nM)[a]	L-TP (uM)[b]	5-HT (nM)	L-TP (uM)	5-HT (nM)[c]	L-TP (uM)[d]
MTLE (patient #)						
1	1.4	2.70	1.5	2.75	1.5	6.40
2	2.5	2.92	NA	NA	1.3	2.98
3	1.3	8.75	1.4	6.25	1.4	17.8
4	1.3	5.28	1.4	1.00	1.2	2.65
5	2.3	4.50	3.0	2.68	1.5	7.00
6	1.6	7.68	2.9	4.30	2.1	4.60
7	2.2	3.98	2.2	2.98	1.8	5.48
8	2.1	11.8	3.2	12.5	1.3	4.35
NTLE (patient #)						
10	1.5	2.67	1.9	2.66	1.3	7.64
11	3.6	2.90	5.1	3.60	2.0	3.80
12	3.6	7.65	4.7	6.05	4.0	5.13
13	3.2	2.77	2.5	2.65	1.5	3.10

Note: Mann–Whitney rank sum test for 5-HT and L-Trp concentration MTLE vs NTLE.

[a]Concentrations of 5-HT were significantly greater for NTLE vs MTLE in granular cells of the DG ($p = 0.055, p < 0.05$).

[b]Concentrations of 5-HT were significantly greater for NTLE vs MTLE in pyramidal cell layer (CA1 Cell Layer [CA1 subfield] ($p = 0.095, p < 0.05$).

[c]Concentrations of L-TP were significantly greater for MTLE vs NTLE in granular cells of the DG ($p = 0.024, p < 0.01$).

[d]Concentrations of L-TP were significantly greater for MTLE vs NTLE in pyramidal cell layer [CA1–4 subfields] ($p = 0.055, p < 0.05$).

Table 13.2 5-HT and L-TP concentrations in temporal lobe neocortex (layer IV)

	5-HT concentration (nM)	L-TP concentration (μM)[a]
MTLE (patient #)		
1	1.5	2.85
2	1.7	2.58
3	1.3	9.30
4	1.3	5.15
5	1.9	3.85
6	2.0	2.74
7	3.5	7.60
8	3.2	5.16
9	3.6	2.66
NTLE (patient #)		
10	2.0	1.20
11	2.1	0.48
12	4.9	2.76
13	1.7	1.30
14	2.2	5.35

Note: Mann–Whitney rank sum test for 5-HT and L-TP concentration MTLE vs NTLE. 5-HT concentrations in MTLE vs NTLE patients were not significantly different from each other ($p = 0.689, p > 0.05$).

[a]L-TP release was significantly higher in temporal neocortex (layer IV) of MTLE vs NTLE patients ($p = 0.06, p < 0.05$).

13.4.1.1 The hippocampal biogenic amine values in MTLE and NTLE patients

Table 13.1 lists the 5-HT and L-TP concentrations for the 8 MTLE and 4 NTLE patients in neuronal substrates of hippocampus.

As shown in Table 13.1, the analysis of the 5-HT concentrations for these 12 patients revealed significantly lower concentrations in the granular cell and pyramidal cell layer (CA1) of the MTLE patients compared to the same regions analyzed in the NTLE group. In the present analysis, for these same 12 patients, L-TP was significantly increased in the granular cell layer and pyramidal layer of the hippocampi in MTLE patients compared to the same region in NTLE patients ($p < 0.05$, Mann–Whitney rank sum test). There was a trend toward higher levels of L-TP in the polymorphic layer, the alveus, of the hippocampi in MTLE patients; the data did not reach significance further confirming the white matter distinctive profiles as recognized by the imaging theragnostic Broderick nanoprobe.

13.4.1.2 The polymorphic layer of the hippocampus

Figure 13.3 shows the Schaffer collaterals of the hippocampal pyramidal cells in CA 3, the mossy fibers of the dentate fascia, the basket cells of the *stratum pyramidate*, as well as the polymorphic layer, the alveus (hippocampal white matter).

13.4.1.3 The neocortical values in MTLE and NTLE patients

Table 13.2 lists the 5-HT and L-TP concentrations in antero-lateral temporal neocortex layer IV for all 14 patients studied (9 MTLE and 5 NTLE).

As shown in Table 13.2, 5-HT concentrations were similar in the two groups and provocatively, and L-TP concentrations were found to be significantly lower in the neocortex of NTLE patients ($p < 0.01$, Mann–Whitney rank sum test).

13.4.1.4 The patient data

To investigate differences in continuous variables between our two groups of patients, the *Mesial Temporal Lobe Epilepsy* (MTLE) and the *Neocortical Temporal Lobe Epilepsy* (NTLE) subtypes, the Mann–Whitney rank sum test was used for nonnormally distributed variables. The Q crit test for small numbers of observations was used to determine outliers. The occurrence of outliers was minimal, being determined for only one value for one patient, i.e., for the L-Tryptophan (L-TP) value in granular cells of the Dentate Gyrus in resected tissue from patient #12. The Q crit values for the rejection of the L-TP concentration in granular cells of the DG for patient #12 was 0.899. Moreover, one MTLE patient, i.e., patient # 9, had surgical specimens that were inadequate for the purposes of this study and one NTLE patient did not require hippocampal resection. Finally, the polymorphic layer of the Dentate Gyrus, that is the alveus from patient #2 could not be identified reliably and was subsequently excluded from the study.

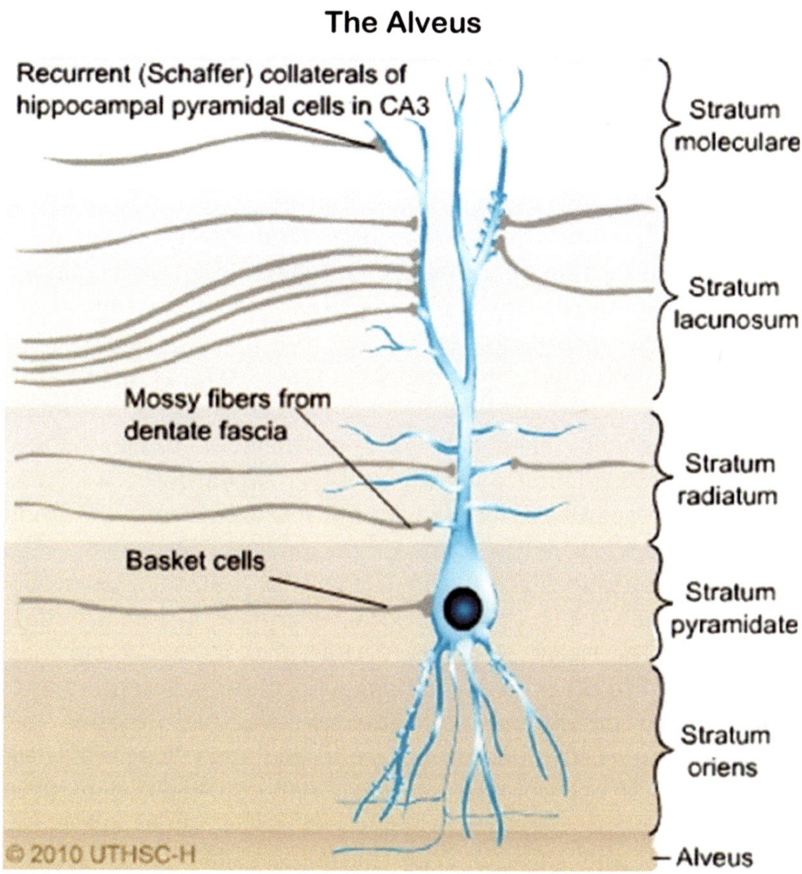

Figure 13.3 The alveus of the hippocampus is comprised of white matter. The response to the nanoprobe is similar to the neocortical white matter in this case, Layer IV, the Band of Baillarger. The hippocampus reconciles information from a variety of sources-such as amygdala and the anterior thalamic nucleus. Outputs or efferents from the hippocampus pass directly from the subiculum to the entorhinal cortex and amygdala or through the fornix to a variety of anterior structures. For further reference, see, https://nba.uth.tmc.edu/neuroscience/m/s4/chapter05.html#:~:text=The%20polymorphic%20layer%20is%20the,from%20a%20variety%20of%20sources.

13.4.1.5 The Broderick nanoprobe identifies electroactive signals: *In situ-in vitro-in vivo*

It is critical to have insight into what really matters for the accurate identification and validation of electrochemical signatures using the BRODERICK PROBE®, that is, the Broderick nanoprobe with innovative Neuromolecular Imaging (NMI). What really matters is that the biogenic amine, serotonin (5-HT) and its precursor, L-Tryptophan (L-TP), contain amine groups that are protonated at neutral pH and therefore, exist as cations, whereas metabolites of the monoamines are deprotonated at neutral pH and exist as anions. The amino group is one of several nitrogen-containing functional groups found in organic molecules. What distinguishes the amino group is that the nitrogen atom is connected by single bonds to either hydrogen or carbon.

L-Tryptophan is not a metabolite of 5-HT; as mentioned before, it is precursor to 5-HT and it too exists as a cation, thereby precluding metabolites of other chemicals to be oxidized at the same oxidation potential as L-TP. Nonetheless, we have studied the possible interference of the oxidation of L-TP from neuropeptides such as somatostatin and/or dynorphin A17. These neuropeptides are signaled separately with the nanoprobe and NMI; the signals are free from interferences at similar redox potentials. The results are reproducible, that is, that found that we can oxidize other neuropeptides with these nanosensors and there is no interference with the electroactive signaling for L-Tryptophan (L-TP). Furthermore, our laboratory has performed a considerable body of work on deciphering the redox potentials of other possible interfering chemicals such as acetic acid, lactic acid, alcohol and glutamic acid; the operationally stable results show that our reported signals are free of such interferences [2, 3, 5, 56].

13.4.1.6 The ratio of a nutrient to a neurotransmitter makes a difference

The data sheds light on therapeutics that might be reasonably and rationally used to help alleviate the neurochemical challenges with which the epilepsy patients meet on a daily basis. After all, the person afflicted with the disease of epilepsy is not seizing every nanosecond! Therefore, we need to find daily nutrients that will shift the course of the excessive neuronal firing in the brain of the epileptic far away from the seizing state.

At any rate, the outcome of the work by the BRODERICK PROBE® for treating the person suffering from epilepsy can be briefly worded as follows:

- Serotonin (5-HT) release was significantly lower whereas L-Tryptophan (L-TP) release was significantly higher in **MTLE** patients within the *hippocampal granular cells of the Dentate Gyrus* ($p < 0.05$, Mann–Whitney rank sum).
- Serotonin (5-HT) release was significantly higher whereas L-Tryptophan (L-TP) release was significantly lower in **NTLE** patients within *the hippocampal granular cells of the Dentate Gyrus* ($p < 0.05$, Mann–Whitney rank sum).
- In CA1 subfield of *hippocampal pyramidal layer,* 5-HT release was significantly lower whereas L-TP release was significantly higher in **MTLE** patients ($p < 0.05$, Mann–Whitney rank sum).
- In CA1 subfield of *hippocampal pyramidal layer,* 5-HT release was significantly higher whereas L-TP release was significantly lower in **NTLE** patients ($p < 0.05$, Mann–Whitney rank sum).
- In the hippocampal **polymorphic layer** of the Dentate Gyrus, *the alveus,* there was a trend toward lower 5-HT release and higher L-TP release in **MTLE** patients ($p < 0.01$, Mann–Whitney rank sum).
- In the hippocampal **polymorphic layer** of the Dentate Gyrus, the alveus, there was a trend toward higher 5-HT release and lower L-TP release in **NTLE** patients ($p < 0.01$, Mann–Whitney rank sum).
- In the neocortex, *Layer IV, Band of Baillarger.* 5-HT release was significantly lower whereas L-TP release was significantly higher in **MTLE** patients ($p < 0.05$, Mann–Whitney rank sum).
- In the neocortex, *Layer IV, Band of Baillarger,* 5-HT release was significantly higher whereas L-TP release was significantly lower in **NTLE** patients ($p < 0.05$, Mann–Whitney rank sum).

The ratio between the Nutrient, levorotary tryptophan, to the Neurotransmitter, serotonin, makes the difference. Thus, the work highly suggests that L-TP, an essential amino acid, may be used as therapy when hippocampal and neocortical neuroanatomic substrates that are comprised of white matter need medical adjustment. Both MTLE and NTLE patients would appear to be eligible for such therapeutic nutrient supplement as long as **white matter** is the entity showing the abnormality as imaged by the Broderick theragnostic nanoprobe. White matter abnormalities are reported [31, 37, 45, 67]. The role of glia in white matter is discussed [70]. The role of myelin in white matter is discussed [69].

13.4.1.6.1 White matter abnormality is treatable with the nutrient, tryptophan: Ratio: 5-HT/TRP

Data derived from the *Mesial Temporal Lobe Epilepsy* (MTLE) patients exhibit a low synaptic release of serotonin (5-HT) while synaptic release of levorotary tryptophan (L-TP) is significantly higher in the mesial temporal lobe patient data. The data lead us to hypothesize that the pathway from levorotary tryptophan to serotonin is hindered in the MTLE patients. On the other hand, our Neocortical *Temporal Lobe Epilepsy* (NTLE) patient data exhibits the reverse brain biochemistry [18, 21, 50, 51]. and these data have brought us to the likely theragnostic value of the tiny nanoprobe in patients with epilepsy. Thus L-Tryptophan may be used as a therapeutic theragnostic in our NTLE patients.

Moreover, realizing that the polymorphic layer is the alveus and is comprised of white matter, the nanoprobe sees **white matter** in the hippocampus and neocortex similarly and in addition, further suggests that levorotary tryptophan (L-TP) as a therapeutic theragnostic for white matter dysfunctions. The data highly indicates that the tryptophan pathway that culminates in the release of the neurotransmitter, serotonin, is the one that is primarily utilized in white matter.

Summary

- The nutrient, L-Tryptophan emerges as a therapeutic theragnostic.

- The Broderick nanoprobe is utilized as a theragnostic.
- The nutrient, L-Tryptophan is a major biomarker to identify epilepsy.

13.4.1.6.2 Enzymatic precursor control maximizes the effect of the nutrient, L-TP as therapy

Indeed, precursor control of neurotransmitters was considered a previously rigidly held process [72] which is considered by Fernstrom [24] specifically for the neurotransmitter, serotonin as a matter of supply and demand; the concept of unsaturated enzymatic tryptophan 2,3-hydroxylase empirically is further supported by Schechter and Wurtman [57]. Enzymatic control for the indoleamine, serotonin, as based on the action of its enzyme, tryptophan, 2,3-hydroxylase, is different from that of the catecholamine, dopamine and its precursor enzyme, tyrosine hydroxylase.

Thus, white matter in the brain of the epilepsy patient enables tryptophan to take its course to synthesize serotonin more readily than does the course of the catecholamine, dopamine, to take its course from its precursor enzyme, tyrosine hydroxylase. This thinking provides the rationale for the nutrient, tryptophan to be used as a therapy for white matter abnormalities in epilepsy patients. The rationale is supported by elegant studies on the mechanism of action of the enzyme, tryptophan hydroxylase (TDO) [24, 57, 72].

Therefore, albeit only 5% of levorotary tryptophan (L-TP) signals to 5-HT, serotonin, Broderick suggests that the value *5%* is maximally effective in the case of white matter signals in the brain's neuronal tissues in the case of the NTLE patient.

13.4.1.6.3 Gray matter can be offered therapy with IDO neuromodulation: Ratio: KYN/TP

The gray matter data from our studies show that L-tryptophan likely catabolizes to the kynurenines and such catabolism is replete in the literature. The data then appears to suggest that L-tryptophan would not be the first treatment for the epilepsy patient who presents with hippocampal or neocortical abnormalities in gray matter. Recent data show that indoleamine 2,3-dioxygenase

(IDO) deficiency delayed *status epilepticus* onset [23]; these data support our hypothesis. Also supporting our hypothesis are data showing macrophages/activated microglia expressing IDO during experimental autoimmune encephalomyelitis (EAE). Kynurenines have been seen in serum and cerebrospinal fluid of patients with complex partial seizures [33] and Scharfman and Ofer were able to block epilepsy seizures with kynurenic acid, one of the IDO catabolites [58].

Summary

- Gray matter may be treated with IDO neuromodulation.
- Gray matter conceptual framework shows that IDO is a biomarker for epilepsy.

13.4.2 The Kynurenines Have a Life of Their Own

13.4.2.1 Tryptophan catabolism in the therapeutics of cancer

Previously thought somewhat as a step child of serotonin, the kynurenines have come into their own rightful and just place within science and medicine. Tryptophan's catabolism to the kynurenines via enzyme, 2,3-indoleamine dioxygenase (IDO) is emerging, exponentially recognizable, as a therapeutic for cancer as a suppressant for tumor immune responses. The literature reveals a novel mechanism by which antigen-presenting cells can regulate T cell immune activation *via* tryptophan catabolism, that is, the expression of IDO by certain antigen–presenting cells *in vivo* allows these antigen-presenting cells to suppress unwanted T cell responses [46, 49]. IDO was measured as the ratio between kynurenine and tryptophan levels [40] in Juvenile myoclonic epilepsy; a type of primary generalized epilepsy was implicated.

L-Tryptophan is an amazing amino acid that is intimately involved in tumor cancer. Indeed, if levorotary tryptophan is deficient, diminished or indeed, missing, a tumor microenvironment comes into play and the tumor microenvironment allows tumor cancer to trigger immune activity with T cell induction. The fascinating immunobiology of tryptophan involves the tryptophan 2,3,-dioxygenase (TDO) step that catabolizes tryptophan to serotonin as well as the indoleamine 2,3-dioxygenase 1 (IDO1) step which is the rate-limiting enzyme that catalyzes the kynurenine pathway of tryptophan metabolism and begins the kynurenine pathway.

Both TDO and IDO 1 play a fundamental role in immunosuppression in the tumor microenvironment. In clinical trials, many IDO inhibitors have been successful antitumor therapies. There was one in a Phase III study that failed unfortunately due to over compensation by TDO in the antitumor IDO microenvironment. Epacadostat failed in 2018 and the cause was likely a compensatory effect offered by tryptophan 2,3-dioxygenase (TDO). Enter then the dual target inhibitors for tumor cancer. The mechanism of action for the new dual target inhibitors of IDO 1 depends on pharmacophore fusion strategy (Zhang et al. 2022 [74]).

In 2020, Meireson et al [47] reported on a cytosolic haem-containing enzyme involved in the catabolism of tryptophan to the kynurenines. IDO 1 expression has gone beyond the level of innate immune responses, classically discussed. Therefore, the discovery was made that the tumor surrounding stroma is another and separate tumor target. The stroma is comprised of endothelial cells, immune cells, fibroblasts, and mesenchymal cells. IDO expression has also been reported in the peripheral blood. Thus, tumor cancer is compartmentalized.

13.4.2.2 Tryptophan's magic: From quinolinic acid neurotoxicity to the therapeutic, niacin

Figure 13.4 shows the tryptophan-nicotinamide pathway. The pathway is comprised of two parts: the first part is from tryptophan to quinolinic acid and the second part is from quinolinic acid to 2-Py and 4-Py and this second part includes the NAD cycle and nicotinamide catabolism.

Thus, niacin is combined with the addicting drug, nicotine, bonded to nitric acid which is nicotinamide. Within this tryptophan-nicotinamide pathway, we see that niacin is synthesized from a neurotoxic kynurenine metabolite, that is *quinolinic acid*. Niacin is a vitamer; it is a form of the vitamin, B3. We see that a toxic substance can become a healthy nutrient in the tryptophan biochemical pathway. We see that niacin which is the vitamer B3 binds with nicotinic acid to make a healthy coenzyme and cofactor called, Nicotinamide Adenine Dinucleotide (NAD). NAD synthesis is reported [36] and the vitamer, B3, is associated with the treatment of epilepsy [44]. Caveat: Vitamin B3 interacts with antiseizure medications [65]. We see that NAD is made from nicotine and yet, we have a coenzyme and cofactor, NAD, that is critical to a healthy metabolism. *This is the magic of the tryptophan pathway.*

The Biochemical Pathway for Tryptophan to Nicotinamide to Niacin-Vitamin B3

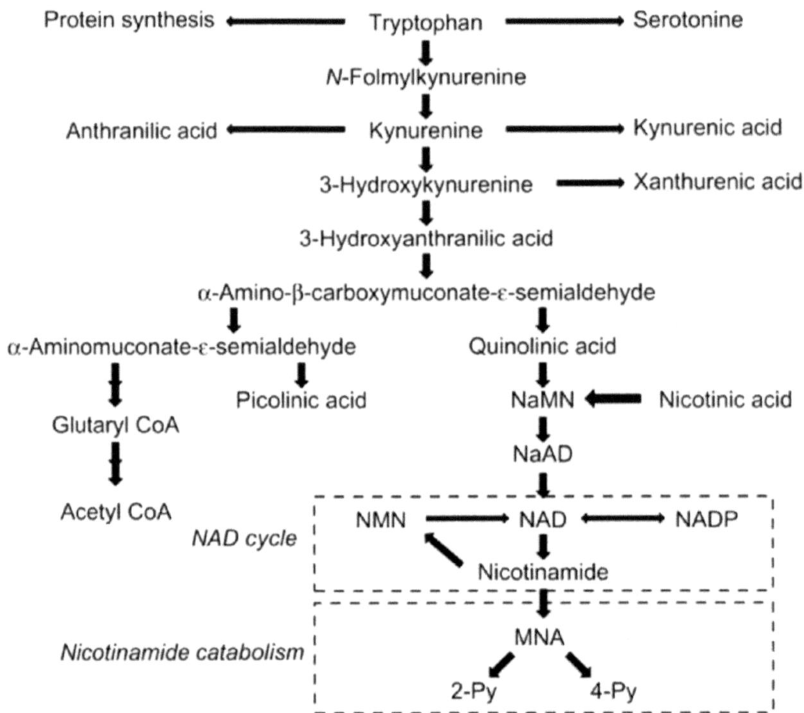

Figure 13.4 The nicotinamide-tryptophan pathway. Thus, niacinamide *via* ingestion of fish, meat, eggs and milk brings us a healthy nutrition. Cofactors are essential factors in the equation for biochemical synthesis of neurotransmitters. *Abbreviations*: NaMN, nicotinic acid mononucleotide; NMN, nicotinamide mononucleotide; MNA, N^1-methylnicotinamide; 2-Py, N^1-methyl-2-pridone-5-carboxamide; 4-Py, N^1-methyl-4-pridone-3 carboxamide [27].

Among the neurotoxic around which niacin must journey, is quinolinic acid, that is unevenly present in different brain regions and, unlike nicotinamide, the molecule enters the blood-brain barrier poorly [26] while still neuroactive in brain, for one example, in neuronal inflammatory disease [32]. Quinolinic acid acts as a *N*-methyl-D-aspartate (NMDA) receptor agonist, thus causing excitotoxic neuronal lesions and oxidative stress and excitation. NMDA is associated with the neurotransmitter, glutamate, which is generally known for pro-convulsant activity, that is, to induce the epileptogenic state. In addition, quinolinic acid concentrations in the cortical brain positively correlate with age, thus contributing to neuron synapsis "dropout" occurring during aging, Finally, neuroinflammation, neurodegeneration and mood changes are accompanied by increased quinolinic acid levels in plasma and/or cerebrospinal fluid. Thus, neurotoxic kynurenines, manufactured endogenously from IDO are known and studied [59, 63, 64]. Indeed, Rohrig et al have figured out the structure activity design for inhibitors of IDO [55]. Nonetheless, there is a fair warning about the caution exercised in regard to the study of the fascinating kynurenines. From Schwartz and Stone:

> "One important problem facing neurobiologists is that the traditionally accepted scheme of tryptophan metabolism *via* kynurenine does not always agree with experimental findings, raising fundamental questions with regard to the origin and disposition of components of the pathway. For example, the concentration of quinolinic acid in the brain does not decline as predicted after pharmacological inhibition or genomic elimination of kynurenine 3-monooxygenase (KMO). Thus, despite large increases in the brain levels of kynurenic acid (10-100-fold), indicating a shift of pathway metabolism towards the formation of the neuroprotective agent, administration of a selective kynurenine 3-monooxygenase KMO inhibitor does not affect brain quinolinic acid levels at all (and even the complete genomic deletion of KMO reduces brain quinolinic acid levels only by around 20%). At the same time, brain levels of anthranilic acid and picolinic acid, two pathway metabolites which have received relatively little attention so far, increase several-fold in these experimental situations" [59].

13.4.2.3 The candidate theragnostic therapeutic niacin

The author has studied electrocatalysis extensively and has studied the cofactor, ascorbate in detail (Chapter 5). In the discussion on the kynurenines then, Patricia feels compelled to mention an important enzyme cofactor, *Nicotinamide Adenine Dinucleotide* (NAD+). NAD+ synthesis occurs *via* the kynurenine pathway and it is an attractive use for the theragnostic Broderick nanoprobe within the tryptophan-nicotinamide pathway. Niacinamide, the generic name, Niacin, is a water-soluble form of Vitamin B3 and niacin serves as an essential nutrient that works *via* sirtuins; these are signaling proteins that regulate metabolic pathways.

Niacin's *modus operandi* is to synthesize Nicotinamide Adenine Dinucleotide (NAD+) by bonding with nicotinic acid. Like the daily need for mammalians to ingest levorotary tryptophan, niacinamide must be ingested or administered and therefore, replenished regularly. Four hundred enzymes are required for Nicotinamide Adenine Dinucleotide to perform necessary beneficial catalysis in the body and this is more than any other vitamin known [54].

Niacin is called nicotinic acid because it is made from the oxidation of nicotine with nitric acid. The name, niacin, came about because the populace could not conceive of the so-called drug-abusing drug of nicotine as one of its constituents. But indeed, the therapeutic medicine, vitamer B3 is made from addicting nicotine. How fascinating are the chemistries!

13.4.2.4 NAD, NAD+, and NADH

Nicotinamide adenine dinucleotide (NAD) is an essential cofactor and coenzyme that participates in crucial biochemical processes in the body. Scientists believe its functions are linked to the aging process and that a decline in NAD levels underlies the aging process, at least in part. NAD is made from nicotinic acid and nicotinamide.

NAD therapy replenishes the amount of NAD in the body, slowing down the aging process and providing cells with the materials they need to operate at a youthful level. The resulting benefits of NAD therapy are manifold, including mood improvement, increased energy, improved mental clarity, and better memory and concentration. Beyond NAD's visible effects, what happens at a cellular level, and what is the difference between NAD, NAD+ and NADH?

13.4.2.5 The Oxidation and Reduction Process

The living organism generates energy through the transfer of electrons, which is also the basis of all chemical reactions. When electrons change hands, the process creates different molecules while generating energy in a process called oxidation and reduction. Think of oxidation and reduction as an exchange of batteries. One molecule has batteries that another molecule needs. But they need an intermediary to help them swap batteries. In this case, cofactors like NAD step in. However, the role of oxidation-reduction extends beyond just power generation. In some cases, the process can help 'disarm' free radicals (charged molecules that can harm the cell) or modulate cellular and metabolic signaling pathways. For a reference for electrochemical principles, see Kissinger [39].

With this simply written piece, NAD fits into the theragnostic framework as follows.

13.4.2.6 The NAD+ NADH cycle

Combining the two processes creates the NAD+/NADH oxidation-reduction chain. Figure 13.5 shows the redox reactions that take place among the essential processes that take place in the formation of nicotinamide adenine dinucleotide (NAD).

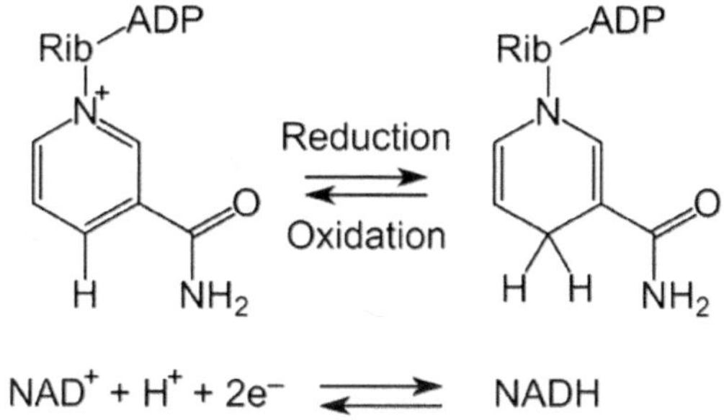

Figure 13.5

Here's how the process works!

NAD+ enters the cell or organelle (like mitochondria), accepts electrons in the form of hydrogen atoms from NADH (remember it has some extra electrons to donate), and becomes NADH. Meanwhile, the NADH that donated electrons becomes NAD+, and the process repeats itself.

NAD+ and NADH are simply different forms of NAD, one with extra electrons and one without, and they pass these electrons back and forth between themselves as a catalytic chain reaction.

This coupled reaction explains why, when NAD+ levels fall, the entire chain reaction starts to fail. Similarly, replenishing NAD+ levels via IV infusion helps restore these reactions to optimum levels.

13.4.2.7 The forms of niacin

Figure 13.6 shows the co-enzymatic (coenzymatic) forms of niacin, after Gasperi [28].

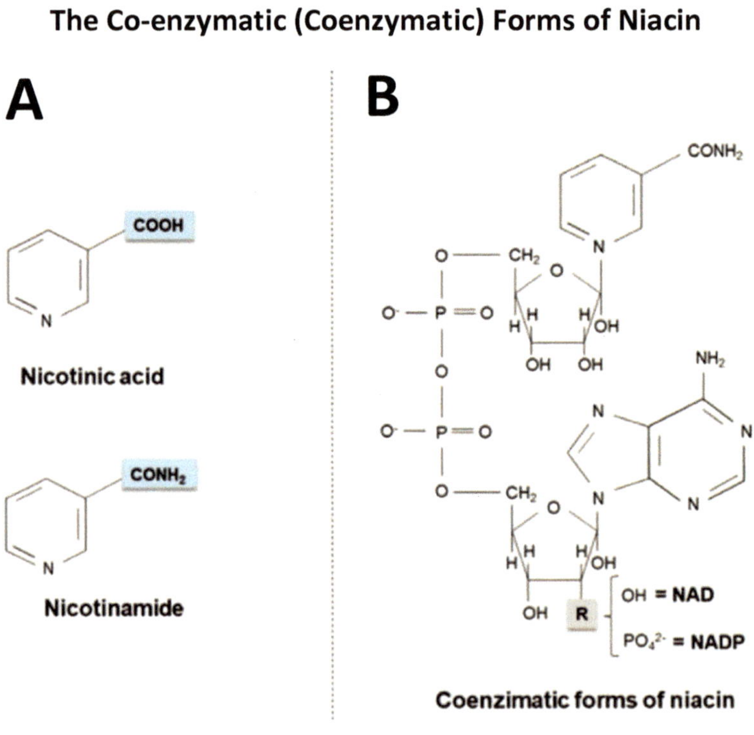

Figure 13.6 Chemical structures of niacin vitamers (**A**) and active coenzimatic forms (**B**). NAD: nicotinamide adenine dinucleotide. NADP: nicotinamide adenine dinucleotide phosphate [28]. Niacin as a theragnostic provides another fertile valley for the use of the Broderick nanoprobe.

The Electrochemistry of Niacin

Niacin (also known as "vitamin B_3" or "vitamin PP") is the generic descriptor for two vitamers, nicotinic acid (pyridine-3-carboxylic acid) and nicotinamide (nicotinic acid amide). The double "p" nomenclature is derived from the pyridine molecules from which it comes. Both bond to the biologically active coenzymes, nicotinamide adenine dinucleotide (NAD) and its phosphate analog, the nicotinamide adenine dinucleotide phosphate (NADP). The two coenzymes take part in redox reactions crucial for energy production: in particular, the pyridinic ring can accept and donate a hydride ion ($:H^-$, the equivalent of a proton and two electrons), thus acting as an electron carrier. Nonetheless, NAD and NADP play different metabolic roles in the cytosol: the NADH/NAD^+ ratio is small (about 8×10^{-4}), thus favoring oxidative catabolism, whereas the NADPH/$NADP^+$ ratio is higher (about 75), thus providing a strongly reducing environment for biosynthetic reactions [28].

13.4.2.8 The Candidate therapeutic theragnostic neopterin (NEO)

13.4.2.8.1 The tryptophan neopterin connection: NEO/L-TP

Tryptophan has really come into its own life, into the life of nutrition, microbiota composition and life style, and this topic is relevant to the work of the author. Thinking about the shift in tryptophan's benefits in the area of tryptophan catabolism, we should look at the neopterin/tryptophan ratio as another novel way to connect to white and gray matter as a candidate therapeutic theragnostic for the Broderick nanoprobe. Neopterin is the catabolic product of guanine triphosphate and its structure is of a purine nucleotide. Neopterin spins off the tryptophan catabolite pathway to provide a dramatic marker of oxidative stress.

Gostner et al., [29] emphasizes the importance of the immunometabolic fate of levorotary tryptophan in terms of neopterin as relevant to neuropsychiatric symptoms. When neopterin is reduced, there is a corresponding decrease in tetrahydrobiopterin, a cofactor of several mono-oxygenases, further linked between tryptophan and the biogenic amines. The psychoneuroimmunological consequences of tryptophan metabolism and catabolism and its relationship to neuromodulaiion by a variety of nutritional and microbiota factors have important implications to the theragnostic of the Broderick nanoprobe. A high neopterin concentration while tryptophan concentration is low, should alert all of us to the presence of disease [29], associated with the tryptophan pathway.

13.5 What the Data Means for the Epilepsy Patient

13.5.1 The Nutrient Way

The person who presents with epileptic seizures is not an epileptic person who seizes every nanosecond: Seizing is intermittent. Thus, the *World Health Organization* describes "epilepsy" as

> One seizure does not signify epilepsy (up to 10% of people worldwide have one seizure during their lifetime). Epilepsy is defined as having two or more unprovoked seizures.
> https://www.who.int

Therefore, the tide is turning as the clinical management for chronic diseases such as, for example, what is seen in the person with epilepsy [30, 38]. Moving forward with a nutrient way is the likely choice to make when treating the person with epilepsy. In fact, the *World Health Organization* describes a person's "health" as,

> Health is a state of complete physical, mental and social well-being and not merely the absence of disease or infirmity. The enjoyment of the highest attainable standard of health is one of the fundamental rights of every human being without distinction of race, religion, political belief, economic or social condition.
> https://www.who.int

13.5.1.1 Short chain fatty acids as Broderick nanoprobe

In Chapter 6 herein, the importance of fatty acids in the diet as well as the author's work regarding biomimetics, biochemically involved, is highlighted.

However, a crucial point to make here is the necessity to emphasize *short-chain fatty acids (SCFAs)* as nutrients in our diet, as these molecules impact seizure susceptibility through modulating (1) excitatory/inhibitory neurotransmitters, (2) oxidative stress and neuroinflammation, and (3) psychosocial stress. These functions are also connected to shared pathologies of epilepsy, that is, for example, depression and anxiety. As the enhancement of SCFA production is enabled through direct administration, as well as probiotics and prebiotics, related dietary treatments may exert antiseizure effects [38].

13.5.1.2 Gut microbes regulate innate immunity and epilepsy

Colonic luminal aromatic amines have been historically considered to be derived from dietary source, especially fermented foods; however, recent studies indicate that the gut microbiota serves as an alternative source of these amines. A method to detect the enzymatic aromatic decarboxylase amines can be found in [35]. Too, the method to directly image the amines can be found in Broderick [4, 7–9, 13–16, 19, 22]. The gastrointestinal flora, when disturbed, weakens the function of the intestinal immune barrier, causing local or systemic inflammation and this inflammation sometimes damages distal blood–brain barrier membranes *via* its surface molecules or metabolites. The destruction of the blood–brain barrier exposes the brain to attack by peripheral inflame immune cells in the brain, thereby lowering seizure threshold to unravel neuronal firing rate and enabling epileptiform discharges and indeed, seizures [73].

13.6 Discovery

- A nutrient, an amino acid, L-tryptophan, is a major biomarker for glial white matter.
- A catabolite of an amino acid, L-tryptophan, is a major biomarker for neuronal gray matter.

- A novel clinical method to discern white matter tumors is marketable.
- Therapeutics for epilepsy patients and patients presenting with tumor cancers are unveiled.
- The BRODERICK PROBE® is a theragnostic imaging device.
- Cost savings for the healthcare system are minimized.

13.6.1 Summary

A theragnostic (imaging plus therapeutics) device is described in this chapter. The device is comprised of a series of reliable and novel, operationally stable biosensors. The Neuromolecular Imaging (NMI) nanotechnology with miniature sensors, known globally as the BRODERICK PROBE®, is presented and discussed in detail. Thus, new state of the art nanoscale technologies is in place as described in this monograph. Other reviews and manuscripts are available for perusal [5, 12, 66, 68].

13.6.2 One More Thought

13.6.2.1 Epilepsy therapy: Electrochemical redox reactions in the brain

Electron Transfer may be just how anticonvulsants and anti-epileptic medications work. Indeed, all neuronal reactions may be redox reactions!

What follows is a quoted redox reaction for the mechanism of action for the anticonvulsants, *progabide* and *zonisamide.*

> Indeed, "reduction potentials were determined for various anticonvulsants, including progabide, SL 75.102, CGS 9896, pyridazines, zonisamide, 1,2,3-triazoles, and copper complexes. The values generally were in the range of about −0.1 to −0.6 V for the protonated drugs and the metal complexes. Reduction potentials provide information on the feasibility of electron transfer *in vivo* if the value is relatively positive (greater than about −0.6 V), the agent can act catalytically as an electron acceptor from an appropriate cellular donor. A concomitant favorable influence on abnormal neuronal processes associated with epilepsy could occur" [1].

References

1. Ames JR, Kovacic P, Kadaba PK, Kiser PF (1992) Electrochemistry of anticonvulsants: electron transfer as a possible mode of action. *Epilepsia* 33(5):936–943.
2. Broderick PA (1987) Striatal neurochemistry of dynorphin-(1-13): in vivo electrochemical semidifferential analyses. *Neuropeptides* 10: 369–386.
3. Broderick PA (1988) Distinguishing in vitro electrochemical signatures for norepinephrine and dopamine. *Neuroscience letters* 95:275–280.
4. Broderick PA, (1989) Characterizing stearate probes in vitro for the electrochemical detection of dopamine and serotonin. *Brain research* 495:115–121.
5. Broderick PA (1990) State-of-the-Art microelectrodes for in vivo voltammetry. *Electroanalysis* 2:241–251.
6. Broderick PA (1991) In vivo voltammetric studies on release mechanisms for cocaine with gamma-butyrolactone. *Pharmacology, biochemistry, and behavior* 40:969–975.
7. Broderick PA (1995) Microelectrodes and their use in cathodic electrochemical circuits.U.S. Patent # 5,433,710.
8. Broderick PA (1996) Microelectrodes and their use in cathodic electrochemical arrangement with telemetric application. European patent # 90914306.7.
9. Broderick PA (1999) Microelectrodes and their use in cathodic electrochemical arrangement with telemetric application. U.S. Patent # 5,938,903.
10. Broderick PA (2015) Noninvasive Electroactive Photonic Sensor with Polymer Memory Transduction Using Organic and Inorganic Elements as Platforms. *US Provisional Application No. 62/273,693.2017.*
11. Broderick PA (2016) Noninvasive Electroactive Photonic Protein Sensor with Polymer Photovoltaic Optics for Memory Transduction Using Organic and Inorganic Elements as Platforms. *US Provisional Application No. PCT/US 16/68879.2020.*
12. Broderick PA (2018) Sensing the epilepsy brain: a unique nanosurgical biomedical device to treat the epilepsy patient. *Med Case Rep* 8(2): 60–65.

13. Broderick PA (2020) Identification, Diagnosis, and Treatment of Tau Pathologies Using Electrodes. *US Application No. 63/130,453.* December 24th.

14. Broderick PA (2021) identifcation, Diagnosis and Treatment of Tau Pathologies Using Electrodes. *US Application No. 63/293,334,* December 23rd.

15. Broderick PA (2021) Noninvasive Electroactive Photonic Protein Sensor with Polymer Photovoltaic Optics for Memory Transduction using Organic and Inorganic Elements as Platforms. *US Non-Provisional Patent. International Application Patent# PCT/US16/68879. US Serial No. 11,197,615, B2, December 14th, 2021,* Issued.

16. Broderick PA (2021) Tau peptide signals are seen in Parkinson subjects by BRODERICK PROBE® sensors *MOJ Clin Med Case Rep.* 11(2):46–50. doi: 10.15406/mojcr.2021.11.00380.

17. Broderick PA, Cofresi SL (2021) Neuroimaging a cytokine storm by transducing IL-1α to hippocampal cornu ammonis: COVID-19 SARS-CoV-2, In: Preedy VR, Patel BP, eds., *Compendium on the Neuroscience of Depression: Genetics, Cell Biology, Neurology, Behaviour and Diet.* Chapter 11, Elsevier, NY and London, pp 107–117. Artigo em Inglês | Scopus | ID: covidwho-1767804.

18. Broderick PA, Doyle WK, Pacia SV, Kuzniecky RI, Devinsky O, Kolodny EH (2009) A clinical trial of an advanced diagnostic biomedical device for epilepsy patients. *J Long-Term Effects of Medical Implants* vol.18, June 1.

19. Broderick PA et al. (2021) Identification, Diagnosis, and Treatment of Neuropathologies, Neurotoxicities, Tumors, and Brain and Spinal Cord Injuries Using Electrodes with Microvoltammetry. *US Patent Serial No10,980,460*, Issued.

20. Broderick PA, Lynch, V (1982) Behavioral and biochemical changes induced by lithium and *l*-tryptophan in muricidal rats. *Neuropharmacology* 21:671–679. doi: 10.1016/0028-3908 (82) 90010-7.

21. Broderick PA, Pacia SV, Doyle WK, Devinsky O (2000) Monoamine neurotransmitters in resected hippocampal subparcellations from neocortical and mesial temporal lobe epilepsy patients: In situ microvoltammetric. *Brain research* 878:48–63.

22. Broderick PA (2022) Methods of Diagnosis and Treatment of the Skin Using Electrodes. *US Application Serial No. 63/349,767, June 30th. PTO on June 30, 2022 at Reel 060366, Frame 0895.*

23. Deng N, Hu J, Xie W (2021) Indoleamine-2,3-Dioxygenase 1 deficiency suppresses seizures in epilepsy. *Front Cell Neurosci* February 18. doi: 10.3389/fncel.2021.638854.

24. Fernstrom JD (1983) Role of precursor availability in control of monoamine biosynthesis in brain. *Physiol Rev* 63(2):484–546.

25. Frangos S, Giannoula E, Buscombe J, Iakovou J (2021) Theragnostics: a neologism that generates schism—The Hellenic Thesis. *J Nucl Med* 62(6):880. doi: https://doi.org/10.2967/jnumed.121.262101.

26. Fukui S, Schwarcz R, Rapoport SI, Takada, Smith QR (1991) Blood-brain barrier transport of kynurenines: implications for brain synthesis and metabolism. *J Neurochem* 56(6):2007–2017. doi: 10.1111/j.1471-4159.1991.tb03460.x.

27. Fukuwatari T, Shibata K (2013) Nutritional aspect of tryptophan metabolism. *Int J Tryptophan Res* 6 (Suppl 1):3–8. doi: 10.4137/IJTR.S11588.

28. Gasperi V, Sibilano M, Savini I, Catani MV (2019) Niacin in the central nervous system: an update of biological aspects and clinical applications. *Int J Mol Sci.* 20(4):974. doi: 10.3390/ijms20040974. Special issue, *Functional Mechanism of B-Vitamins and Their Metabolites.*

29. Gostnera JM et al., (2020) Tryptophan metabolism and related pathways in psychoneuroimmunology: the impact of nutrition and lifestyle. *Neuropsychobiology* 79:89–99. doi: 10.1159/000496293.

30. Hachinski V, Avan A, Gilliland J, Oveisgharan S. (2021) A new definition of brain health. *Lancet Neurol.* 20:335–336.

31. Hammers A, Koepp M J, Hurlemann R, Thom M, Richardson MP, Brooks DJ, Duncan JS (2002) Abnormalities of grey and white matter [11C]flumazenil binding in temporal lobe epilepsy with normal MRI. *Brain J Neurol* 125:2257–2271.

32. Heyes MP, Saito K, Major EO, Milstien S, Markey SP, Vickers JH. (1993) A mechanism of quinolinic acid formation by brain in inflammatory neurological disease. Attenuation of synthesis from L-tryptophan by 6-chlorotryptophan and 4-chloro-3-hydroxyanthranilate. *Brain* 116(Pt 6):1425–1450. doi: 10.1093/brain/116.6.1425.

33. Heyes M P, Saito K, Devinsky O, Nadi N S (1994) Kynurenine pathway metabolites in cerebrospinal fluid and serum in complex partial seizures. *Epilepsia* 35:251–257.

34. Hoglund E, Overli O, Winberg S (2019) Tryptophan metabolic pathways and brain serotonergic activity: a comparative review.

Front Endocrinol, Sec. Exp Endocrinol 10:158. doi: https://doi.org/10.3389/fendo.2019.00158.

35. Hyland K, Clayton PT (December 1992) Aromatic L-amino acid decarboxylase deficiency: diagnostic methodology. *Clin Chem* 38(12):2405–2410.

36. Ikeda M, Tsuji H, Nakamura S, Ichiyama A, Nishizuka Y, Hayaishi O (1965) Studies on the biosynthesis of nicotinamide adenine dinucleotide II. A role of picolinic carboxylase in the biosynthesis of nicotinamide adenine dinucleotide from tryptophan in mammals. *J Biol Chem* 240(3):1395–1401.

37. Kendal C, Everall I, Polkey C, Al-Sarraj S (1999) Glial cell changes in the white matter in temporal lobe epilepsy. *Epilepsy Res* 36:43–51.

38. Kim S, Park S, Choi TG, Kim SS (2022) Role of short chain fatty acids in epilepsy and potential benefits of probiotics and prebiotics: targeting "health" of epileptic patients. *Nutrients* 14(14):2982. doi: 10.3390/nu14142982.

39. Kissinger PT, Preddy CR, Shoup RE, Heineman WR (1984) Fundamental concepts of analytical electrochemistry. In: Kissinger PT, Heineman WR, eds., *Laboratory Techniques in Electroanalytical Chemistry*, Marcell Dekker Inc., New York, pp. 9–49.

40. Kwidzinski E et al. (2005) Indolamine 2,3-dioxygenase is expressed in the CNS and down-regulates autoimmune inflammation. *FASEB J* 19(10):1347–1349. doi: 10.1096/fj.04-3228fje.

41. Lauren A, Erland E, Saxena P (2019) Auxin driven indoleamine biosynthesis and the role of tryptophan as an inductive signal in Hypericum perforatum L. *PLoS One* 14(10): e0223878. doi: 10.1371/journal.pone.0223878.

42. Liimatainen S, Lehtimaki K, Raitala A, Peltola M, Oja SS, Peltola J, Hurme MA (2011) Increased indoleamine 2,3-dioxygenase (IDO) activity in idiopathic generalized epilepsy. *Epilepsy Res* 94:206–212.

43. Lu KT, Gean PW (1998) Endogenous serotonin inhibits epileptiform activity in rat hippocampal CA1 neurons via 5-hydroxytryptamine1A receptor activation. *Neuroscience* 86:729–737.

44. Mariusz J et al., (2016) Protective action of nicotinic acid benzylamide in a variety of chemically-induced seizures in mice. *Pharmacol Rep* 68(2):297–300. doi: 10.1016/j.pharep.2015.09.010. Epub 2015 Oct 9.

45. Marshal L, Sullivan EV, Moreell M, Lim KO, Pfefferbuam A (2001) Structural brain abnormalities in patients with schizophrenia

epilepsy and epilepsy with chronic interctal psychosis. *Psychiatry Res* 108:1–15.

46. Mellor AL, Munn DH (2004) IDO expression by dendritic cells: tolerance and tryptophan catabolism. *Nat Rev Immunol* 4(10): 762–774. doi: 10.1038/nri1457.
47. Meireson, A, Devos, M, Brochez L (2020) IDO expression in cancer: different compartment, different functionality? *Front Immunol* 11:531491. doi: 10.3389/fimmu.2020.531491. eCollection 2020.
48. Moroni F (1999) Tryptophan metabolism and brain function: focus on kynurenine and other indole metabolites. *Eur J Pharmacol* 375: 87–100.
49. Munn DH, Shafizadeh E, Attwood JT, Bondarev I, Pashine A, Mellor AL (1999) Inhibition of T cell proliferation by macrophage tryptophan catabolism. *J Exp Med* 189(9):1363–1372. doi: 10.1084/jem.189.9.1363.
50. Pacia SV, Broderick PA (2005) Bioimaging L-tryptophan in human hippocampus and neocortex: subtyping temporal lobe epilepsy. In: Broderick PA, Rahni, DN, Kolodny EH, eds., *Bioimaging in Neurodegeneration* (Humana Press, Totowa, NJ, USA) Springer Nature, London, UK, Berlin, Germany, NY, USA. Chapter 13; pp 141–147.
51. Pacia S V, Doyle W K, Broderick P A (2001) Biogenic amines in resected human neocortex in patients with neocortical and mesial temporal lobe epilepsy: Identification with in situ microvoltammetry. *Brain Res* 899:106–111.
52. Palme K, Nagy F (2008) A new gene for auxin synthesis. *Cell* 133(1): 31–32.
53. Pene F, Courtine E, Cariou A, Mira, J-P (2009) Toward theragnostics. *Crit Care Med* 37(1 Suppl):S50–558. doi: 10.1097/CCM.0b013e3181921349.
54. Penberthy WT, Kirkland JB (2012) Niacin. In: Erdman JW, et al., eds., *Present Knowledge in Nutrition*, 10th ed., Chapter 19. Oxford-Wiley Blackwell, NJ, USA, Singapore. doi: https://doi.org/10.1002/9781119946045.ch19.
55. Rohrig UF, Majjigapu SR, Grosdidier A, Bron S, Strooba TV, Pilotte L, Colan D, Vogel P, Van den Eynde BJ, Zoete V, Michielin O (2012) Rational design of 4-aryl-1,2,3-triazoles for indoleamine 2,3-dioxygenase 1 inhibition. *J Med Chem* 55(11):5270–5290.
56. Saleem W, Broderick PA (2013) Biomarkers for brain disorders electrochemically detected by BRODERICK PROBE® micro-electrodes/biosensors. *J Biosens Bioelectron* S2:003.

57. Schaechter JD, Wurtman RJ (1990) Serotonin release varies with brain tryptophan levels. *Brain Res* 532(1–2):203–210.

58. Scharfman H E, Ofer A (1997) Pretreatment with L-kynurenine, the precursor to the excitatory amino acid antagonist kynurenic acid, suppresses epileptiform activity in combined entorhinal/hippocampal slices. *Neurosci Lett* 224:115–118.

59. Schwarcz R, Stone TW (2017) The kynurenine pathway and the brain: challenges, controversies and promises. *Neuropharmacology* 112(Pt B):237–247. Special issue, *The Kynurenine Pathway in Health and Disease.* doi: 10.1016/j.neuropharm.2016.08.003.

60. Simmons DA, Broderick, PA (2005) Cytokines, stressors, and clinical depression: augmented adaptation responses underlie depression pathogenesis. *Prog Neuropsychopharmacol Biol Psychiatr* 29(5): 793–807. doi: 10.1016/j.pnpbp.2005.03.009.

61. Stachowiak MK, Stricker EM, Jacoby JH, Zigmond MJ (1986) Increased tryptophan hydroxylase activity in serotonergic nerve terminals spared by 5,7-dihydroxytryptamine. Biochem Pharmacol 35(8): 1241–1248.

62. Statnick M A, Dailey J W, Jobe PC, Browning R A (1996) Abnormalities in brain serotonin concentration, high-affinity uptake, and tryptophan hydroxylase activity in severe-seizure genetically epilepsy-prone rats. *Epilepsia* 37:311–321.

63. Stone T W (2001) Kynurenic acid antagonists and kynurenine pathway inhibitors. *Expert Opin Investig Drugs* 10:633–645.

64. Stone T W (2001) Kynurenines in the CNS: from endogenous obscurity to therapeutic importance. *Prog Neurobiol* 64:185–218.

65. Synnøve E et al. (2021) Vitamin B status and association with antiseizure medication in pregnant women with epilepsy. *Epilepsia* 62(12):2968–2980. doi: 10.1111/epi.17076.

66. Thijs RD, Surges R, O'Brien TJ, Sander JW (2019) Epilepsy in adults. *Lancet* 393:689–701.

67. Thomas P, Hazif-Thomas C, Saccardy F, Vandermarq P (2004) Loss of motivation and frontal dysfunction. Role of the white matter change. *L'Encephale* 30:52–59.

68. Tiwari S, Sharma V, Mujawar M, Mishra YK, Kaushik A, Ghosal A (2019) Biosensors for epilepsy management: state-of-art and future aspects. *Sensors* 19:1525.

69. van der Knaap MS, Valk J (1995) *Magnetic Resonance of Myelin, Myelination, and Myelin Disorders*, 2 ed., Springer, Berlin.

70. Walz W (1989) Role of glial cells in the regulation of the brain ion microenvironment. *Prog Neurobiol* 33:309–333.

71. Wieser, H (1983) *Electroclinical Features of the Psychomotor Seizure: A Stereo-Encephalographic Study of Ictal Symptoms and Chronotopographical Seizure Patterns Including Clinical Effects of Intracerebral Stimulation*. Gustav Fisher Vertag, New York.

72. Wurtman RJ, Hefti F, Melamed E (1980) Precursor control of neurotransmitter synthesis. *Pharmacol Rev* 32(4):31.

73. Zhang L, Li S, Tai Z, Yu C, Xu Z (2022) Gut microbes regulate innate immunity and epilepsy. *Front Neurosci* 16:870197. doi: 10.3389/fnins.2022.870197.

74. Zhang Y, Hu Z, Zhang J, Ren C, Wang Y (2022) Dual-target inhibitors of indoleamine 2, 3 dioxygenase 1 (Ido1): a promising direction in cancer immunotherapy. *Eur. J. Med. Chem.* 238:114524. doi: 10.1016/j.ejmech.2022.114524.

75. Munn DH, Shafizadeh E, Attwood JT, Bondarev I, Pashine A, Mellor AL (1999) Inhibition of T cell proliferation by macrophage tryptophan catabolism *J Exp Med* 189(9):1363–1372. doi: 10.1084/jem.189.9.1363.

Chapter 14

The Next Generation of White Matter Imaging Techniques Is a Nanotechnology

The author predicts that the next generation of white matter imaging techniques will move away from magnets and x-rays to electrons and photons to Artificial Intelligence with and without particle physics. The advancement is universally inherent in the ease of sensing that is manifested with increased surface area for electron and photon transfer using tiny pliant devices in the safety nook of near infrared. The move is in nanotech, nanoprobe delivery and nanoprobe sensing. The journey is made by the tiniest probe, the nanoprobe that has trafficked from in vivo electrochemistry to Neuromolecular Imaging to voltaic Quantum Photonics. The nanotechnology is the BRODERICK PROBE®. *Since white matter in the brain and spinal cord of patients predicts aging and neurodegeneration as in the dementias and the epilepsies, this chapter focuses on white matter in the patient with epileptic seizures. Seizures are known to evolve into peritumoral epilepsy and to accompany gliomas. White matter diseases and inherited metabolic disorders include multiple sclerosis, cerebral ischemia, leukodystrophies, adrenoleukodystrophies, leukoaraiosis, leukoencephalopathies, as well as mitochondrial disorders.*

14.1 Abstract of the Discovery

We studied 15 temporal lobe epilepsy patients who presented with either neocortical or mesial temporal lobe epilepsy. Resected

Neuroimaging: Sensing Biochemistry in the Brain
Patricia A. Broderick

ISBN 978-981-4613-49-1 (Hardcover), 978-0-429-02974-5 (eBook)
www.jennystanford.com

tissue from these patients was studied with Neuromolecular Imaging (NMI) using miniature carbon-based BRODERICK PROBE® stearic and lauric acid sensors. In separate studies, a sensor was inserted into neocortex or hippocampus, specifically into individual layers of neocortical temporal gyrus as well as temporal stem and hippocampal pyramidal layer, dentate gyrus, alveus, and subiculum. The catecholamines, dopamine and norepinephrine and the indoleamine, serotonin as well as ascorbic acid, an enzyme catalyst in the dopamine metabolic pathway, were detected in separate electroactive signals by these sensors in real time. Results showed that gray matter and white matter were readily distinguished both by characteristic microvoltammetric waveforms and by concentration differences, that is, (1) electroactive species for catecholamine signals in white matter were diffusion waveforms whereas the indoleamine waveform was an adsorption waveform and (2) concentrations of neurotransmitters were significantly lower in white matter than in gray matter. White matter signals were detected and distinguished from gray matter signals in both neocortical and hippocampal neuroanatomic substrates. Results showed that an average of 80% of patient electroactive signals for white matter in the three white matter structures were positive both for waveforms and for concentration characteristics. Another 21.4% and 10% were partially positive for white matter in alveus and subiculum, respectively, that is, positive for concentration characteristics but not positive for both catecholamine and indoleamine waveforms. An average of 90% of patient electroactive signals for gray matter were positive both for waveforms and for concentrations. Thus, NMI with these specialized sensors provides a unique technology for detecting, monitoring, and measuring neurotransmitters and related neurochemicals in white matter *vis-à-vis* gray matter in temporal lobe epilepsy patients.

14.1.1 Background for the Discovery

White matter demyelination with and/or without the destruction of white matter causes devastating consequences to neuronal conduction. Since the discovery that white matter can be imaged with computed tomography, correlations between white matter and neuronal pathologies have been under intense examination. One study has found that age-related changes in white matter in frontal lobe were significantly associated with decreased cognitive status whereas leukoaraiosis was associated with loss of motivation [40]. In addition, using diffusion tensor magnetic resonance imaging (MRI), O'Sullivan et al. [37] confirmed a role for white matter damage and disruption of white matter connections in the pathogenesis of cognitive impairment in cerebral small vessel disease. Van der Knapp et al. have reported a distinct encephalopathic nosologic entity which presents in children as H-ABC. This entity involves white matter atrophy in basal ganglia and the cerebellum as shown by MRI and spectroscopy; H-ABC is associated with extrapyramidal symptoms, movement dysfunction likely related to basal ganglia nigrostriatal neuronal systems [41]. This same group also has discovered genetic links to vanishing white matter encephalopathy [42].

White matter abnormalities have been reported in temporal lobe epilepsy (TLE) patients by using glial fibrillary acidic protein analysis. The results from this study showed that tissue resected from hippocampal sclerotic patients and tumor/TLE patients exhibited significantly reduced densities for cells immunopositive for glial fibrillary acidic protein when compared with control tissue as well as tissue from idiopathic epilepsy patients [25]. Another study, an immunohistochemical study, showed that temporal lobe white matter in normal tissue existed in higher levels than white matter resected from epilepsy patients [20]. Positron emission tomography imaging with [11C] flumazenil, a γ-aminobenzoic acid antagonist, also has been used to show discrepancies in white matter in epilepsy patients, that is, increased flumazenil binding in temporal lobe white matter was observed in 11 of 18 epilepsy patients [23].

Using single-voxel MRI technology, temporal lobe white matter of patients with hippocampal sclerosis was found to be reduced by using *N*-acetyl-aspartate/choline ratios as a marker. The data implicate a decreased number of axons and a reduction in myelin density [34]. Interestingly, coronal spin-echo MRI sequence was used to study a possible differentiation in white matter function between patients with TLE and those with chronic interictal psychosis; the results showed that only TLE patients had white matter deficits [33]. Volumetric studies further showed a corpus callosum involvement in childhood onset of TLE [24].

The importance of neurotransmitters in white matter is becoming more and more critical to our understanding of neurologic disorders and white matter abnormalities. Indeed, white matter glia facilitates both the redox-cycling of dopamine (DA) to free radicals and the reverse transport of these free radicals to DA in neuronal gray matter. This oxidation/reduction reaction helps to suppress the yield of toxic free radicals in gray matter [18]. This laboratory has been engaged in detecting neurotransmitters in white matter and in gray matter in neocortical and hippocampal tissue from epilepsy patients by using NMI. With such devices, based on the presence or absence of specific neurotransmitters and/or concentration differences among neurotransmitters, TLE was neurochemically subtyped [14, 38]. In the present edition of this book, correlations between serotonin (5-HT) and its precursor, L-tryptophan (L-Trp) are reported that further differentiate subtypes of TLE in human epilepsy both neocortically and mesially [39].

The purpose of this chapter is (1) to present a unique technology to bioimage white matter distinct from gray matter in TLE patients, and (2) to compare NMI with magnetic resonance imaging (MRI).

14.1.2 The Science of Advanced Materials

We used NMI with BRODERICK PROBE® stearic acid and lauric acid sensors to detect ascorbic acid (AA), DA, norepinephrine (NE), and serotonin (5-HT) in resected temporal lobe tissue (n = 15) from patients with TLE. Neurotransmitters were detected in separate signals within the same recording cycle in seconds, on line, and in real time. Three gray matter structures (neocortical temporal gyrus, hippocampal pyramidal layer and hippocampal dentate gyrus, [granular layer] tissue) and three white matter structures (neocortical temporal stem, hippocampal alveus, and hippocampal subiculum) were investigated. Experimentally derived oxidative potentials were determined in Ringer's Lactate and/or phosphate buffer. Sensors were placed as neuroanatomically appropriate.

Methods are published in detail elsewhere [4–16, 38, 39]. Criteria for distinguishing electroactive species for white matter vs gray matter were (1) characteristic waveform differences and (2) characteristic concentrations.

14.1.2.1 The quintessential Cottrell equation named after Cottrell

$$i_t = \frac{nFAC_0 D_0^{1/2}}{3.14^{1/2} t^{1/2}}$$

where i = current at time t

n = number of electrons, eq/mol

F = Faraday's constant, 96,486 C/eq

A = electrode area, cm^2

C = concentration of O, mol/cm^3

D = diffusion coefficient of O, cm^2/s

Figure 14.1 The Cottrell equation: the proportionality, between charge and mass of an electrochemical reaction, describes the relationship between the charge of each monoamine and AA being oxidized and/or reduced and the concentration of each monoamine and AA, respectively. In electrochemistry, the Cottrell equation describes the change in electric current with respect to time in a controlled potential experiment. Specifically, it describes the current response when the potential is a step function in time. It was derived by Frederick Gardner Cottrell in 1903.

In Vivo Voltammetry

In vivo voltammetry, also known as *in vivo* electrochemistry, has consisted of placing an indicator electrode into brain, and after triggering a potential difference, which causes a redox reaction, a current is produced. One can derive information about an analyte, a neurotransmitter, or its metabolite, from an electrochemical current which is a function of potential difference as this activates the surface of an electrochemical electrode. Therefore,

- Voltammetry involves the measurement of current in an electrochemical cell as a function of applied potential.
- Electroactive species undergo oxidation or reduction at a characteristic half wave potential, according to the formula:

$O + ne^- \rightarrow R$,

wherein O = oxidation, ne = number of electrons R = reduction.

- The amount of current is proportional to concentration according to the Cottrell equation (Figure 14.1).
- Chemically modified conventional electrodes utilize more specific chemical reactions.

14.1.2.2 The novel nanotechnology named after Patricia's father

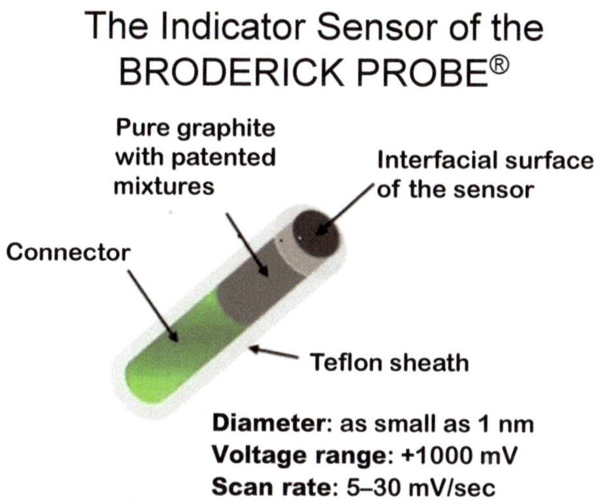

Figure 14.2 The BRODERICK PROBE®: A schematic diagram depicting a versatile nanoprobe for science and medicine.

Neuromolecular Imaging

BRODERICK PROBE® inventions (depicted schematically in Figure 14.2) relate to a variety of unique, patented, and trademarked carbon sensors that are composed selectively of a series of compounds that include, among others, classes of molecules in the biochemical categories of lipid, glycolipid, lipoprotein, saturated, and unsaturated fatty acid. These inventions are able to detect electrochemical signals for a vast number of neurotransmitters, neuromodulators, and metabolites, including the neuropeptides, hormones and vitamins In this laboratory, we routinely and selectively detect, in discrete neuroanatomic substrates of living human and animal brain, the monoamines, DA, NE, and 5-HT, in addition to L-Trp (the precursor to 5-HT), AA, and uric acid [4–16, 38, 39]. This laboratory also has differentiated the catecholamines, DA and NE, electrochemically. The electrochemical detection of somatostatin and dynorphin A are among our latest discoveries [7]. Nitrous as well as nitric oxide detection during intraoperative brain surgery for treatment of TLE epilepsy is another recent discovery from this laboratory (see present edition of this book; ref. 15).

Each new formulation of the sensor reliably detects selective properties for detection of specific analytes, neurotransmitters and metabolites. Moreover, the size of each new formulation is directly related to capacitance at the surface of this sensor, this nanotechnology (as low as 1 nm in surface reliably reduces excess charging current).

These nano- and microdevices detect basal (normal or steady state) concentrations *in vivo*, *in situ*, and *in vitro* and any alterations in neurochemistry in brain, body, beaker, or body fluids before and after pharmacologic manipulation with drugs and other compounds. Neurochemicals during actual, induced, or mimicked brain diseases can be detected. Neurochemicals in the brain and body of animals and humans can be detected.

14.1.2.3 MEMS (micro-electromechanical systems)

One recording selectively recorded several analytes, each at distinct and separate signature oxidation potentials. Each analyte was detected on line within 10 s. Repetitive scanning with this nano- and microtechnology is also reliable. Charging (background) current was recorded and eliminated in the first 20 s of each recording. Scans were recorded for at least one half an hour in each neocortical and hippocampal subparcellation. For these studies, a semiderivative circuit was used; this circuit provides the first half derivative of the linear analog signal. A semiderivative voltammetric circuit combines an additional series of resistors and capacitors called a "ladder network" [36] with the traditional linear scanning technology. The addition of the "ladder network" to the original linear scanning circuit allows for sharper and more clearly defined peaks than those previously obtained from the linear scanning methodology. Early on, the detection of electrochemical signals was computerized and operated with manual control in this laboratory. Recently, we have updated our computerized control detection and measurement of electrochemical signals with an Autolab (manufactured in the Netherlands and distributed by Brinkmann, Long Island, NY).

14.1.2.4 The study protocol

In separate experiments, each layer of neocortex and each subparcellation of hippocampus was studied. As recordings were taken from each neocortical or hippocampal specimen O_2 was intermittently bubbled through the Ringer's lactate medium to diminish possible effects of hypoxia; electrochemical signals for monoamines have been studied and are easily recognizable when hypoxia is present [7].

After the indicator sensor was placed in specific neuroanatomic substrates, scanning potentials (E_{app}) in mV were applied to the indicator, *versus* Ag/AgCl referenced stainless steel auxiliary electrodes, *via* a redesigned CV37 potentiostat (BAS) at a scan rate of 10 mV/s, with an initial applied potential of 0.2 V and an ending applied potential of 0.4 V. Sensitivity was set at 5 nA/V; time constant was set at a 5 and/or 1 s time constant. The CV37 detector was electrically connected to a Minigard surge suppressor (Jefferson Electric, Magnetek, NY), which was then connected to an electrical ground in isolation.

14.1.2.5 The patient presurgical evaluation

After intracranial EEG seizure localization and functional mapping of language for dominant temporal lobe patients, the patients underwent anterior and mesial temporal lobe resections. There was a variable extent to lateral resection, which depended on the location of ictal onset and its proximity to language function.

Tissue resection limitations were established by chronic and intraoperative electrocorticography, and these included the epileptogenic zone. The amygdala and hippocampus were removed in both neocortical TLE patients (with the exception of one) and mesial TLE patients. This procedure was performed in neocortical patients despite the lack of involvement of the mesiobasal temporal regions at ictal onset because of the known rapid spread of seizures into mesial temporal regions in many patients with neocortical foci. Data from animal studies have revealed that the hippocampus is a pace-maker or an amplifier of seizures beginning in temporal lobe [44]. Additionally, neocortical TLE patients demonstrated memory dysfunction on presurgical WADA testing.

14.1.2.6 The patient classification

Fifteen consecutive patients who had temporal resection surgeries for intractable seizures were studied. Neocortical tissue was resected in all 15 patients and hippocampal tissue was resected in 14 of the 15 patients. Five patients were diagnosed with neocortical TLE, and ten patients were diagnosed with mesial TLE. Neocortical TLE patients were classified as such and based on MRI scans that showed no atrophy. The classification was further based on tissue examination pathologically as well as seizure onset seen in temporal neocortex during chronic intracranial EEG studies. These studies were performed by using lateral temporal subdural grid electrodes and multiple basomesial temporal subdural strip electrodes. Ten patients were classified as having mesial TLE if pathologic examination of the resected temporal lobe revealed severe hippocampal neuronal loss and gliosis and if examination of the neocortical tissue revealed no other etiology for the patient's epilepsy.

14.2 The Outcome of the Discovery

Catecholamine (DA, NE) as well as AA electroactive signals in white matter were comprised of (1) representative waveforms known as diffusion (broad) waveforms and (2) lower concentrations when compared with gray matter. Indoleamine (5-HT) electroactive signals in white matter were (a) representative waveforms known as adsorption (sharp) waveforms and (b) lower concentrations when compared with gray matter.

There were 79 tissues studied. Of the 41 temporal lobe gray matter tissues sampled, 37 gray matter tissues were positive for gray matter, that is, these tissues exhibited electroactive signals consistent with the above-defined criteria for identification of gray matter. Only four of these tissues did not meet criteria. Of the 38 white matter tissues sampled, 30 white matter tissues were positive for white matter, that is, these tissues exhibited electroactive signals consistent with the above-defined criteria for identification of white matter. Moreover, there were three tissues in alveus and one in subiculum that exhibited partial positive identification, that is, these tissues were positive for low concentrations of catecholamines and indoleamines but did not reveal characteristic white matter waveforms for the catecholamines, although they did exhibit characteristic white matter waveforms for the indoleamine, 5-HT.

White matter signals were detected and distinguished from gray matter in both neocortical and hippocampal neuroanatomic structures. An average of 80% of patient electroactive signals for white matter in the three white matter structures were positive for both catecholamine and indoleamine waveforms and characteristic concentrations as well. Another 21.4% and 10% were partially positive for white matter in alveus and subiculum, respectively, that is, positive for concentration characteristics but not positive for both catecholamine and indoleamine wave forms. An average of 90% of patient electroactive signals for gray matter were positive for both characteristic catecholamine and indoleamine waveforms and characteristic concentration values (approximately three orders of magnitude less in white matter as compared with gray matter, that is, pM concentrations for white matter as opposed to nM concentrations for gray matter; unpublished data). Recordings drawn from raw data, from human TLE patients, are shown in Figures 14.3–14.5. Figure 14.6 shows results from white versus gray matter studies in histogram form.

There appeared to be no significant difference in the distinction between subtypes of TLE as to characterization of white matter *vs* gray matter electroactive signals, although this aspect remains under study. The possible exception to this may only be in concentrations differences, that is, that catecholamines in temporal stem white matter exhibit higher amplitude waveforms associated with higher concentrations than do hippocampal white matter.

Patient Care: Clinical Case 1

Neocortical Theragnostics: Gray Matter in Neocortex *vs* White Matter in Neocortex

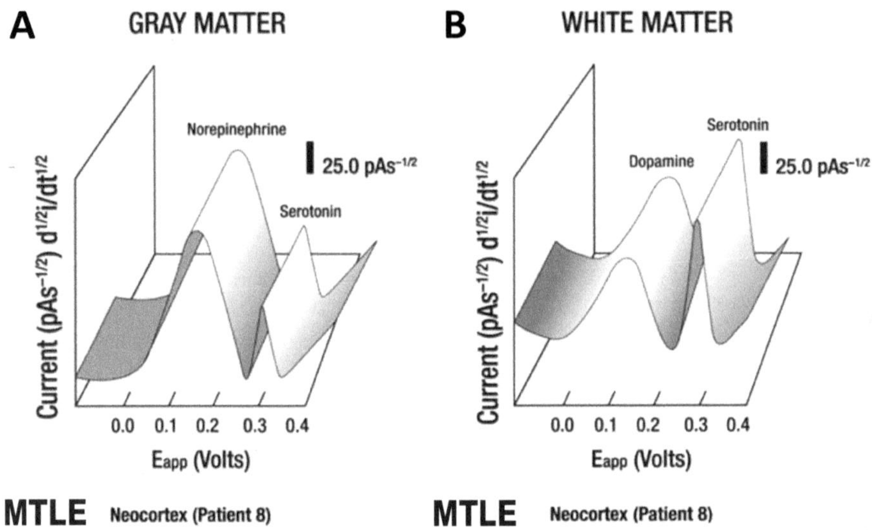

Figure 14.3 **(A)** Gray matter recording from resected neocortical tissue from mesial temporal lobe epilepsy (MTLE) patient 8. Recording, taken with NMI and BRODERICK PROBE® sensors, show the real-time electro chemical detection of norepinephrine and serotonin at experimentally derived oxidation potentials. *x-axis*: oxidation potentials in millivolts. *y-axis*: current in picoamperes (pA) per semidifferentiation of the second. Temporal gyrus gray matter signals were derived from separate studies in each of the neocortical layers 1 through 6 with the exception of layer 4, the Baillarger's band. Electroactive signals from Baillarger's band, which is composed of white matter, were previously published [39]. It is noteworthy that the terms semidifferentiation and semiderivative are used interchangeably, although they are not electronically equal. The difference can reside in the feedback circuit to operational amplifiers [36]. **(B)** White matter recording from resected neocortical tissue from mesial temporal lobe epilepsy (MTLE) patient 8. Recording, taken with NMI and BRODERICK PROBE® sensors, shows the real-time electrochemical detection of dopamine and serotonin at experimentally derived oxidation potentials. *x-axis*: oxidation potentials in millivolts. *y-axis*: current in picoamperes (pA) per semidifferentiation of the second. White matter signals in patients were studied in the neuroanatomic substrate, temporal stem.

Patient Care: Clinical Case 2

Hippocampal Theragnostics: Gray Matter in Pyramidal Layer *vs* White Matter in Subiculum

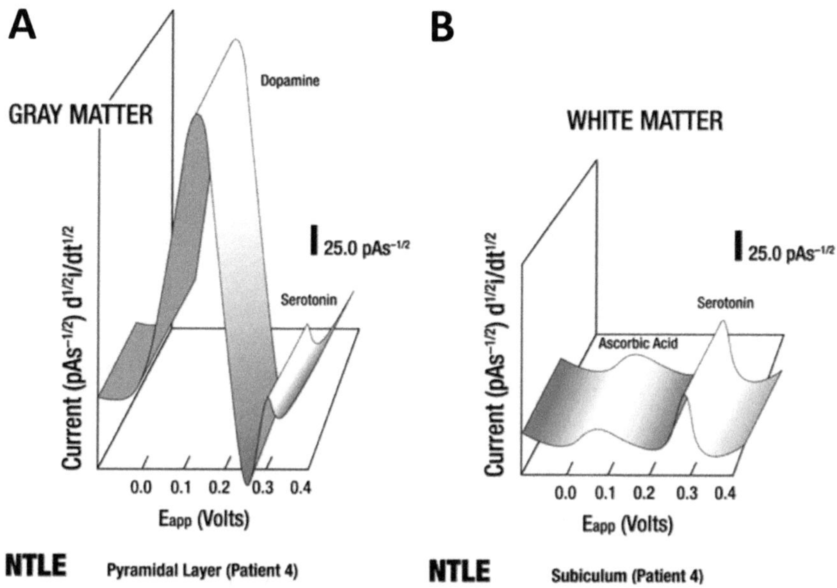

Figure 14.4 **(A)** Gray matter recording from resected hippocampal tissue from neocortical temporal lobe epilepsy (NTLE) patient 4. Recording, taken with NMI and BRODERICK PROBE® sensors, shows the real-time electrochemical detection of dopamine and serotonin at experimentally derived oxidation potentials. *x-axis*: oxidation potentials in millivolts. *y-axis*: current in picoamperes (pA) per semidifferentiation of the second. Gray matter signals in patients were studied in the neuroanatomic substrate, pyramidal layer. **(B)** White matter recording from resected hippocampal tissue of neocortical temporal lobe epilepsy (NTLE) patient 4. Recording, taken with NMI and BRODERICK PROBE® sensors, shows the real-time electrochemical detection of ascorbic acid and serotonin at experimentally derived oxidation potentials. *x-axis*: oxidation potentials in millivolts. *y-axis*: current in picoamperes (pA) per semidifferentiation of the second. White matter signals in patients were studied in the neuroanatomic substrate, subiculum.

White matter exhibited substantially lower concentrations of catecholamines and the indoleamine, 5-HT (approximately three orders of magnitude less in white matter as compared with gray matter, that is, pM concentrations for white matter as opposed to nM concentrations for gray matter; unpublished data).

Patient Care: Clinical Case 3

Hippocampal Theragnostics: Gray Matter in Dentate Gyrus *vs* White Matter in Alveus

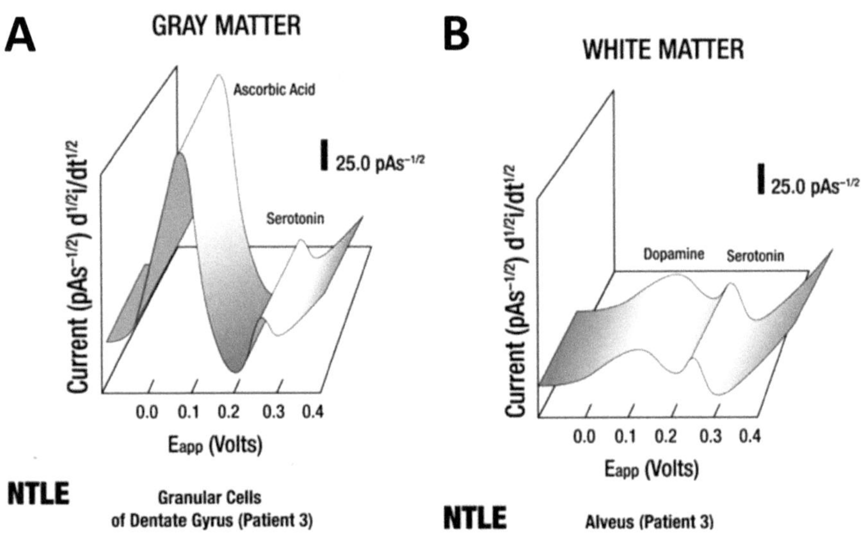

Figure 14.5 (A) Gray matter recording from resected hippocampal tissue from neocortical temporal lobe epilepsy (NTLE) patient 3. Recording, taken with NMI and BRODERICK PROBE® sensors, shows the real-time electrochemical detection of ascorbic acid and serotonin at experimentally derived oxidation potentials. *x-axis*: oxidation potentials in millivolts. *y-axis*: current in picoamperes (pA) per semidifferentiation of the second. Gray matter signals in patients were studied in the neuroanatomic substrate, granular cells of the dentate gyrus. **(B)** White matter recording from resected hippocampal tissue of neocortical temporal lobe epilepsy (NTLE) patient 3. Recording, taken with NMI and BRODERICK PROBE® sensors, shows the real-time electrochemical detection of ascorbic acid and serotonin at experimentally derived oxidation potentials. *x-axis*: oxidation potentials in millivolts. *y-axis*: current in picoamperes (pA) per semidifferentiation of the second. White matter signals in patients were studied in the neuroanatomic substrate, alveus.

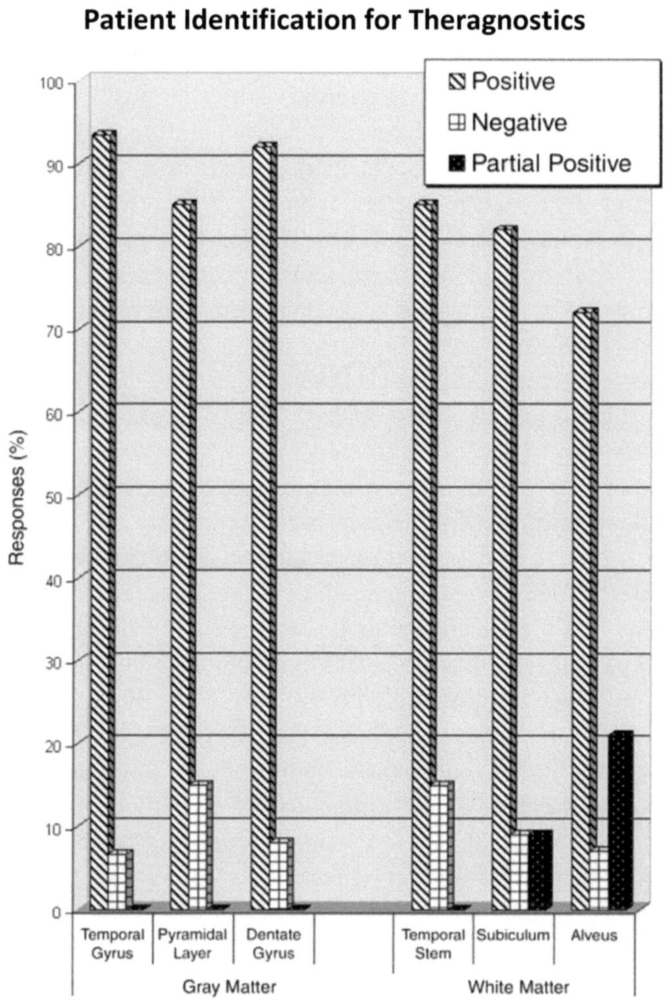

Figure 14.6 Patient percentages. On the left is a histogram representation of percentages of patients who met criteria for positive identification of gray matter waveforms; negative percentages are also shown. In the temporal gyrus, pyramidal layer, and dentate gyrus, 14 of 15 patients, 11 of 13 patients, and 12 of 13 patients, respectively, exhibited waveforms positive for gray matter. On the right is a histogram representation of percentages of patients who met criteria for positive identification of white matter waveforms; negative percentages are also shown. In temporal stem, alveus, and subiculum, 11 of 13 patients, 10 of 14 patients and 9 of 11 patients, respectively, exhibited waveforms positive for white matter. In alveus and subiculum, 3 of 14 patients and 1 of 11 patients respectively, met criteria for partially positive identification of white matter waveforms.

14.3 Matters of White vs Gray Matter

White matter has inherent differences in chemical composition from gray matter. Furthermore, white matter has inherent neuroanatomic, physiologic, neuronal, structural, and functional differences from gray matter. White and gray matter of the central nervous system significantly differ in gross morphology, in water content, and macromolecular components, for example, membrane lipids. White matter is composed of myelinated bundles of axons, whereas gray matter primarily is composed of neurons and neuronal processes. Several lipids are found in the lipid bilayer of the membrane of white matter, for example, phospholipid, sphingomyelin, galactocerebroside, and cholesterol. These are lipids that account for myelin membrane strength [30, 41], and demyelination of the lipid bilayer may occur from any number of missteps such as trauma, injury, and/or chemical denaturation [2]. Indeed, it has been reported that the ionic environment of glial cells is a significant factor in the regulation of white matter [43]. Importantly, a change in the ionic environment of glial cells has recently been associated with the pathophysiology of epileptic seizures [3] albeit one of the first discoveries of the glial role in epilepsy patients was contributed by Bordey and Sontheimer.[1]

These differences in neuroanatomy, physiology, as well as structural, functional, and chemical composition differences, lend explanation for the representative electroactive waveforms observed here for white matter *vs* gray matter. These differences may also lend explanatory notes for the observed differences in concentrations of neurotransmitters in white matter vs gray matter in TLE.

Simply stated, white matter is composed of fewer neuronal processes than is gray matter, and this difference provides a rational explanation for the observed lower concentrations of neurotransmitters and neurochemicals in white matter as compared with gray matter. However, the waveform differences observed in the detection of white matter vs gray matter very likely depend on chemical compositional differences. In fact and *a priori*, a lipid environment is an important determinant in

---

[1]Bordey A, Sontheimer H (1998) Properties of human glial cells associated with epileptic seizure foci. *Epilepsy Res.* 32:286–303.

signal amplification and in the mechanism of action of BRODERICK PROBE® sensors. Several studies, *in vitro* and *in vivo*, have validated these data [4–16]. Other technologies such as Raman resonance technology and surface-enhanced Raman spectroscopy confirm these results [21].

Moreover, the exact chemical modification of each specific formulation of these miniature sensors, provides a further rational explanation for the ability of these miniature sensors to distinguish electroactive signals for white matter from gray matter in terms of oxidative waveforms. Electron transfer is not only enhanced by lipids and suppressed by protein [4–16, 38, 39] but electron transfer is affected by the hydrophobicity, hydrophilicity, carbon chain length, and polar end groups of saturated and unsaturated fatty acids, simple lipids, and complex lipids.

Therefore, relevant to the present study is the difference in electron transfer properties between the BRODERICK PROBE® stearic acid and lauric acid sensors in that electron transfer for lauric acid, a shorter chain, 12-carbon saturated fatty acid is greater than that of stearic acid, a longer chain, 18-carbon saturated fatty acid. Thus, differences in electron transfer, depending on the chemical modification of the surface, factor into the specific detection properties of sensors as each sensor interacts with ionic differences in white vs gray matter. Therefore, it is suggested that electron transfer may well be a key mechanism for detection of analytes with this nano-microtechnology. In addition, neuroanatomic structure, ionic, and chemical composition and such of white matter *vs* gray matter, as these relate to detection properties on the surface of the microelectrode sensor is another key by which distinct signals for white *vs* gray matter are provided by this unique sensor technology.

14.3.1 About the NMI Inventive Art

In NMI, action potentials are not involved; it is not a depolarization technology. It is important to note that these miniature sensors do not sense membrane potentials. These sensors pass small but finite currents while neurotransmitters close to the surface undergo oxidation and/or reduction [1]. The current, which is formed from this flow of electrons, is dependent on voltage according to Ohm's law. Thus, the detection of electrochemical or electroactive signals from neuroanatomic brain sites is termed *faradaic* because the amount of the oxidative and/or reductive species detected at the surface of the microelectrode surface is calculated by a derivation of Faraday's law, which is the Cottrell equation [26]. Faradaic electrochemistry using conventional electrodes has been described [19]. The NMI electrochemical art that sees white vs gray matter in brain and spinal cord of patients, is invented by Broderick [13].

14.3.2 About the MRI Inventive Art

In MRI, action potentials are not involved; MRI is not a depolarization technology. It is important to note that as myelination occurs, there is a loss of water within the myelin sheath which decreases proton density, thus, enhancing the T_1-weighted image and reducing the T_2-weighted image (*see* ref. [29] for overview). Too, signal changes on T_1-weighted MRI patterns parallel increases in lipids that occur during myelin formed from oligodendrocytes [17, 22, 28]. Signal changes on T_2-weighted MRI patterns may correlate with the period of maturation of the myelin sheath [2]. In MRI detection, hydrogen atoms of lipids do not contribute appreciably to the MR signal because hydrogen atoms of lipids are immobile and bound tightly to long-chain fatty acids [28]. Instead, lipid protons affect MRI by interacting with the mobile water protons by, for example, chemical exchange with transiently immobilized water protons [30]. Chemicals such as myelin-bound cholesterol [17, 22, 27] and galactose cerebroside [30] are responsible for shorter T_1/T_2-weighted relaxation times for normal myelinated white matter on MRI (see ref. [35] for review).

14.3.3 About the NMI and MRI Technologies

It is of interest that neither NMI nor the MRI technologies use action potentials. Neither technology is a depolarization technique. Whereas T_1/T_2-weighted images in MRI signals white matter changes and the presence or absence of white matter, it is electron transfer in NMI with the BRODERICK PROBE® sensors that signals white matter changes and the presence or absence of white matter. In addition, in MRI detection, the hydrogen atoms of lipids do not contribute appreciably to the MRI signal. Lipid protons affect the MR signal in MRI indirectly by communicating with water protons. In NMI with BRODERICK PROBE® sensors, lipids act to enhance signals indirectly as well, by enhancing electron transfer for electroactive species. Finally, in both technologies, structural and chemical changes in axons, in addition to the length of carbon chain in fatty acids, and properties such as hydrophobicity and hydrophilicity, e.g., are significant factors, affecting and influencing the image and waveform patterns of white matter.

In summary, NMI with BRODERICK PROBE® sensors reliably differentiates temporal lobe gray from white matter in separate subparcellations of neocortex and hippocampus in TLE patients. These results have important implications for enabling direct *in vivo* measurement of neurotransmitters and critical analytes in diverse neuroanatomic substrates during surgery. Thus, one can envision, with numerous sensors, the production of a temporospatial image of synaptic changes in TLE neuronal circuitry, which may well enable more precise diagnosis and pharmacotherapy for TLE, in addition to enabling more precise intraoperative interventions. These data hopefully will lead to human trials and be helpful also in the areas of tumor, leukodystrophies, leukoencephalopathies, stroke and spinal cord injury.

14.4 Acknowledgement

The protocol for human studies was approved by the CUNY and NYU Investigational Review Boards. We thank the FACES Campaign, Parents Against Childhood Epilepsy (PACE Foundation) and the National Institute of Health, NIH/NIGMS SCORE AWARD # SO 6 GM 08168 for partial financial support. The author gratefully acknowledges Bridget T. O'Sullivan, OP, MA, *in memoriam*; Msgr. Scanlan, HS; and Karen Schulz, Humana Press, for secretarial and artistic assistance, respectively. The author thanks Fang Zhou and Ratna Medicherla, second year, CUNY Medical School students, for their participation in this project.

Adapted from Broderick PA, Rahni DN, Kolodny EH, (eds.) Bioimaging in Neurodegeneration: Humana Press, Totowa, NJ, Springer Nature Publishing Group, London, UK (Global), Berlin, Germany, NY, NY, USA, Chapter 17, pp. 199–208, with permission under the terms of the Creative Commons Attribution License.

References

1. Adams RN, Marsden CA (1982) New techniques in psychopharmacology. In: Iversen LL, Snyder SH, eds., *Handbook of Psychopharmacology*, Plenum Press, New York.
2. Barkovich AJ, Lyon G, Evrard P (1992) Formation, maturation, and disorders of white matter. *Am J Neuroradiol* 13:447–461.
3. Bordey A, Spencer DD (2004) Distinct electrophysiological alterations in dentate gyrus versus glial cells from epileptic humans with temporal lobe sclerosis. *Epilepsy Res* 59:107–122.
4. Broderick PA (1987) Striatal neurochemistry of dynorphin-(1-13): *in vivo* electrochemical semidifferential analyses. *Neuropeptides* 10: 369–386.
5. Broderick PA (1988) Distinguishing *in vitro* electrochemical signatures for norepinephrine and dopamine. *Neurosci Lett* 95:275–280.
6. Broderick PA (1989) Characterizing stearate probes *in vitro* for the electrochemical detection of dopamine and serotonin. *Brain Res* 495: 115–121.
7. Broderick PA, Gibson GE (1989) Dopamine and serotonin in rat striatum during *in vivo* hypoxic-hypoxia. *Metabol Brain Dis* 4: 143–153.
8. Broderick PA (1990) State-of-the-art microelectrodes for *in vivo* voltammetry. *Electroanalysis* 2:241–251.
9. Broderick PA (1991) *In vivo* voltammetric studies on release mechanisms for cocaine with gamma-butyrolactone. *Pharmacol Biochem Behav* 40:969–975.
10. Broderick PA (1995) Microelectrodes and their use in cathodic electrochemical circuits. U.S. Patent # 5,433,710, Issued.
11. Broderick PA (1996) Microelectrodes and their use in cathodic electrochemical arrangement with telemetric application. European Patent # 90914306.7, Issued.

12. Broderick PA (1999) Microelectrodes and their use in cathodic electrochemical arrangement with telemetric application. U.S. Patent # 5,938,903, Issued.

13. Broderick PA, et al. (2012) Identification, diagnosis and treatment of neuropathologies, neurotoxicities, tumors, and brain and spinal cord injuries using microelectrodes with microvoltammetry; US Patent Application Serial No. 10/118,571.

14. Broderick PA, Pacia SV, Doyle WK, Devinsky O (2000) Monoamine neurotransmitters in resected hippocampal subparcellations from neocortical and mesial temporal lobe epilepsy patients: *In Situ* microvoltammetric studies. *Brain Res* 878:48–63.

15. Broderick PA, Rahni DN, Pacia SV (2005) Nano- and microimaging surgical anesthesia in epilepsy patients. In: Broderick PA, Rahni DN, Kolodny EH, eds., *Bioimaging in Neurodegeneration*, Humana Press Inc., Totowa, NJ, pp. xiii–xvi.

16. Broderick PA, Pacia SV (2006) Identification, diagnosis, and treatment of neuropathologies, neurotoxicities, tumors and brain and spinal cord injuries using microelectrodes with microvoltammetry. U.S. Patent # 7,112,319, Issued.

17. Brody BA, Kinney HC, Kloman AS, Gilles FH, et al. (1987) Sequence of central nervous system myelination in human infancy. I. An autopsy study of myelination. *J Neuropathol Exp Neurol* 46:283–301.

18. Cadet JL, Brannock C (2002) Mechanism of methamphetamine-induced neurotoxicity. In: Massaro EJ, Broderick PA, Mattsson JL, Schardin JL, Schaepfer TE, eds., *Handbook of Neurotoxicology*, Humana Press,Totowa, NJ, Vol. 2, pp. 259–168.

19. Dayton MA, Brown JC, Stutts KJ, Wightman RM (1980) Faradaic electrochemistry at microvoltammetric electrodes. *Anal Chem* 52: 948–950.

20. Emery JA, Roper SN, Rojiani AM (1997) White matter neuronal heterotopia in temporal lobe epilepsy: a morphometric and immunohistochemical study. *J Neuropathol Exp Neurol* 56:1276–1282.

21. Foucault R, Broderick PA, Rahni DN, Lombardi JR, Birke RL (2002) Neurotransmitter signatures: a correlation between Raman spectroscopy and microvoltammetric BRODERICK PROBE® sensors. *NIH/NIGMS Symposium*, New Orleans, LA.

22. Fralix TA, Ceckler TL, Wolff SD, Simon SA, Balaban RS (1991) Lipid bilayer and water proton magnetization transfer: effect of cholesterol. *Magn Reson Med* 18:214–223.

23. Hammers A, Koepp MJ, Hurlemann R, Thom M, Richardson MP, Brooks DJ, Duncan JS (2002) Abnormalities of grey and white matter [11C] flumazenil binding in temporal lobe epilepsy with normal MRI. *Brain* 125:2257–2271.

24. Hermann B, Hansen R, Seidenberg M, Magnotta V, O'Leary D (2003) Neurodevelopmental vulnerability of the corpus callosum to childhood onset localization-related epilepsy. *Neuroimage* 18:284–292.

25. Kendal C, Everall I, Polkey C, Al-Sarraj S (1999) Glial cell changes in the white matter in temporal lobe epilepsy. *Epilepsy Res* 36:43–51.

26. Kissinger PT, Preddy CR, Shoup RE, Heineman WR (1996) Fundamental concepts of analytical electrochemistry. In: Kissinger PT, Heineman WR, eds., *Laboratory Techniques in Electroanalytical Chemistry*, Marcell Dekker Inc., New York, pp. 11–50.

27. Koenig SH, Brown RD 3rd, Spiller M, Lundbom N (1990) Relaxometry of brain: why white matter appears bright in MRI. *Magn Reson Med* 14:482–495.

28. Koenig SH (1991) Cholesterol of myelin is the determination of grey-white matter contrast in MRI of the brain. *Magn Reson Med* 20:285–296.

29. Kolodny EH (2005) Overview of leukoencephalopies: an MRI point of view. In: Broderick PA, Rahni DN, Kolodny EH, eds., *Bioimaging in Neurodegeneration*, Humana Press Inc., Totowa, NJ, pp. 209–214.

30. Kucharczy W, Macdonald PM, Stanisz GJ, Henkelman RM (1994) Relaxivity and magnetization transfer of white matter lipids at MR imaging: importance of cerebrosides and pH. *Radiology* 192:521–529.

31. Leegwater PA, Pronk JC, van der Knaap MS (2003) Leukoencephalopathy with vanishing white matter: from magnetic resonance imaging pattern to five genes. *J Child Neurology* 18:639–645.

32. Lu KT, Gean PW (1998) Endogenous serotonin inhibits epileptiform activity in rat hippocampal CA1 neurons *via* 5-hydroxytryptamine 1A receptor activation. *Neuroscience* 86:729–737.

33. Marsh L, Sullivan EV, Morrell M, Lim KO, Pfefferbuam A (2001) Structural brain abnormalities in patients with schizophrenia, epilepsy, and epilepsy with chronic interctal psychosis. *Psychiatry Res* 108:1–15.

34. Meiners LC, van der Grond J, van Rijen PC, Springorum R, de Kort GAP, Jansen GH (2000) Proton magnetic resonance spectroscopy of temporal lobe white matter in patients with histologically proven hippocampal sclerosis. *J Magn Reson Imaging* 11:25–31.

35. Nusbaum AO, Fung K, Atlas SW (2002) White matter dieases and inherited metabolic disorder. In: Atlas SW, ed., *Magnetic Resonance Imaging of the Brain and Spine*, 3rd ed., Lipincott Williams & Wilkins, Philadelphia, pp. 457–563.
36. Oldham K (1973) Semi-integral electroanalysis: analog implementation. *Anal Chem* 45:39–50.
37. O'Sullivan M, Morris RG, Huckstep B, Jones DK, Williams SC, Markus HS (2004) Diffusion tensor MRI correlates with executive dysfunction in patients with ischaemic leukoaraiosis. *J Neurol Neurosurg Psychiatry* 75:441–447.
38. Pacia SV, Doyle WK, Broderick PA (2001) Biogenic amines in resected human neocortex in patients with neocortical and mesial temporal lobe epilepsy: identification with *in situ* microvoltammetry. *Brain Res* 899:106–111.
39. Pacia SV, Broderick PA (2005) Bioimaging L-Tryptophan in human hippocampus and neocortex: subtyping temporal lobe epilepsy. In: Broderick PA, Rahni DN, Kolodny EH, eds., *Bioimaging in Neurodegeneration*, Humana Press, Totowa, NJ, pp. 141–147.
40. Thomas P, Hazif-Thomas C, Saccardy F, Vandermarq P (2004) Loss of motivation and frontal dysfunction: Role of the white matter change. *L'Encephale* 30:52–59.
41. van der Knaap MS, Valk J (1995) *Magnetic Resonance of Myelin, Myelination, and Myelin Disorders*. 2nd ed.; Springer, Berlin, pp. 1–17.
42. van der Knaap MS, Naidu S, Pouwels PJ, Bonavita S, van Coster R, Lagae L, Sperner J, Surtees R, Schiffmann R, Valk J (2002) New syndrome characterized by hypomyelination with atrophy of the basal ganglia and cerebellum. *AJNR. Amer J Neuroradiol* 23:1466–1474.
43. Walz W (1989) Role of glial cells in the regulation of brain microenvironment. *Prog Neurobiol* 33:309–333.
44. Wieser H (1983) *Electroclinical Features of the Psychomotor Seizure: A Stereo-Encephalographic Study of Ictal Symptoms and Chronotopographical Seizure Patterns Including Clinical Effects of Intracerebral Stimulation*. Gustav Fisher Verlag, New York.

Chapter 15

Aging and the Dementias

This *"short but sweet"* clinical paper is about a next generation "uptick" nanotechnology that demonstrates a unique, real time imaging inventive art enabling a different kind of look at the brain, actually, to see inside neuronal and glial circuitry in the brain and spinal cord of the living human being and animal. "*Lewy body dementia is characterized by the abnormal buildup of proteins into masses known as Lewy bodies. This protein is also associated with Parkinson's disease. People who have Lewy bodies in their brains also have the plaques and tangles associated with Alzheimer's disease. Apr 26, 2019, Mayo clinic*" [1]. This is the problem and thus far, the problem is addressed primarily after autopsy, called *post mortem*, also problematic. Therefore, the purpose of this paper is to introduce the online and real time and spatial sensitive voltaic image of the Tau peptide complex video-tracked by the BRODERICK PROBE® biomedical sensors in striatum of the living Parkinson subject. This nanoprobe enables studies of the striking consequences among intensities of phosphorylated Tau. Identifying Tau on line is unmistakably relevant to longevity, both individual and societal, and this relevance is inexorably critical to and on behalf of humanity *per se*.

15.1 Tau Peptide Signals Are Imaged in Parkinson Subjects by Sensors

About this next generation "uptick" nanotechnology, it is further specified as Neuromolecular Imaging, abbreviated NMI.

Neuroimaging: Sensing Biochemistry in the Brain
Patricia A. Broderick

ISBN 978-981-4613-49-1 (Hardcover), 978-0-429-02974-5 (eBook)
www.jennystanford.com

The BRODERICK PROBE® biomedical library of sensors are the principal part of this LIVE and online novel imaging nanotechnology. The original formulae of these sensors are comprised of carbon/lipid/phosphotidyl polymers and this dramatic advance has transformed spectrometry into a spectral analysis in the form of LIVE electrochemistry using carbon allotropes in lipid matrices [3–5] to video-track time sensitive movement of neuromolecules while at the same time, the sensors image from movement nuclei and further while the subject is actually walking and/or in altered states of consciousness and further to another Broderick sensing device, a semiconductor carbon allotrope, fullerene [13] and further to a photoluminescent fiber optic protein probe [8, 11]. Each and every sensor is inventive art formalized under the umbrella of the BRODERICK PROBE® trademark.

15.1.1 The Present Theragnostics

When we look at the diagnosis of Parkinson and Alzheimer disease, the conventional empirical and theoretical art virtually remains at a loss for adequate treatment given the caveat that positron emission tomography (PET), images Tau and Aβ plaques to assess cognitive deficits. Nonetheless, when we discern Parkinson disease, Lewy Body Dementia (LBD) and Alzheimer disease including other neurodegenerative disorders of asynchrony similar to epileptogenic seizures, we are met with a conundrum to either diagnose or treat because central distinctions among these deadly diseases are beset with formerly known exact criteria but these criteria are now confused. In fact, we thought we had this down to a dance step as the Lewy body seemed to be a suitable solution for a comfortable distinction. As Lewy bodies so true of Parkinsonian patients and animals are comprised of alpha(α)synuclein, we were able to say in the past that Parkinsonian subjects present with Lewy bodies, Alzheimer subjects present with Aβ plaques plus Tau and epilepsy subjects exhibit Tau.

15.1.2 Enter- the Prionopathies and the Synucleinopathies

What has happened? These essentially untreatable diseases present with horrendous complexities such that Parkinsonian patients now present with Lewy bodies plus Aβ plaques and Tau; Alzheimer subjects present with Aβ plaques, plus Lewy bodies plus Tau. To compound the felony, the trio of deadly diseases come from within, infections from within, a type of protein unfolding, derived from prions, called prionopathies and derived from alpha synuclein, called synucleinopathies. To this point, *post mortem* studies have assayed Lewy bodies, alpha synuclein, the amyloid precursor proteins and Tau but the studies are performed *post mortem* which means after death and the limitations of "life after death" may simply add to already mis-diagnoses.

15.1.3 What Is Tau?

Tau is a group of six highly soluble protein isoforms produced by alternative splicing from the gene *MAPT* (microtubule-associated protein tau). Microtubules are tubular polymers, comprised of tubulin that self-assemble on a scaffold-like design to make up the infrastructure of a cell; these tubular polymers act as a kind of railway to transport organelles throughout the cytoplasm. These intracellular structures have a stable region, as well as a region that remains dynamic, both important to playing their role in a cell. It is noteworthy that the conventional role of Tau resided in the maintenance of microtubule *stability* in axons, an established thesis that is highly questioned in current times [21, 25, 31, 32]. Tau peptides and proteins are abundant in the neurons of the central nervous system (CNS). It is generally known that Tau is less common elsewhere but are also expressed at very low levels in CNS astrocytes and oligodendrocytes. As defined herein, Tau protein/peptide includes but is not limited to phosphorylated forms. **Tau is a healthy protein! It is healthy until it is unhealthy by dysregulation.**

15.1.4 The Broderick Nanoprobe: The Art of Phosphorylation

Thus, Tau, when unhealthy, is the biomarker for Tau pathologies or tauopathies. Tauopathies are neurodegenerative disorders characterized by the deposition of *abnormal* Tau in the brain. Pathologies and dementias of the nervous system such as, for example, Alzheimer disease, Parkinson disease and Tau-associated Epilepsy are associated with Tau that have become hyperphosphorylated insoluble aggregates called neurofibrillary *tangles*. The path from phosphorylation to hyperphosphorylation to aggregation is puzzling; this is why the BRODERICK PROBE® polymer is important. The nanoprobe acts to image through mechanisms of phosphorylation, enabling precise images showing differential degrees of phosphorylation [10, 12]. Therefore, we can study degrees of phosphorylation to the point at which aggregation occurs, the proverbial "hitting the nail on the head." ***The* BRODERICK PROBE® *enables the online view, that is, actually the ability to see the point at which there is little or no return from Tau phosphorylation as it proceeds to aggregation.*** Such a feat then allows the view of the image of the Tau peptide complex and moreover, the image of the selective Tauopathies.

15.1.5 The Rationale

Therefore, the purpose of this paper is to introduce the online and real time voltaic image of the Tau peptides by the BRODERICK PROBE® biomedical sensors in the striatum of the living Parkinsonian animal model. Identifying Tau on line brings real time diagnostic, therapeutic and prognostic answers to a world that is daily increasing in cases of dementia not at all material to its derivation. The quest is imperative as the consequences to not pursuing this line of "research questions" may be our proverbial "Waterloo." Indeed, this unwise alternative may well be the demise of a cognitive human race at least as it is recognized presently as a kingdom domain, inclusive of separate and distinct species of plants and animals [24] albeit there are now six kingdoms classified. Humans are in the animal kingdom [2].

15.2 The Broderick Brand: The Nanoprobe's Salience

15.2.1 About the Brand

The BRODERICK PROBE® series of biosensors work by electrochemical detection and are comprised of carbon and/or carbon allotropes in one example and fiber optic proteins in another of the many examples. In the present study, the original carbon sensors, the carbon/lipid/phosphotidyl polymers were used. This biomedical device is unique from other techniques and methodologies in that this series of biosensors can image signals *in vivo, in vitro or in situ* within sub-seconds in real time and the sensors are so tiny that they can be used anywhere and for any length of time and controlled studies have found these operationally stable biomedical biosensors to *be void of causing* gliosis (scar tissue) or producing infection (bacterial growth) [9].

Further details and examples of Broderick probes may be found in [3–18, 23], including circuit diagrams and methods of making Broderick probes. Broderick probes may be in electrical contact with an auxiliary electrode and/or a reference electrode. It will be apparent to those of ordinary skill in the art particularly in view of the cited patent documents, that "Broderick probe" is a term that relates to a number of electrodes that vary by composition and the type of circuit in which it is employed and that these variations give rise to differences in detection properties.

15.2.1.1 At the nanoscale

This inventive art relates to the recent leaps in areas not only in spectroscopy but in microscopy and this inventive art gives scientists a nanotechnology that takes advantage of LIVE phenomena that occur naturally in the brain and spinal cord where matter is organized at the nanoscale. The science of advanced nanomaterials allows these phenomena to be studied based on quantum mechanics and allows focus on an expanded surface area. Indeed, the majority of biological processes do occur at the nanoscale, bringing us to brand new constructs and designs and the like in medicine, imaging, computing, printing, chemical catalysis, electrocatalysis and materials synthesis. We cannot think of "Nanotechnology" as just working at ever smaller dimensions. This would be a mistake.

Rather, imagine working at the nanoscale that engages scientists in a brave new world, talking with each other to utilize unique platforms to cure disease. Imagine uniting diagnostic and therapeutic platforms such as physical, biochemical, chemical, mechanical, electrical, piezoelectric, photoelectric, stimulating and optical properties that naturally occur at nanoscale. The BRODERICK PROBE® works at the nanoscale, enabling tunability of carbon allotropes [3–6] and opsin laser photodiodes [8, 11] that see inside the brain without opening or minimally touching the brain as well as semiconductors such as fullerene [13] to fulfill the hope of a better quality of life beyond degeneration. In this clinical report, the original carbon allotropic polymer expands the realm of the uses of the BRODERICK PROBE® to advance the quest for diagnoses and treatment of Tau-related physiologies and pathologies in the healthy and the unhealthy living subject.

15.2.1.2 Advantages over the prior and the present art

Advantages over other methodologies are clear. The literature cites that the dialysis (microdialysis) method has disadvantages relative to sampling time resolution. For one example, dialysis offers somewhat limited time resolution. The use of highly sensitive analytical techniques (e.g. neuromolecular imaging with the Broderick probe) images in temporal resolutions of seconds and less, spatial resolutions of nanospheres and current resolution of nanoamperes. Dialysis samples may not provide real time information regarding changes in the neurochemical environment. Also, dialysis is an invasive technique and therefore causes some damage to the tissue under study and critically, the dialysis process creates an area around the membrane in which all solutes capable of crossing the membrane are depleted [19].

Post-Mortem studies are daily used to distinguish between Alzheimer disease and the age-old expression of "dementia" as these diseases relate to the pathological presence of Aβ plaques and neurofibrillary tangles versus Lewy bodies respectively. Tau and Aβ plaques are seen as a double prion disease *post mortem* by Prusiner who discovered the prion [28, 29]. Alzheimer patients are often seen *post mortem* to exhibit alpha synucleated Lewy bodies in addition to Aβ plaques, thus resulting in confusion, compounded with the innate hindrance of studying any disease *post mortem*. Physiological and pathological processes may not be translatable to life *per definitionem*. Tau is a physiological and pathological process.

15.2.1.3 The BRODERICK PROBE® provides solutions

The BRODERICK PROBE® provides solutions since Tau is imaged LIVE. NMI of Tau's anionic redox dynamic movement is perhaps synaptic, likely away from the microtubule and is comprised of, in at least one embodiment as, but not limited to, one of a complex, comprised of metal (Broderick patent claims [10]) as well as in another patent by Broderick.[1] Tau exists in axonal microtubules as a microtubule associated protein (MAP) and MAP-1 and MAP-2 are commonly used. Metal complex forms of Tau are rare in the contemporary field of Tau theragnostics and the field is appraised [22]. When Tau is detached from the microtubule, this form may assist in seeing Tau as a neuroprotector [30].

15.2.1.4 The BRODERICK PROBE® provides solutions to the disease of Parkinson

Indeed, imaging the neurotransmitter facets of the disease called, Parkinsonian, is a nanotechnology and a nanobiotechnology which is studied extensively in the Broderick laboratory. The laboratory is well known for delineating the neuroprotective transmitter elements of the disease as well as its degenerative elements in the living animal model of this movement and cognitive disorder, named after Parkinson. Often beset with depression and consequent to treatment, *per se*, to psychosis, the disease is a Lewy body disease of alpha (α) synuclein; the disease is a prionopathy and a synucleinopathy. Details of the experimental design are provided [16–18].

[1]Broderick PA (2021) Identification, diagnosis, and treatment of tau-associated pathologies using electrodes. US Provisional Application No. 63/293,334, December 23.

15.2.1.5 The precision of the nanoprobe

The precision of this nanoprobe is enhanced further beyond the previously used techniques because the molecular transmission imaging is derived from the actual basal ganglia neurons responsible for motor function. The BRODERICK PROBE® sensed striatal neurons in order to study the functional activity of efferent dorsal striatal neuronal circuitry from the substantia nigra pars compacta (SNPC). Parkinson model subjects were lesioned at the SNPC using 6-hydroxydopamine in order to inhibit neurons in the dorsal striatal area originating from the SNPC from producing dopamine; surgical technicians at the North Carolina facility, the International Charles River Laboratories, performed the expert lesion technology.

15.3 The Path to Seeing the Tau LIVE in the Parkinson Subject

The neuroprobe/nanosensor signals from the brain are remembered by the Broderick nanoprobe polymer, the carbon polymer, the carbon/lipid/phosphotidyl polymer. The following drawings are provided by way of example only and without limitation, wherein like reference numerals (when used) indicate corresponding elements throughout the several views. The reference numerals are described in detail within the patents referenced in [3–6, 8, 10–13].

The dopamine (DA) neurotransmitter was absent in lesioned group (PD). In fact, the neurotransmitter, serotonin, is absent at times, especially when there is corresponding motor incoordination as is the case with degenerative disorders, specifically, the disease of Parkinson in this example. The direct cause and effect discovery between release of a serotonin molecule with a specific classified functional movement as it happens in real time, is critical to the meaning of the data presented since *Temporal Synchrony*, shown between serotonin and movement, may well define the healthy brain [7]. The importance of such real time correlations between brain and behavior has reached the textbook [18].

The Operation of the BRODERICK PROBE® nanoprobe polymer

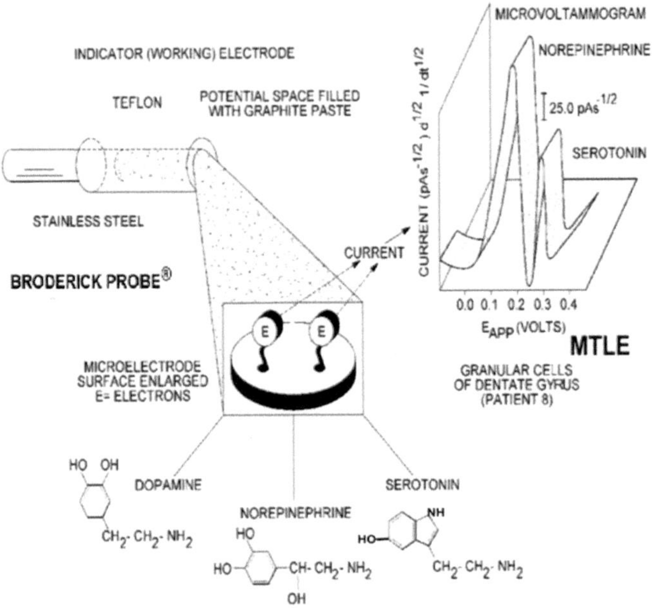

Figure 15.1 The operation of a Broderick nanoprobe polymer in imaging neurotransmitters and a scan resulting there from.

The BRODERICK PROBE® Photoluminescent Fiber Optic Protein Probe

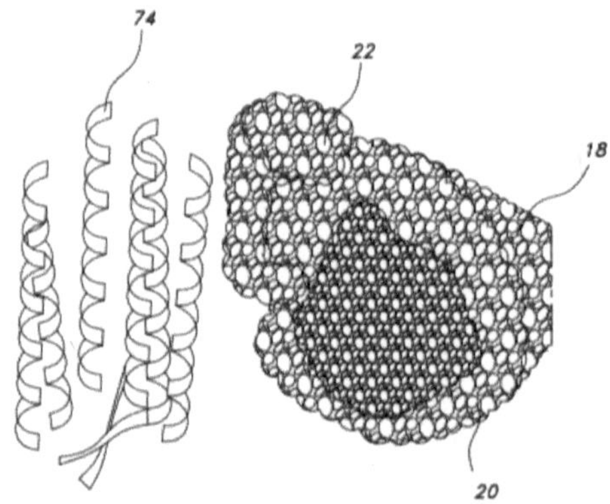

Figure 15.2 The photoluminescent fiber optic protein probe. A schematic of the opsin, rhodopsin is shown to the left (74). The Broderick Semiconductor Fullerene Probe is shown to the right in the center of the schematic as encased in the Broderick probe original carbon/lipid/phosphotidyl polymer. The biopolymer is numbered (18, 20, 22). Each of these sensing devices are trademarked, the BRODERICK PROBE®.

The Sensing Device, BRODERICK PROBE® Photoluminescent Fiber Optic Protein Probe, Operating within the Lobes of the Brain

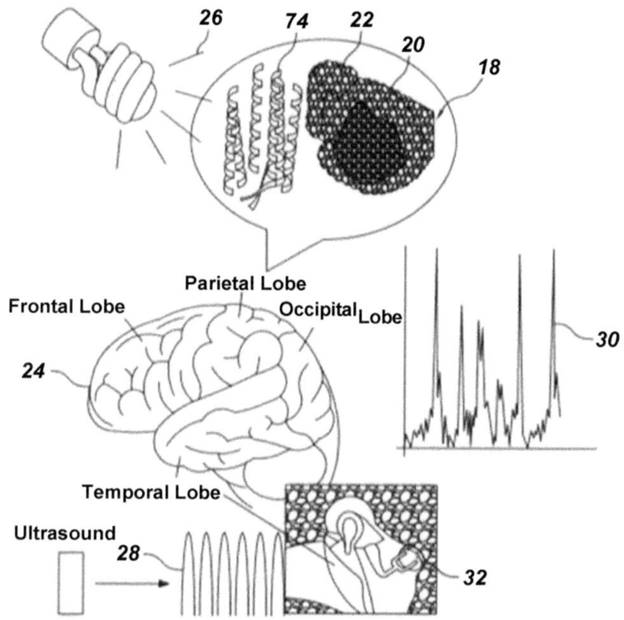

Figure 15.3 The sensing device operating within the BRODERICK PROBE® photoluminescent fiber optic protein probe as the sensing device interacts with the lobes of the brain that connect *via* the radio frequency light spectrum. Note that the number "30" refers to the intellectual property of Broderick, comprised of neuromolecular images for neurotransmitters and metabolites. Note that the number "32" refers to the auditory canal. Although Ultraviolet Light is shown as an example, the patents do not limit the photons of light chosen. However, the light selected for the experimental design is preferably Near Infrared, it is not limited to the Near Infrared Spectrum that the author has chosen for the Tau epilepsy patients and the patients presenting with Alzheimer disease and/or the disorder of Parkinson.

About Figure 15.3: In the bottom left, the application of piezoelectric sensing devices is shown as proposed and patented to act in concert with the BRODERICK PROBE®. Data output derived from the calcium channel blocker, *nimodipine*, is shown to the center, right side. Same subject design is an advantage of our nanotechnology, NMI. The data were gleaned from a hypoxic-hypoxia animal model designed in our laboratory to introduce the mechanism of action of memory defects that commonly occur in post-operative surgery [23].

The LIVE Tau Peptide

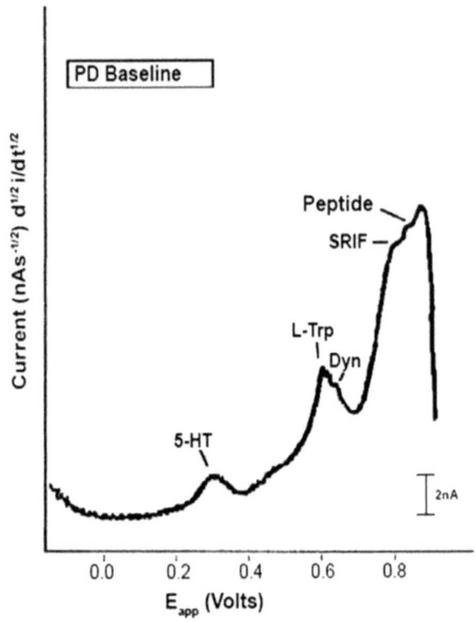

Figure 15.4 NMI results from lesioned group (PD animal model) with endogenous neurochemical profile. On the x-axis: oxidation potential in term of applied voltage in volts (V). On the y-axis: current in nanoamperes (nA). NMI was conducted using the BRODERICK PROBE® laurate biosensor. Serotonin (5-HT), L-tryptophan (L-Trp) and neuropeptides (dynorphin and somstostatin) were present. Note that dopamine (DA) neurotransmitter was absent in lesioned group (PD).

About Figure 15.4: The signal marked "peptide" is the Tau-associated peptide/protein. This is the first report of Tau signaling LIVE in electroactive images in a LIVE subject presenting with and exhibiting the cardinal symptomatology of the brain disease called Parkinson. In one embodiment, the electrode or sensor, nanoprobe detects the activities of Tau proteins/peptides (i.e., "Tau") [10, 13].

15.3.1 Let's Talk About It

15.3.1.1 Tau in the epilepsies

Tauopathies are a group of neurodegenerative diseases in which aggregates of the hyperphosphorylated form of the microtubule-associated protein (MAP), tau, are believed to play a critical role in the pathogenesis. Many of the tauopathies are associated with an increased incidence of development of epilepsy. Deposits of hyperphosphorylated tau have also been documented pathologically in animal models of epilepsy, and in patients with drug-resistant focal epilepsy who have undergone epilepsy surgery. The precise mechanisms by which hyperphosphorylated tau contributes to epileptogenesis are currently uncertain, but there is increasing evidence from animal models that targeting tau-based mechanism, either genetically or pharmacologically, can have both antiseizure and antiepileptogenic effects. Indeed, animal models of epilepsy-associated tauopathies, biomarkers of tau pathology, experimental evidence implicating hyper-phosphorylated tau in the pathophysiology of acquired epilepsies, in addition to therapies for Tau epilepsies are described in [26] and critical sensing epilepsy work is contained in [9, 33].

15.3.1.2 Tau in the disease of Alzheimer

The Tau gene and its reported six Tau isoforms have been found in the brain of Alzheimer's patients [27]. In Tau-related prionopathies, for example but not limited to mad cow, people and or patients who are affected by this disease experience unusual sensations that include prickling or tingling feelings in the face, hands, legs, and feet while sometimes sores appear on the feet for example. Sensations on these body parts can transform into burning sensations and permeate other body parts in a "pins and needles" sensation. Paresthesia can occur, causing a persistent lack of feeling in these body parts, producing clinical numbness.

15.3.1.3 Tau in the disorder of Parkinson

Dugger et al. [20] provide data that

> "Dementia is a frequent complication of Parkinson's disease (PD) and that about half of PD dementia (PDD) is hypothesized to be due to progression of the underlying Lewy body pathology

into limbic regions and the cerebral cortex while the other half is thought to be due to coexistent Alzheimer disease (AD). Clinically, however, these are indistinguishable. Since the spread of amyloid plaques to the striatum has been reported to be a sensitive and specific indicator of dementia due to AD, these authors analyzed neuropathologically-confirmed cases of PD without dementia (PDND, $N = 31$), PDD without AD (PDD, $N = 31$) and PD with dementia meeting clinicopathological criteria for AD (PDAD, $N = 40$). The minimum diagnostic criterion for AD was defined as including a clinical history of dementia, moderate or frequent CERAD cortical neuritic plaque density and Braak neurofibrillary stage III-VI. Striatal amyloid plaque densities were determined using Campbell-Switzer and Thioflavine S stains. Striatal plaque densities were significantly higher in PDAD compared to PDD ($p < 0.001$). The presence of striatal plaques was approximately 80% sensitive and 80% specific for predicting AD. In comparison, the presence of cerebral cortex plaques alone was highly sensitive (100%) but had poor specificity (48% to 55%). The results suggest that striatal amyloid imaging may be clinically useful for making the distinction between PDD and PDAD."

Dugger et al.'s work is an important contribution to distinguishing Parkinsonism by amyloid concentration. Positron Emission Tomography in cerebrospinal fluid is new and hopeful. Nuclear magnetic resonance (NMR) for flexible Tau forms and Cryo-EM for the most rigid forms or fluid immuno-diagnosis [21] coupled to brain imaging research such as the BRODERICK PROBE® [3–18, 23] bring promise. Video microscopy enables another useful tool in this highly active and necessary research arena, both clinical and pre-clinical. Antibody labeling is daily used *antemortem* and *post mortem*; it has brought forth valuable data on tauopathies despite its *post mortem* status [21].

15.3.1.4 DNA telomere length gives us a clue

Telomere length on the ends of strands of deoxyribose nucleic acid (DNA) is associated with deciphering correlative cognitive enhancement or decline, another avenue of strategic cures for tauopathies [26].[2]

[2]In 2009, Elizabeth H. Blackburn, PhD, won the Nobel Prize for her original work in the area of DNA, the telomere and telomerase. https://www.nobelprize.org/prizes/medicine/2009/blackburn/facts/.

15.3.1.5 In press

In press, from Singapore, is the author's authentic imaging textbook, a monograph. The innovative scientific monograph explains with original experimental protocols in detail, the know-how of NMI along with describing the discoveries made by the BRODERICK PROBE®. The author interjects her own person and her personality as she interacts with competitors and colleagues alike within her riveting travel to her original discoveries [15].

15.4 A Message to Take Home

It is the Broderick Laboratory that identifies herein the LIVE electroactive image for Tau in the precise striatal basal neurons of the living subject as a metal complex phosphorylated and vital, presenting with Parkinsonian disease as portrayed in this clinical paper and her patent [10, 12] Mueller et al., [25] posit: "tau functions as a signaling hub, regulating a wide set of cellular processes through the modulation of selected kinases and phosphatases." Thus, the elusive nature of Tau is no longer elusive as Tau is showcased herein as a dynamic redox reaction occurring LIVE in the brain of the living Tau subject *via* the phosphorylated BRODERICK PROBE®.

15.5 Permissions

Adapted with permission from **Broderick, PA (2021) Tau peptide signals are seen in Parkinson subjects by BRODERICK PROBE® sensors. *MOJ Clin Med Case Rep 11(2)* 46–50**. Medcrave: *Communications*: 2524 N. Broadway Edmond OK 73034.USA; *Editorial:* 1051 Budapest, 1st floor, Roosevelt 7/ 8, Széchenyi István tér 7–8 C tower. *The Medical Professor, Patricia A. Broderick, Ph.D. owns the copyright to her article published by Medcrave*, ©2021 Broderick.

References

1. Alzheimer's disease (2020) Symptoms and causes. *Mayo Clinic.*
2. Bailey RB (2019) The Six Kingdoms of Life. *Thought Co.* Nov. 28.

3. Broderick PA (1989) Cathodic Electrochemical Current Arrangement with Telemetric Application. US Patent Serial No. 4,883,057, Issued.
4. Broderick PA (1995) Microelectrodes and Their Use in Cathodic Electrochemical Current Arrangement with Telemetric Application. US Patent Serial No. 5,433,710, Issued.
5. Broderick PA (1999) Microelectrodes and Their Use in an Electrochemical Arrangement with Telemetric Application. US Patent Serial No. 5,938,903, Issued.
6. Broderick PA et al. (2006) Identification, Diagnosis, and Treatment of Neuropathologies, Neurotoxicities, Tumors, and Brain and Spinal Cord injuries using Microelectrodes with Microvoltammetry. US Patent Serial No. 7,112,319, Issued.
7. Broderick PA (2013) Neuromolecular Imaging Shows Temporal Synchrony Patterns between Serotonin and Movement within Neuronal Motor Circuits in the Brain. *Brain Sci. 3* (2):992–1012.
8. Broderick PA (2017) Noninvasive Electroactive Photonic Sensor with Polymer Memory Transduction Using Organic and Inorganic Elements as Platforms. US Provisional Application No. 62/273,693.
9. Broderick PA. (2018) Sensing the epilepsy brain: a unique nanosurgical biomedical device to treat the epilepsy patient. *MOJ Clin Med Case Rep* 8(2):60–65.
10. Broderick, PA (2020) Identification, Diagnosis, and Treatment of Tau-Associated Pathologies Using Electrodes. US Provisional Application No. 63/130,453, December 24.
11. Broderick PA (2021) Noninvasive Electroactive Photonic Protein Sensor with Polymer Photovoltaic Optics for Memory Transduction Using Organic and Inorganic Elements as Platforms. US Non–Provisional International Application No. PCT/US 16/68879, Issued.
12. Broderick, PA (2021) Tau peptide signals are seen in Parkinson subjects by Broderick Probe® sensors. *MOJ Clin Med Case Rep* 11(2) 46–50.
13. Broderick, PA, et al. (2021) Identification, Diagnosis and Treatment of Neuropathologies, Neurotoxicities Tumors and Brain and Spinal Cord Injuries Using Electrodes with Microvoltammetry. US Patent Serial No. 10,980,460. Issued, April, 20th.
14. Broderick PA (2022) A New Approach to Tumor Cancer with a Novel Imaging Profile for White matter abnormalities Including Leukodystrophies: Sensing the Human Brain. *Med Res Arch* [online] 10(11), https://doi.org/10.18103/mra.v10i11.

15. Broderick PA (2022) *Neuroimaging: Sensing Biochemistry in the Brain,* Jenny Stanford Publishing, Pte Ltd., Singapore, December, in press. ISBN 978-981-4613-49-1 (Hardcover), 978-0-429-02974-5 (eBook).
16. Broderick PA, Kolodny EH (2009) Real time imaging of biomarkers in the Parkinson's brain using mini–implantable biosensors. II. Pharmaceutical therapy with bromocriptine. *Pharmaceuticals (Basel, Switzerland)* 2(3):236–249.
17. Broderick PA, Wenning L, Li YS (2016) Neuromolecular imaging, a nanobiotechnology for Parkinson's disease: advancing pharmacotherapy for personalized medicine. *J Neural Transm (Vienna, Austria*) 124(1):57–78.
18. Broderick PA, Wenning L (2020) Neuromolecular imaging in Parkinson's disease. In: Martin CR, Preedy VR, eds, *Genetics, Neurology, Behavior, and Diet in Parkinson's Disease,* The Neuroscience of Parkinson's Disease, Elsevier Academic Press, United Kingdom, Vol. 2, Chapter 16, pp. 251–265.
19. Chefer VI, Thompson AC, Zapata A, et al. (2009) Overview of brain microdialysis. *Curr Protoc Neurosci* Chapter 7: Unit 7.1.
20. Dugger BN, Serrano GE, Sue LI et al. (2012) Presence of striatal amyloid plaques in Parkinson's disease dementia predicts concomitant Alzheimer's disease: usefulness for amyloid imaging. *J Parkinson's Dis* 2(1):57–65.
21. Fichou Y, Al–Hilaly YK, Devred F, et al. (2019) The elusive tau molecular structures: can we translate the recent breakthroughs into new targets for intervention? *Acta Neuropathol Commun.* (7)31.
22. Gorantla NV, Landge VG, Nagaraju PG, et al. (2019) Molecular cobalt (II) complexes for tau polymerization in Alzheimer's disease. *ACS Omega* 4(16):16702–16714.
23. Haile M, Broderick P, Li Y, et al. (2011) Nimodipine reverses the elevation of synaptic striatal dopamine and serotonin during *in vivo* hypoxia. *Open Neurosci J* 5:1–7.
24. Linnaeus C (2013) *Systemanaturae, sive regna trianaturae systematic eproposita per classes.* ordines, genera, & species. Leiden: Haak. October 3, 1735; p. 1–12.
25. Mueller RL, et al. (2021) Tau: a signaling hub protein. *Front Mol Neurosci* 14.
26. Mahoney ER, Dumitrescu L, Seto M, et al. (2019) Telomere length associations with cognition depend on Alzheimer's disease biomarkers. *Alzheimer's Demen* 5(1):883–890.

27. Noble W, Hanger DP, Miller C, et. al. (2013) The importance of tau phosphorylation for neurodegenerative diseases. *Front Neurol* 4:83.

28. Prusiner SB (1998) Nobel Lecture: Prions. *Proc Natl Acad Sci* 95(23):13363–13383.

29. Prusiner SB, Woerman AL, Mordes DA, et al. (2015) Evidence for α-synuclein prisons causing multiple system atrophy in human with parkinsonism. *Proc Natl Acad Sci U S A* 112(38): E5308–E5317.

30. Schneider A, Biernat J, von Bergen M, et al. (1999) Phosphorylation that detaches tau protein from microtubules (Ser262, Ser214) also protects it against aggregation into Alzheimer paired helical filaments. *Biochemistry* 38(12):3549–3558.

31. Scholl M, Maass A, Mattson N, et al. (2019) Biomarkers for tau pathology. *Mol Cell Neurosci* 97:18–33.

32. Trinczek B, Biernat J, Baumann K, et al. (1995) Domains of tau protein, differential phosphorylation, and dynamic instability of microtubules. *Mol Biol Cell* (2):1887–1902.

33. Zheng, P, Shultz S, Hovens C, et al. (2017) Animal models of acquired epilepsy and tauopathies. *Models Seizures Epilepsy* 1031–1041.

Chapter 16

The Living Legacy

Photograph of Dr. Broderick by Jeff Smith for *Patient Preferred Physician,* Preferred Health Magazine, 2021.

Reproduced with permission from *National Digest.*

Neuroimaging: Sensing Biochemistry in the Brain
Patricia A. Broderick

ISBN 978-981-4613-49-1 (Hardcover), 978-0-429-02974-5 (eBook)
www.jennystanford.com

Trademarked, the BRODERICK PROBE®, the tiny nanoprobe was created by a tiny but mighty five-foot girl, Dr. Patricia Ann Broderick, PhD. The name was to honor her father, Patrick Broderick, Patricia felt a strong desire to name the nanoprobe sensor after her father as her father was killed and taken from her in a shocking oxyacetylene torch welding accident on Christmas Eve when Patricia had just turned the age of seven. The abandonment that Patricia feels every Christmas Eve is formidable and the complexity of the brain always propelled her to elucidate how the brain works. What happened to her father?

Patricia is a first-generation Irish American. She is the child of Patrick Broderick and Margaret Daly, immigrants to America from County Clare, Ireland, and Bantry Bay, County Cork, Ireland. Her family took root in the Throggs Neck Section in the Bronx, moved to Philadelphia and then back to the Bronx when Patricia was five. This time it was the South Bronx, a Catholic, blue-collar, poor and happy Irish neighborhood where everyone was family.

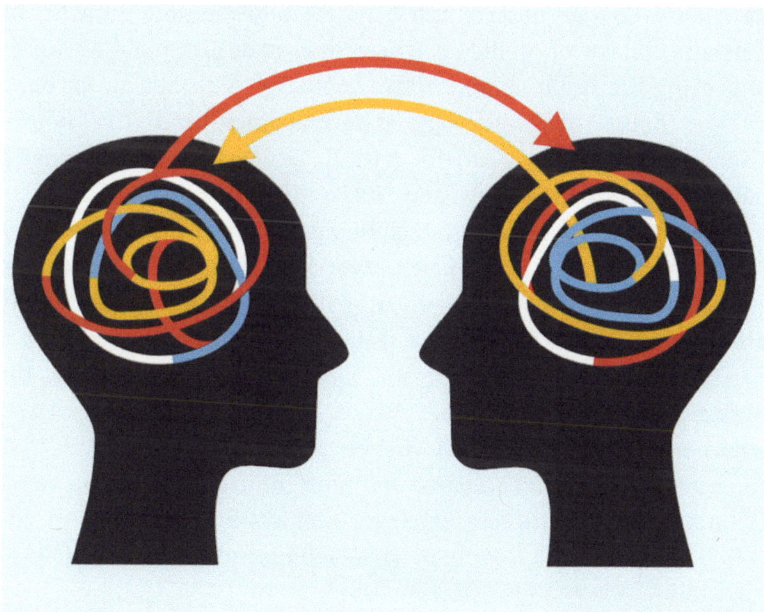

The environment was supporting and loving which helped the family to be strong in the face of such horror. Patricia was always bright and extremely perceptive. Quietly, she helped all her peers at every level to study and pass exams smoothly with the manner in which she taught. Everyone waited for Patricia to synthesize the course, no matter what the course was, and break it down into understanding. To this day, Patricia makes complexity into simplicity and she is affectionately called a "super power brain."

The nanoprobe just may make it to the Nobel Prize despite her humble beginnings in her happy, yet poor Irish family in a tenement neighborhood in the South Bronx, New York, NY, USA. So many people of all walks in life believe that she will make it in honor of her father and her courageous mother, her family and dear colleagues. Patricia has spent a lifetime figuring out how the brain works to make the world better. Humble herself, she asks to this day for so little financial support and seeks less. It is said that her work is maximal on such minimal requests.

Patricia began her innovative spectroscopy work at St John's University College of Arts and Sciences and continued on to the Einstein College of Medicine where she made dramatic advances in sensing the brain using electrochemistry. With her background in the fields of physiology, pharmacology and the neurosciences, Patricia has made critical advances in the sensing field of electrochemistry and its application to the field of nanotechnology, an imaging technology, *Neuromolecular Imaging*. She then continued to Cornell University School of Medicine and currently serves as a full Medical Professor at the City University of New York School of Medicine at the City College of New York.

Dr. Broderick, both herself and her work embody the phraseology from the *Nobel Peace Presentation Speech,* 1977, wherein the Irish spirit of justice wins the award. The Irish courageous spirit rails against indignity and promotes the cause of human dignity instead. Patricia's philosophy for her entire life is to promote human dignity in her teaching and her research.

But, what Patricia has done that virtually no one else has done is to promote animal dignity along with human dignity. She has changed the way to study animals by using her nanoprobe and her Neuromolecular Imaging nanotechnology.

Her philosophy is that all are created equally and all should be treated equally and with the highest respect. She believes that people in power should have the grace of office as described in theology.

"Power of authority should not be used as power of abuse."

Rational behavior in the academe and everywhere needs to be honored. One level need not destroy another because of jealousy or ill will. Her students and her research technicians feel her philosophy of equal power based on resolute and solid direction toward the good of the institution and the person. Her philosophy is embedded in the following phrase; it is paraphrased as it regards the Irish American spirit in Patricia herself and it relates and transfers to all people of nationality, age and each and every gender.

> Patricia is an individual who holds the deep and firmly rooted conviction that the ordinary man and woman is capable of making a meaningful contribution to peace. The peace that Patricia has made is in the field of *science and medicine* where she is courageous in promoting the dignity of animals as well as humans. Patricia believes in working diligently and being fair and just in her teaching and research and everyday life.

One part of her work that may well deserve the Nobel Award is her translational work for the epilepsy patient to bring about Personalized LIVE Medicine. The nanoprobe is the first to see the neurochemistry of the human epilepsy patient brain during epilepsy surgery exactly in the operating room side by side with the neurosurgeon and in fact, collegially directing the neurosurgeon in the mechanics of the BRODERICK PROBE® sensing nanoprobe.

The nanoprobe is the first of its kind. It is a polymer made of non-toxic biocompatible scientifically advanced formulas designed by Dr. Broderick. This probe is actually a series of sensors/ nanoprobes, each of which is smaller than one human hair, it has been tested for bacterial infection by the microbiologists at NYU Langone and the results show that it does not promote bacterial formation. The tiny probe does not cause scarring as this has been tested and the report by the pathologist at NYU Langone Medical Center is published in the report from the clinical trial.[1]

[1]Broderick PA et al. (2009) A clinical trial of an advanced diagnostic biomedical device for epilepsy patients. *J Long Term Eff Med Implants* 18(1), June 1.

Indeed, Dr. Broderick has pioneered nanotechnology to be fair to animals wherein she has changed the science and medicine community away from the after-death, *post-mortem* work into LIVE imaging. In fact, her work translates to the clinic, where she is the first, with her team, to change the way that physicians think about science and bedside medicine.

From what anyone can see is that certainly a considerable amount of insight and engineering aplomb has gone into the translation of the sensor from lab to clinic. It is extraordinary, original and the first of its kind. Although the work and Patricia's dynamic personality are loved by many around the globe, there are the misanthropes.

Her breakthrough and cutting-edge research is sought after and she personally lays prey to the tyranny and terror that does exist when one's genius is original and leans toward being lucrative. Yes, her professional life is productive and her work is appreciated globally; however, there are times that this particular injustice of jealousy even from power besets her due to her genius in the work of the tiny probe and due to her principles in engendering dignity to animal research and dignity for all, no matter what level of power.

> Patricia continues to be strong in her belief in honesty and justice in science and medicine in the academe despite the suffering that is put upon her for promoting animal dignity as well as human dignity and for devoting her life to making and sharing the tiniest sensor in the world capable of LIVE studies in humans and animals.

It is true that many are willing to die for a cause and some are willing to live a lifetime fighting for a cause. Patricia is willing to suffer, live and die for the cause.

With this philosophy in mind, Patricia has gone on to change the face of science and medicine because her biomedical sensors link to the brain to see inside the animal and the human brain. In fact, the nanoprobes are the smallest biomedical sensors for LIVE Imaging in the world and they have gained tremendous recognition in the academic as well as in the medical press.

What Patricia has done differently from the majority of the professional scientists is this: The nanoprobe is available to and for the layperson. She has made it a product and it is an affordable product. Yes, the sensors reach the layperson in particular in the inventor's quest to share medicine and science with all people around the globe.

The sensors reach undergraduate students in the City College of New York. The Professor teaches a novel interdisciplinary course in the study of the *Neurobiology of Drug Abuse.* Her 200 interested and interesting students meet *every week, each semester* and she has taught the medical students, who are now the practicing physicians of the world. She has taught the physician assistants and served as a former director. The students enjoy her style of creative techniques in teaching. The students report that Professor Broderick, a full and tenured Medical Professor, makes complex material understandable for life's tasks.

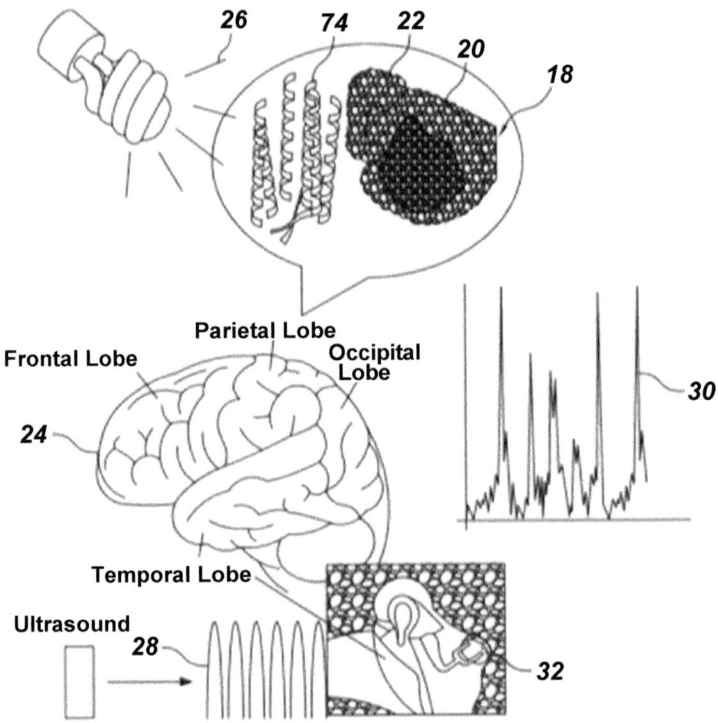

Her enlightening research reaches the health magazines for the layperson. A person can pick up the knowledge at the newsstand as Patricia is published in *Preferred Health Magazine.*

She is known for her in-depth contribution to the COVID-19 pandemic. She was the first to publish her deeply thought-out empirical and theoretical discovery that COVID-19 affects the brain of the healthy and the unhealthy.

The Broderick laboratory has configured the algorithms for the LIVE cytokine mobility into the hippocampal neurons to damage memory. Adaptation and stress mechanisms as these relate to suicide, depression, schizophrenia and bipolar affect represent just part of her contributions.

Brain reward and temporal Synchrony are everyday vocabulary for the world to understand as coming from her work!

Dr. Broderick is a force to bring the new science and medicine to the lay world for understanding.

Along these lines then, Dr. Broderick is the host of the syndicated and global BBM radio show, *The Easysense Show*, remains available on many networks and her *TV Eazysense Show* launched in November 2022; her new-generation nanoprobe is the focus. In fact, that is not all. The work and she are no stranger at all to the airways.

Dr. Broderick has been invited to *Close-Up TV radio with Doug Llewelyn* and *Close-Up TV radio with Jim Masters*, the *Dr. Robi Ludwig Show*, the *Jill Nicolini show*, the *Donna Drake Show*—all are on YouTube. She appeared on the *Melissa Billy Clark Show* on November 17, 2022, and is invited by Doug Llewelyn, Close-up TV-blog radio once again on April 3, 2023, to discuss her unique brain and spinal cord inventive art and her gift of LIVE Neuroimaging for the world.

Patricia's major discovery is the BRODERICK PROBE®. What the series of nanoprobe sensors can do is awesome! The advanced materials and preclinical and clinical applications are considered by many as Nobel material. Patricia has promoted the concept that the brain itself is an electrochemical device.

"The brain is electrochemical and molecular; it matches the BRODERICK PROBE® electrochemical and molecular brain matrix."

> The nanoprobe actually meets the brain, attaches to it and then leaves the brain intact and whole. There is no other technology or nanotechnology that can do it. The nanoprobe is not one probe; it is a series of probes, sensing devices, sensors that see inside the brain whatever you need, wherever you need it.

The nanoprobe sees chemicals and neurotransmitters inside the brain. This nanotechnology is ready for market manufacturers and with the help of partners, for financial funding by sponsors, investors and philanthropists, a response in shares in the company *Eazysense Nanotechnologies Inc.* is in order.

The BRODERICK PROBE® sensors relate to several formulations of unique, patented, and trademarked miniature carbon sensors comprising a series of compounds which include, among others, classes of molecules in the biochemical categories of lipid, glycolipid, lipoprotein, saturated and unsaturated fatty acid.

These inventions are able to detect electrochemical signals in the living animal brain and the human brain for a vast number of neurotransmitters, neuromodulators, and metabolites, including neuropeptides.

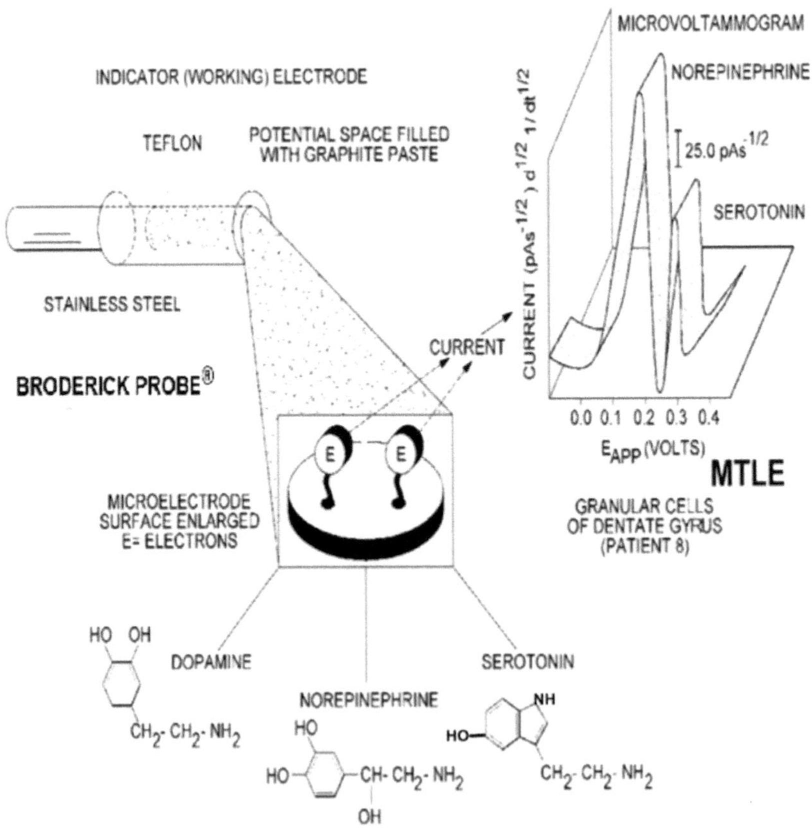

Diagnosis of Tumors

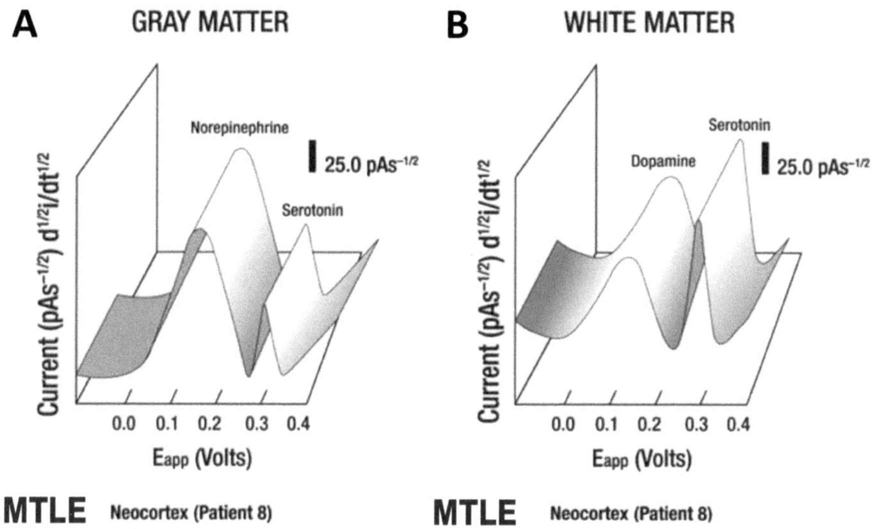

The MRI cannot see white matter **deep in gray**. When the surgeon is tracking the tumor, they can see the tumor in the gray matter. Then, suddenly, the tumor is lost to the eye of the surgeon and to the MRI as white matter is hidden. Water is in the way. Diffusion tensor can help, but still MRI is needed first and with contrast media that causes allergies. Enter *Neuromolecular Imaging (NMI)* to see the white matter. The nanoprobe can see the white matter in subseconds. The BRODERICK PROBE® has shown, for the first time, a distinct difference, in tissue resected from epilepsy patients, between hippocampal white versus gray matter.

The Non-Invasive Optical Nanoprobe-Electrons to Photons

Using Neuromolecular Imaging, the BRODERICK PROBE® is patented preclinically and clinically as carbon and protein-based semiconductor nanoprobes with integrated interfacial surface algorithms for programming online unique neurobehavior. Biosensing recognizable neuromolecular elements in anion and cation ranges of neurotransmitters LIVE, while neurobehavior is imaged online, further advances material science and dramatically moves discoveries in neurotransmission forward.

The disclosed embodiments of the non-invasive patent by Broderick relate to a miniature photosensitive, electroactive polymer sensing device that operates by converting photonic energy into electrical and electrochemical energy that generates a photocurrent in the brain without opening the brain. The output is provided in units of voltage. Laser diodes then enable the electrochemical waveform to be seen as an electrochemical image in color separated into specific spectral frequencies. Thus, the photocurrent provides an image of neurochemicals in real time, noninvasively and/or minimally invasively. The embodiments provide a photoelectrochemical conductance device for nanodiagnostics, nanotherapies, and nanotheranostics. These embodiments are directly applicable to specific neurodegenerative and psychiatric disorders in humans and animals including, but not limited to, diseases of the motor and sensory systems in the brain, body, and blood.

We are poised to read empirically derived signal processing-based algorithms on the BRODERICK PROBE® from multiple interconnected sites in brain, upper and lower limbs, these programmed sensing neuromolecular devices further equip the paralyzed person for agile mobility. The epidural spinal cord patient will benefit. The Parkinson patient will benefit. Exquisite detail in information processing is essential and selective frequency of neurotransmission plus afferent or efferent signal dynamics from one neuroanatomic site to another is paramount to deciphering the significance of neurotransmission in concert with mobility. Along with pure carbon polymers, the BRODERICK PROBE® fiber optical-based programmable vector biosensing devices enable recognizable molecular elements for entropy in artificial intelligence (AI). The dementias, Alzheimer, Parkinson, epilepsy, stroke and brain reward circuits of orgasm and the like suit the tiny probes all the more.

Patricia is the first to sense the brain as an expert in both fields. There is no one else who has accomplished this feat.

In one of Dr. Broderick's recent books, an Imaging Textbook that is indeed a monograph, the evolution of In Vivo Electrochemistry into Neuromolecular Imaging into Non-Invasive Voltaic Photonics: LIVE IMAGING, her inventions that make her work original and far above the fray, is told in detail. The book is in press, on course this year, from Singapore, **Jenny Stanford Publishing Pte Ltd**. For a scientist to, in fact, write a monograph on how she discerns her lifetime work is most unusual and this is an understatement. These major discoveries are delineated by the master herself and precise patented protocols for her imaging techniques are told with unique delivery. It is amazing how she intersperses her personal life into the professional as she writes in her book the exact patented protocols for her discoveries. She tells of her interrelationships, how her dedication, wisdom and sense of humor propelled her through her path that was less chosen as in the poetry of Robert Frost.

In her book, she courageously takes on the fight over electrocatalysis!

The problem of electrocatalysis, that is, the mix-up between the detection of dopamine (DA) and ascorbic acid (AA) electrochemically, was indeed a problem. This problem was the very bane of the existence of the neuroscientist who wished to be the "one" to be the expert "in vivo electrochemist" and vice versa. But these scientists, neuroscientist or electrochemist, apparently could not do it! Each and every scientist involved began to author manuscripts in which the authors could detect ascorbic acid or dopamine but none could detect both molecules separately and clearly. What could be done? Someone had to do it! Then, Broderick stepped up to the plate, so to speak. What did Broderick do? Broderick accomplished the necessary! What was accomplished between 1983 and 2008 was the saving grace for in vivo electrochemistry for growth in the electrochemistry field in vivo and preparation for the advent of neuromolecular imaging and cerebral polymeric, photosensitive, photoelectric and piezoelectric live imaging. The future of in vivo electrochemistry in spectral analysis now has its place, humbly, via the contribution of the neuroscientist electrochemist, Broderick and her team.

Patricia's Legacy

Patricia's advancements to science and medicine continue. Dr. Broderick is honored once again as the sole editor of the second edition of her successful Humana Press book, *Bioimaging in Neurodegeneration*; her next new book is entitled, *Biomedical Imaging in Neurodegeneration*, under contract with the prestigious Springer Nature Publishing Group.

Conclusion

There is one unmistakable chord throughout her closely intertwined work and life that she has a true genius in uncovering the truth. Her brilliant childlike innocent approach to discernment is enchanting and her passion envelops one's very soul. Perhaps, she will humbly meet with the Nobel Prize! Indeed, there are peaceful clusters of scientists around the world discussing the desire that a woman scientist should be one of the recipients. Perhaps she has even been nominated already!

Paragraph from Chapter 3 of her book, *Neuroimaging Sensing Biochemistry in the Brain*:

> *Nathaniel Hawthorne wrote: "Time flies over us, but leaves its shadow behind" Twenty-five years—one quarter of one century, a century of shadows, of joyful shadows, of sorrowful shadows, each and every nanosecond remembered. Each and every scientist, colleague, family, friend and acquaintance, remembered. Each and every student remembered. Each and every discovery remembered. Each and every kind greeting, encouragement and each and every hurtful word and action remembered. Each and every supporter and each and every detractor and/demoralizer, all limbic-related in emotion and cortically analytically ordered to help Patricia's mind/brain/soul grow in depth, mentally and spiritually, is deeply remembered. William Shakespeare wrote: "Time is very slow for those who wait, very fast for those who are scared, very long for those who lament, very short for those who celebrate"—But for those who love, time is eternal. For Patricia, time is eternal.*

Index